STUDENT'S SOLUTIONS MANUAL

to accompany

FINITE MATHEMATICS

SIXTH EDITION

STUDENT'S SOLUTIONS MANUAL

to accompany

FINITE MATHEMATICS
SIXTH EDITION

LIAL • GREENWELL • MILLER

PREPARED WITH THE ASSISTANCE OF

August Zarcone
Gerald Krusinski

College of DuPage

 **ADDISON-WESLEY**

An imprint of Addison Wesley Longman, Inc.

Reading, Massachusetts • Menlo Park, California • New York • Harlow, England
Don Mills, Ontario • Sydney • Mexico City • Madrid • Amsterdam

ISBN Package: 0-321-01625-4
ISBN Manual: 0-321-02864-3

3 4 5 6 7 8 9 10 VG 0099

PREFACE

This book provides solutions for many of the exercises in *Finite Mathematics*, Sixth Edition, by Margaret L. Lial, Raymond N. Greenwell, and Charles D. Miller. Solutions are included for odd–numbered exercises. Solutions are not provided for exercises with open–response answers. Sample tests are provided at the end of each chapter to help you determine if you have mastered the concepts in a given chapter.

This book should be used as an aid as you work to master your coursework. Try to solve the exercises that your instructor assigns before you refer to the solutions in this book. Then, if you have difficulty, read these solutions to guide you in solving the exercises. The solutions have been written so that they are consistent with the methods used in the textbook.

You may find that some of the solutions are presented in greater detail than others. Thus, if you cannot find an explanation for a difficulty that you encountered in one exercise, you may find the explanation in the solution for a similar exercise elsewhere in the exercise set.

Solutions that require graphs will refer you to the answer section of the textbook. These graphs are not included in this book.

In addition to solutions, you will find a list of suggestions on how to be successful in mathematics. A careful reading will be helpful for many students.

The following people have made valuable contributions to the production of this *Student's Solutions Manual:* Terry McGinnis and Abby Tanenbaum, editors; Judy Martinez and Sheri Minkner, typists; Karen Pedigo and Ev Potempa, proofreaders; and Darryl Nester, artist.

We also want to thank Tommy Thompson of Cedar Valley Community College for the essay "To the Student: Success in Mathematics."

CONTENTS

CHAPTER 2 SYSTEMS OF LINEAR EQUATIONS AND MATRICES

CHAPTER 3 LINEAR PROGRAMMING: THE GRAPHICAL METHOD

CHAPTER 4 LINEAR PROGRAMMING: THE SIMPLEX METHOD

CHAPTER 5 MATHEMATICS OF FINANCE

CHAPTER 6 SETS AND PROBABILITY

CHAPTER 7 COUNTING PRINCIPLES; FURTHER PROBABILITY TOPICS

CHAPTER 8 STATISTICS

CHAPTER 9 MARKOV CHAINS

CHAPTER 10 GAME THEORY

TO THE STUDENT: SUCCESS IN MATHEMATICS

The main reason students have difficulty with mathematics is that they don't know how to study it. Studying mathematics *is* different from studying subjects like English or history. The key to success is regular practice.

This should not be surprising. After all, can you learn to play the piano or to ski well without a lot of regular practice? The same thing is true for learning mathematics. Working problems nearly every day is the key to becoming successful. Here is a list of things you can do to help you succeed in studying mathematics.

1. *Attend class regularly.* Pay attention in class to what your instructor says and does, and make careful notes. In particular, note the problems the instructor works on the board and copy the complete solutions. Keep these notes separate from your homework to avoid confusion when you read them over later.

2. Don't hesitate to ask questions in class. It is not a sign of weakness, but of strength. There are always other students with the same question who are too shy to ask.

3. *Read your text carefully.* Many students read only enough to get by, usually only the examples. Reading the complete section will help you to be successful with the homework problems. Most exercises are keyed to specific examples or objectives that will explain the procedures for working them.

4. Before you start on your homework assignment, rework the problems the instructor worked in class. This will reinforce what you have learned. Many students say, "I understand it perfectly when you do it, but I get stuck when I try to work the problem myself."

5. Do your homework assignment only *after* reading the text and reviewing your notes from class. Check your work with the answers in the back of the book. If you get a problem wrong and are unable to see why, mark that problem and ask your instructor about it. Then practice working additional problems of the same type to reinforce what you have learned.

6. Work as neatly as you can. Write your symbols clearly, and make sure the problems are clearly separated from each other. Working neatly will help you to think clearly and also make it easier to review the homework before a test.

7. After you have completed a homework assignment, look over the text again. Try to decide what the main ideas are in the lesson. Often they are clearly highlighted or boxed in the text.

8. Use the chapter test at the end of each chapter as a practice test. Work through the problems under test conditions, without referring to the text or the answers until you are finished. You may want to time yourself to see how long it takes you. When you have finished, check your answers against those in the back of the book and study those problems that you missed. Answers are referenced to the appropriate sections of the text.

9. Keep any quizzes and tests that are returned to you and use them when you study for future tests and the final exam. These quizzes and tests indicate what your instructor considers most important. Be sure to correct any problems on these tests that you missed, so you will have the corrected work to study.

10. Don't worry if you do not understand a new topic right away. As you read more about it and work through the problems, you will gain understanding. Each time you look back at a topic you will understand it a little better. No one understands each topic completely right from the start.

ALGEBRA REFERENCE

R.1 Polynomials

1. $(2x^2 - 6x + 11) + (-3x^2 + 7x - 2)$
$= 2x^2 - 6x + 11 - 3x^2 + 7x - 2$
$= (2 - 3)x^2 + (7 - 6)x + (11 - 2)$
$= -x^2 + x + 9$

2. $(-4y^2 - 3y + 8) - (2y^2 - 6y - 2)$
$= (-4y^2 - 3y + 8) + (-2y^2 + 6y + 2)$
$= -4y^2 - 3y + 8 - 2y^2 + 6y + 2$
$= (-4y^2 - 2y^2) + (-3y + 6y)$
$\quad + (8 + 2)$
$= -6y^2 + 3y + 10$

3. $-3(4q^2 - 3q + 2) + 2(-q^2 + q - 4)$
$= -12q^2 + 9q - 6 - 2q^2 + 2q - 8$
$= -14q^2 + 11q - 14$

4. $2(3r^2 + 4r + 2) - 3(-r^2 + 4r - 5)$
$= (6r^2 + 8r + 4) + (3r^2 - 12r + 15)$
$= (6r^2 + 3r^2) + (8r - 12r)$
$\quad + (4 + 15)$
$= 9r^2 - 4r + 19$

5. $(.613x^2 - 4.215x + .892)$
$\quad - .47(2x^2 - 3x + 5)$
$= .613x^2 - 4.215x + .892$
$\quad - .94x^2 + 1.41x - 2.35$
$= -.327x^2 - 2.805x - 1.458$

6. $.83(5r^2 - 2r + 7) - (7.12r^2 + 6.423r - 2)$
$= (4.15r^2 - 1.66r + 5.81)$
$\quad + (-7.12r^2 - 6.423r + 2)$
$= (4.15r^2 - 7.12r^2)$
$\quad + (-1.66r - 6.423r) + (5.81 + 2)$
$= -2.97r^2 - 8.083r + 7.81$

7. $-9m(2m^2 + 3m - 1)$
$= -9m(2m^2) - 9m(3m) - 9m(-1)$
$= -18m^2 - 27m^2 + 9m$

8. $(6k - 1)(2k - 3)$
$= (6k)(2k) + (6k)(-3) + (-1)(2k)$
$\quad + (-1)(-3)$
$= 12k^2 - 18k - 2k + 3$
$= 12k^2 - 20k + 3$

9. $(5r - 3s)(5r + 4s)$

Use the FOIL method to find this product.

$(5r - 3s)(5r + 4s)$
$= (5r)(5r) + (5r)(4s) + (-3s)(5r) + (-3s)(4s)$
$= 25r^2 + 20rs - 15rs - 12s^2$
$= 25r^2 + 5rs - 12s^2$

10. $(9k + q)(2k - q)$
$= (9k)(2k) + (9k)(-q) + (q)(2k)$
$\quad + (q)(-q)$
$= 18k^2 - 9kq + 2kq - q^2$
$= 18k^2 - 7kq - q^2$

11. $\left(\frac{2}{5}y + \frac{1}{8}z\right)\left(\frac{3}{5}y + \frac{1}{2}z\right)$
$= \left(\frac{2}{5}y\right)\left(\frac{3}{5}y\right) + \left(\frac{2}{5}y\right)\left(\frac{1}{2}z\right) + \left(\frac{1}{8}z\right)\left(\frac{3}{5}y\right)$
$\quad + \left(\frac{1}{8}z\right)\left(\frac{1}{2}z\right)$
$= \frac{6}{25}y^2 + \frac{1}{5}yz + \frac{3}{40}yz + \frac{1}{16}z^2$
$= \frac{6}{25}y^2 + \left(\frac{8}{40} + \frac{3}{40}\right)yz + \frac{1}{16}z^2$
$= \frac{6}{25}y^2 + \frac{11}{40}yz + \frac{1}{16}z^2$

12. $\left(\frac{3}{4}r - \frac{2}{3}s\right)\left(\frac{5}{4}r + \frac{1}{3}s\right)$
$= \left(\frac{3}{4}r\right)\left(\frac{5}{4}r\right) + \left(\frac{3}{4}r\right)\left(\frac{1}{3}s\right) + \left(-\frac{2}{3}s\right)\left(\frac{5}{4}r\right)$
$\quad + \left(-\frac{2}{3}s\right)\left(\frac{1}{3}s\right)$
$= \frac{15}{16}r^2 + \frac{1}{4}rs - \frac{5}{6}rs - \frac{2}{9}s^2$
$= \frac{15}{16}r^2 - \frac{7}{12}rs - \frac{2}{9}s^2$

13. $(12x - 1)(12x + 1)$
$= (12x)^2 - 1^2$
$= 144x^2 - 1$

14. $(6m + 5)(6m - 5)$
$= (6m)(6m) + (6m)(-5) + (5)(6m)$
$\quad + (5)(-5)$
$= 36m^2 - 30m + 30m - 25$
$= 36m^2 - 25$

15. $(3p-1)(9p^2+3p+1)$
$$= (3p-1)(9p^2) + (3p-1)(3p)$$
$$\qquad + (3p-1)(1)$$
$$= 3p(9p^2) - 1(9p^2) + 3p(3p)$$
$$\qquad - 1(3p) + 3p(1) - 1(1)$$
$$= 27p^3 - 9p^2 + 9p^2 - 3p + 3p - 1$$
$$= 27p^3 - 1$$

16. $(2p-1)(3p^2-4p+5)$
$$= (2p)(3p^2) + (2p)(-4p) + (2p)(5)$$
$$\qquad + (-1)(3p^2) + (-1)(-4p) + (-1)(5)$$
$$= 6p^3 - 8p^2 + 10p - 3p^2 + 4p - 5$$
$$= 6p^3 - 11p^2 + 14p - 5$$

17. $(2m+1)(4m^2-2m+1)$
$$= 2m(4m^2-2m+1) + 1(4m^2-2m+1)$$
$$= 8m^3 - 4m^2 + 2m + 4m^2 - 2m + 1$$
$$= 8m^3 + 1$$

18. $(k+2)(12k^3-3k^2+k+1)$
$$= k(12k^3) + k(-3k^2) + k(k) + k(1)$$
$$\qquad + 2(12k^3) + 2(-3k^2) + 2(k) + 2(1)$$
$$= 12k^4 - 3k^3 + k^2 + k + 24k^3 - 6k^2$$
$$\qquad + 2k + 2$$
$$= 12k^4 + 21k^3 - 5k^2 + 3k + 2$$

19. $(m-n+k)(m+2n-3k)$
$$= m(m+2n-3k) - n(m+2n-3k)$$
$$\qquad + k(m+2n-3k)$$
$$= m^2 + 2mn - 3km - mn - 2n^2 + 3kn$$
$$\qquad + km + 2kn - 3k^2$$
$$= m^2 + mn - 2n^2 - 2km + 5kn - 3k^2$$

20. $(r-3s+t)(2r-s+t)$
$$= r(2r) + r(-s) + r(t) - 3s(2r)$$
$$\qquad - 3s(-s) - 3s(t) + t(2r) + t(-s)$$
$$\qquad + t(t)$$
$$= 2r^2 - rs + rt - 6rs + 3s^2 - 3st$$
$$\qquad + 2rt - st + t^2$$
$$= 2r^2 - 7rs + 3s^2 + 3rt - 4st + t^2$$

21. $(x+1)(x+2)(x+3)$
$$= [x(x+2) + 1(x+2)](x+3)$$
$$= \left[x^2 + 2x + x + 2\right](x+3)$$
$$= \left[x^2 + 3x + 2\right](x+3)$$
$$= x^2(x+3) + 3x(x+3) + 2(x+3)$$
$$= x^3 + 3x^2 + 3x^2 + 9x + 2x + 6$$
$$= x^3 + 6x^2 + 11x + 6$$

22. $(x-1)(x+2)(x-3)$
$$= [x(x+2) + (-1)(x+2)](x-3)$$
$$= (x^2 + 2x - x - 2)(x-3)$$
$$= (x^2 + x - 2)(x-3)$$

$$= x^2(x-3) + x(x-3) + (-2)(x-3)$$
$$= x^3 - 3x^2 + x^2 - 3xx - 2x + 6$$
$$= x^3 - 2x^2 - 5x + 6$$

R.2   Factoring

1. $8a^3 - 16a^2 + 24a$
$$= 8a \cdot a^2 - 8a \cdot 2a + 8a \cdot 3$$
$$= 8a(a^2 - 2a + 3)$$

2. $3y^3 + 24y^2 + 9y$
$$= 3y \cdot y^2 + 3y \cdot 8y + 3y \cdot 3$$
$$= 3y(y^2 + 8y + 3)$$

3. $25p^4 - 20p^3q + 100p^2q^2$
$$= 5p^2 \cdot 5p^2 - 5p^2 \cdot 4pq + 5p^2 \cdot 20q^2$$
$$= 5p^2(5p^2 - 4pq + 20q^2)$$

4. $60m^4 - 120m^3n + 50m^2n^2$
$$= 10m^2 \cdot 6m^2 - 10m^2 \cdot 12mn$$
$$\qquad + 10m^2 \cdot 5n^2$$
$$= 10m^2(6m^2 - 12mn + 5n^2)$$

5. $m^2 + 9m + 14 = (m+2)(m+7)$

since $2 \cdot 7 = 14$ and $2 + 7 = 9$.

6. $x^2 + 4x - 5 = (x+5)(x-1)$

7. $z^2 + 9z + 20 = (z+4)(z+5)$

since $4 \cdot 5 = 20$ and $4 + 5 = 9$.

8. $b^2 - 8b + 7 = (b-7)(b-1)$

9. $a^2 - 6ab + 5b^2 = (a-b)(a-5b)$

since $(-b)(-5b) = 5b^2$ and
$-b + (-5b) = -6b$.

10. $s^2 + 2st - 35t^2 = (s-5t)(s+7t)$

11. $y^2 - 4yz - 21z^2 = (y+3z)(y-7z)$

since $(3z)(-7z) = -21z^2$ and
$3z + (-7z) = -4z$.

12. $6a^2 - 48a - 120$
$$= 6(a^2 - 8a - 20)$$
$$= 6(a - 10)(a + 2)$$

13. $3m^3 + 12m^2 + 9m$
$$= 3m(m^2 + 4m + 3)$$
$$= 3m(m + 1)(m + 3)$$

14. $2x^2 - 5x - 3$

The possible factors of $2x^2$ are $2x$ and x and the possible factors of -3 are -3 and 1, or 3 and -1. Try various combinations until one works.

$$2x^2 - 5x - 3 = (2x + 1)(x - 3)$$

15. $3a^2 + 10a + 7$

The possible factors of $3a^2$ are $3a$ and a and the possible factors of 7 are 7 and 1. Try various combinations until one works.

$$3a^2 + 10a + 7 = (a + 1)(3a + 7)$$

16. $2a^2 - 17a + 30 = (2a - 5)(a - 6)$

17. $15y^2 + y - 2 = (5y + 2)(3y - 1)$

18. $21m^2 + 13mn + 2n^2$
$$= (7m + 2n)(3m + n)$$

19. $24a^4 + 10a^3b - 4a^2b^2$
$$= 2a^2(12a^2 + 5ab - 2b^2)$$
$$= 2a^2(4a - b)(3a + 2b)$$

20. $32z^5 - 20z^4a - 12z^3a^2$
$$= 4z^3(8z^2 - 5za - 3a^2)$$
$$= 4z^3(8z + 3a)(z - a)$$

21. $x^2 - 64 = x^2 - 8^2$
$$= (x + 8)(x - 8)$$

22. $9m^2 - 25 = (3m)^2 - (5)^2$
$$= (3m + 5)(3m - 5)$$

23. $121a^2 - 100$
$$= (11a)^2 - 10^2$$
$$= (11a + 10)(11a - 10)$$

24. $9x^2 + 64$ is the *sum* of two perfect squares. It cannot be factored. It is prime.

25. $z^2 + 14zy + 49y^2$
$$= z^2 + 2 \cdot 7zy + 7^2y^2$$
$$= (z + 7y)^2$$

26. $m^2 - 6mn + 9n^2$
$$= m^2 - 2(3mn) + (3n)^2$$
$$= (m - 3n)^2$$

27. $9p^2 - 24p + 16$
$$= (3p)^2 - 2 \cdot 3p \cdot 4 + 4^2$$
$$= (3p - 4)^2$$

28. $a^3 - 216$
$$= a^3 - 6^3$$
$$= (a - 6)[(a)^2 + (a)(6) + (6)^2]$$
$$= (a - 6)(a^2 + 6a + 36)$$

29. $8r^3 - 27s^3$
$$= (2r)^3 - (3s)^3$$
$$= (2r - 3s)(4r^2 + 6rs + 9s^2)$$

30. $64m^3 + 125$
$$= (4m)^3 + 5^3$$
$$= (4m + 5)[(4m)^2 - (4m)(5) + (5)^2]$$
$$= (4m + 5)(16m^2 - 20m + 25)$$

31. $x^4 - y^4 = (x^2)^2 - (y^2)^2$
$$= (x^2 + y^2)(x^2 - y^2)$$
$$= (x^2 + y^2)(x + y)(x - y)$$

32. $16a^4 - 81b^4$
$$= (4a^2)^2 - (9b^2)^2$$
$$= (4a^2 + 9b^2)(4a^2 - 9b^2)$$
$$= (4a^2 + 9b^2)[(2a)^2 - (3b)^2]$$
$$= (4a^2 + 9b^2)(2a + 3b)(2a - 3b)$$

R.3   Rational Expressions

1. $\dfrac{7z^2}{14z} = \dfrac{7z \cdot z}{2(7)z} = \dfrac{z}{2}$

2. $\dfrac{25p^3}{10p^2} = \dfrac{5 \cdot 5 \cdot p \cdot p \cdot p}{2 \cdot 5 \cdot p \cdot p} = \dfrac{5p}{2}$

3. $\dfrac{8k + 16}{9k + 18} = \dfrac{8(k + 2)}{9(k + 2)} = \dfrac{8}{9}$

4. $\dfrac{3(t + 5)}{(t + 5)(t - 3)} = \dfrac{3}{t - 3}$

5. $\dfrac{8x^2 + 16x}{4x^2} = \dfrac{8x(x + 2)}{4x^2}$
$$= \dfrac{2(x + 2)}{x}$$

6. $\dfrac{36y^2 + 72y}{9y} = \dfrac{36y(y + 2)}{9y}$
$$= \dfrac{9 \cdot 4 \cdot y(y + 2)}{9 \cdot y}$$
$$= 4(y + 2)$$

7. $\dfrac{m^2 - 4m + 4}{m^2 + m - 6}$

$\qquad = \dfrac{(m-2)(m-2)}{(m-2)(m+3)}$

$\qquad = \dfrac{m-2}{m+3}$

8. $\dfrac{r^2 - r - 6}{r^2 + r - 12} = \dfrac{(r-3)(r+2)}{(r+4)(r-3)}$

$\qquad\qquad\qquad = \dfrac{r+2}{r+4}$

9. $\dfrac{x^2 + 3x - 4}{x^2 - 1}$

$\qquad = \dfrac{(x-1)(x+4)}{(x-1)(x+1)}$

$\qquad = \dfrac{x+4}{x+1}$

10. $\dfrac{z^2 - 5z + 6}{z^2 - 4} = \dfrac{(z-3)(z-2)}{(z+2)(z-2)}$

$\qquad\qquad\qquad = \dfrac{z-3}{z+2}$

11. $\dfrac{8m^2 + 6m - 9}{16m^2 - 9}$

$\qquad = \dfrac{(4m-3)(2m+3)}{(4m-3)(4m+3)}$

$\qquad = \dfrac{2m+3}{4m+3}$

12. $\dfrac{6y^2 + 11y + 4}{3y^2 + 7y + 4}$

$\qquad = \dfrac{(3y+4)(2y+1)}{(3y+4)(y+1)}$

$\qquad = \dfrac{2y+1}{y+1}$

13. $\dfrac{9k^2}{25} \cdot \dfrac{5}{3k} = \dfrac{3 \cdot 3 \cdot 5k^2}{5 \cdot 5 \cdot 3k} = \dfrac{3k^2}{5k} = \dfrac{3k}{5}$

14. $\dfrac{15p^3}{9p^2} \div \dfrac{6p}{10p^2}$

$\qquad = \dfrac{15p^3}{9p^2} \cdot \dfrac{10p^2}{6p}$

$\qquad = \dfrac{150p^5}{54p^3}$

$\qquad = \dfrac{25 \cdot 6p^5}{9 \cdot 6p^3}$

$\qquad = \dfrac{25p^2}{9}$

15. $\dfrac{a+b}{2p} \cdot \dfrac{12}{5(a+b)}$

$\qquad = \dfrac{(a+b)12}{2p(5)(a+b)} = \dfrac{12}{10p}$

$\qquad = \dfrac{6}{5p}$

16. $\dfrac{a-3}{16} \div \dfrac{a-3}{32} = \dfrac{a-3}{16} \cdot \dfrac{32}{a-3}$

$\qquad\qquad\qquad = \dfrac{a-3}{16} \cdot \dfrac{16 \cdot 2}{a-3}$

$\qquad\qquad\qquad = \dfrac{2}{1} = 2$

17. $\dfrac{2k+8}{6} \div \dfrac{3k+12}{2}$

$\qquad = \dfrac{2(k+4)}{6} \cdot \dfrac{2}{3(k+4)}$

$\qquad = \dfrac{2(k+4)(2)}{2 \cdot 3 \cdot 3(k+4)}$

$\qquad = \dfrac{2}{9}$

18. $\dfrac{9y - 18}{6y + 12} \cdot \dfrac{3y + 6}{15y - 30}$

$\qquad = \dfrac{9(y-2)}{6(y+2)} \cdot \dfrac{3(y+2)}{15(y-2)}$

$\qquad = \dfrac{27}{90} = \dfrac{3 \cdot 3}{10 \cdot 3} = \dfrac{3}{10}$

19. $\dfrac{4a + 12}{2a - 10} \div \dfrac{a^2 - 9}{a^2 - a - 20}$

$\qquad = \dfrac{4(a+3)}{2(a-5)} \cdot \dfrac{(a-5)(a+4)}{(a-3)(a+3)}$

$\qquad = \dfrac{2(a+4)}{a-3}$

20. $\dfrac{6r - 18}{9r^2 + 6r - 24} \cdot \dfrac{12r - 16}{4r - 12}$

$\qquad = \dfrac{6(r-3)}{3(3r^2 + 2r - 8)} \cdot \dfrac{4(3r-4)}{4(r-3)}$

$\qquad = \dfrac{6(r-3)}{3(3r-4)(r+2)} \cdot \dfrac{4(3r-4)}{4(r-3)}$

$\qquad = \dfrac{6}{3(r+2)}$

$\qquad = \dfrac{2}{r+2}$

21. $\dfrac{k^2 - k - 6}{k^2 + k - 12} \cdot \dfrac{k^2 + 3k - 4}{k^2 + 2k - 3}$

$\quad = \dfrac{(k-3)(k+2)(k-1)(k+4)}{(k+4)(k-3)(k+3)(k-1)}$

$\quad = \dfrac{k+2}{k+3}$

22. $\dfrac{m^2 + 3m + 2}{m^2 + 5m + 4} \div \dfrac{m^2 + 5m + 6}{m^2 + 10m + 24}$

$\quad = \dfrac{m^2 + 3m + 2}{m^2 + 5m + 4} \cdot \dfrac{m^2 + 10m + 24}{m^2 + 5m + 6}$

$\quad = \dfrac{(m+1)(m+2)}{(m+4)(m+1)} \cdot \dfrac{(m+6)(m+4)}{(m+3)(m+2)}$

$\quad = \dfrac{m+6}{m+3}$

23. $\dfrac{2m^2 - 5m - 12}{m^2 - 10m + 24} \div \dfrac{4m^2 - 9}{m^2 - 9m + 18}$

$\quad = \dfrac{2m^2 - 5m - 12}{m^2 - 10m + 24} \cdot \dfrac{m^2 - 9m + 18}{4m^2 - 9}$

$\quad = \dfrac{(2m+3)(m-4)(m-6)(m-3)}{(m-6)(m-4)(2m-3)(2m+3)}$

$\quad = \dfrac{m-3}{2m-3}$

24. $\dfrac{6n^2 - 5n - 6}{6n^2 + 5n - 6} \cdot \dfrac{12n^2 - 17n + 6}{12n^2 - n - 6}$

$\quad = \dfrac{(2n-3)(3n+2)}{(2n+3)(3n-2)} \cdot \dfrac{(3n-2)(4n-3)}{(3n+2)(4n-3)}$

$\quad = \dfrac{2n-3}{2n+3}$

25. $\dfrac{a+1}{2} - \dfrac{a-1}{2}$

$\quad = \dfrac{(a+1)-(a-1)}{2}$

$\quad = \dfrac{a+1-a+1}{2}$

$\quad = \dfrac{2}{2} = 1$

26. $\dfrac{3}{p} + \dfrac{1}{2}$

Multiply the first term by $\frac{2}{2}$ and the second by $\frac{p}{p}$.

$\qquad \dfrac{2 \cdot 2}{2 \cdot p} + \dfrac{p \cdot 1}{p \cdot 2} = \dfrac{6}{2p} + \dfrac{p}{2p}$

$\qquad\qquad\qquad\qquad = \dfrac{6+p}{2p}$

27. $\dfrac{2}{y} - \dfrac{1}{4} = \left(\dfrac{4}{4}\right)\dfrac{2}{y} - \left(\dfrac{y}{y}\right)\dfrac{1}{4}$

$\qquad\quad = \dfrac{8-y}{4y}$

28. $\dfrac{1}{6m} + \dfrac{2}{5m} + \dfrac{4}{m}$

$\quad = \dfrac{5 \cdot 1}{5 \cdot 6m} + \dfrac{6 \cdot 2}{6 \cdot 5m} + \dfrac{30 \cdot 4}{30 \cdot m}$

$\quad = \dfrac{5}{30m} + \dfrac{12}{30m} + \dfrac{120}{30m}$

$\quad = \dfrac{5 + 12 + 120}{30m}$

$\quad = \dfrac{137}{30m}$

29. $\dfrac{1}{m-1} + \dfrac{2}{m}$

$\quad = \dfrac{m}{m}\left(\dfrac{1}{m-1}\right) + \dfrac{m-1}{m-1}\left(\dfrac{2}{m}\right)$

$\quad = \dfrac{m + 2m - 2}{m(m-1)}$

$\quad = \dfrac{3m - 2}{m(m-1)}$

30. $\dfrac{6}{r} - \dfrac{5}{r-2}$

$\quad = \dfrac{6(r-2)}{r(r-2)} - \dfrac{5r}{r(r-2)}$

$\quad = \dfrac{6(r-2) - 5r}{r(r-2)}$

$\quad = \dfrac{6r - 12 - 5r}{r(r-2)}$

$\quad = \dfrac{r - 12}{r(r-2)}$

31. $\dfrac{8}{3(a-1)} + \dfrac{2}{a-1}$

$\quad = \dfrac{8}{3(a-1)} + \dfrac{3}{3}\left(\dfrac{2}{a+1}\right)$

$\quad = \dfrac{8 + 6}{3(a-1)}$

$\quad = \dfrac{14}{3(a-1)}$

32. $\dfrac{2}{5(k-2)} + \dfrac{3}{4(k-2)}$

$= \dfrac{4\cdot 2}{4\cdot 5(k-2)} + \dfrac{5\cdot 3}{5\cdot 4(k-2)}$

$= \dfrac{8}{20(k-2)} + \dfrac{15}{20(k-2)}$

$= \dfrac{8+15}{20(k-2)}$

$= \dfrac{23}{20(k-2)}$

33. $\dfrac{2}{x^2-2x-3} + \dfrac{5}{x^2-x-6}$

$= \dfrac{2}{(x-3)(x+1)} + \dfrac{5}{(x-3)(x+2)}$

$= \left(\dfrac{x+2}{x+2}\right)\dfrac{2}{(x-3)(x+1)}$

$\quad + \left(\dfrac{x+1}{x+1}\right)\dfrac{5}{(x-3)(x+2)}$

$= \dfrac{2x+4+5x+5}{(x+2)(x-3)(x+1)}$

$= \dfrac{7x+9}{(x+2)(x-3)(x+1)}$

34. $\dfrac{2y}{y^2+7y+12} - \dfrac{y}{y^2+5y+6}$

$= \dfrac{2y}{(y+4)(y+3)} - \dfrac{y}{(y+3)(y+2)}$

$= \dfrac{2y(y+2)}{(y+4)(y+3)(y+2)}$

$\quad - \dfrac{y(y+4)}{(y+3)(y+2)(y+4)}$

$= \dfrac{2y(y+2) - y(y+4)}{(y+4)(y+3)(y+2)}$

$= \dfrac{2y^2+4y - y^2 - 4y}{(y+4)(y+3)(y+2)}$

$= \dfrac{y^2}{(y+4)(y+3)(y+2)}$

35. $\dfrac{3k}{2k^2+3k-2} - \dfrac{2k}{2k^2-7k+3}$

$= \dfrac{3k}{(2k-1)(k+2)} - \dfrac{2k}{(2k-1)(k-3)}$

$\left(\dfrac{k-3}{k-3}\right)\dfrac{3k}{(2k-1)(k+2)}$

$\quad - \left(\dfrac{k+2}{k+2}\right)\dfrac{2k}{(2k-1)(k-3)}$

$= \dfrac{(3k^2-9k) - (2k^2+4k)}{(2k-1)(k+2)(k-3)}$

$= \dfrac{k^2-13k}{(2k-1)(k+2)(k-3)}$

$= \dfrac{k(k-13)}{(2k-1)(k+2)(k-3)}$

36. $\dfrac{4m}{3m^2+7m-6} - \dfrac{m}{3m^2-14m+8}$

$= \dfrac{4m}{(3m-2)(m+3)} - \dfrac{m}{(3m-2)(m-4)}$

$= \dfrac{4m(m-4)}{(3m-2)(m+3)(m-4)}$

$\quad - \dfrac{m(m+3)}{(3m-2)(m-4)(m+3)}$

$= \dfrac{4m(m-4) - m(m+3)}{(3m-2)(m-4)(m+3)}$

$= \dfrac{4m^2-16m - m^2 - 3m}{(3m-2)(m+3)(m-4)}$

$= \dfrac{3m^2-19m}{(3m-2)(m+3)(m-4)}$

$= \dfrac{m(3m-19)}{(3m-2)(m+3)(m-4)}$

37. $\dfrac{2}{a+2} + \dfrac{1}{a} + \dfrac{a-1}{a^2+2a}$

$= \dfrac{2}{a+2} + \dfrac{1}{a} + \dfrac{a-1}{a(a+2)}$

$= \left(\dfrac{a}{a}\right)\dfrac{2}{a+2} + \left(\dfrac{a+2}{a+2}\right)\dfrac{1}{a} + \dfrac{a-1}{a(a+2)}$

$= \dfrac{2a+a+2+a-1}{a(a+2)}$

$= \dfrac{4a+1}{a(a+2)}$

38. $\dfrac{5x+2}{x^2-1} + \dfrac{3}{x^2+x} - \dfrac{1}{x^2-x}$

$= \dfrac{5x+2}{(x+1)(x-1)} + \dfrac{3}{x(x+1)} - \dfrac{1}{x(x-1)}$

$= \left(\dfrac{x}{x}\right)\left(\dfrac{5x+2}{(x+1)(x-1)}\right) + \left(\dfrac{x-1}{x-1}\right)\left(\dfrac{3}{x(x+1)}\right)$

$\quad - \left(\dfrac{x+1}{x+1}\right)\left(\dfrac{1}{x(x-1)}\right)$

$= \dfrac{x(5x+2) + (x-1)(3) - (x+1)(1)}{x(x+1)(x-1)}$

$= \dfrac{5x^2+2x+3x-3 - x - 1}{x(x+1)(x-1)}$

$= \dfrac{5x^2+4x-4}{x(x+1)(x-1)}$

R.4 Equations

1. $.2m - .5 = .1m + .7$

$10(.2m - .5) = 10(.1m + .7)$

$2m - 5 = m + 7$

$m - 5 = 7$

$m = 12$

The solution is 12.

2. $\dfrac{5}{6}k - 2k + \dfrac{1}{3} = \dfrac{2}{3}$

Multiply both sides of the equation by 6.

$$6\left(\frac{5}{6}k\right) - 6(2k) + 6\left(\frac{1}{3}\right) = 6\left(\frac{2}{3}\right)$$

$$5k - 12k + 2 = 4$$

$$-7k + 2 = 4$$

$$-7k = 2$$

$$k = -\frac{2}{7}$$

The solution is $-\frac{2}{7}$.

3. $3r + 2 - 5(r + 1) = 6r + 4$

$3r + 2 - 5r - 5 = 6r + 4$

$-3 - 2r = 6r + 4$

$-3 = 8r + 4$

$-7 = 8r$

$-\dfrac{7}{8} = r$

The solution is $-\frac{7}{8}$.

4. $2[m - (4 + 2m) + 3] = 2m + 2$

$2[m - 4 - 2m + 3] = 2m + 2$

$2[-m - 1] = 2m + 2$

$-2m - 2 = 2m + 2$

$-2m = 2m + 4$

$-4m = 4$

$m = -1$

The solution is -1.

5. $|3x + 2| = 9$

$3x + 2 = 9$ or $-(3x + 2) = 9$

$3x = 7$ $-3x - 2 = 9$

$-3x = 11$

$x = \dfrac{7}{3}$ or $x = -\dfrac{11}{3}$

Substituting in the original equation shows that both $\frac{7}{3}$ and $-\frac{11}{3}$ are solutions.

6. $|4 - 7x| = 15$

$4 - 7x = 15$ or $-(4 - 7x) = 15$

$-7x = 11$ $-4 + 7x = 15$

$7x = 19$

$x = -\dfrac{11}{7}$ or $x = \dfrac{19}{7}$

The solutions are $-\frac{11}{7}$ and $\frac{19}{7}$.

7. $|2x + 8| = |x - 4|$

$2x + 8 = x - 4$ or $-(2x + 8) = x - 4$

$-2x - 8 = x - 4$

$-3x = 4$

$x = -12$ or $x = -\dfrac{4}{3}$

The solutions are -12 and $-\frac{4}{3}$.

8. $|5x + 2| = |8 - 3x|$

Rewrite the equation as follows.

$5x + 2 = 8 - 3x$ or $-(5x + 2) = 8 - 3x$

$5x = 6 - 3x$ $-5x - 2 = 8 - 3x$

$8x = 6$ $-5x = 10 - 3x$

$x = \dfrac{6}{8}$ $-2x = 10$

$x = \dfrac{3}{4}$ or $x = -5$

The solutions are $\frac{3}{4}$ and -5.

9. $x^2 + 5x + 6 = 0$

$(x + 3)(x + 2) = 0$

$x + 3 = 0$ or $x + 2 = 0$

$x = -3$ or $x = -2$

The solutions are -3 and -2.

10. $x^2 = 3 + 2x$

$x^2 - 2x - 3 = 0$

$(x - 3)(x + 1) = 0$

$x - 3 = 0$ or $x + 1 = 0$

$x = 3$ or $x = -1$

The solutions are 3 and -1.

11. $m^2 + 16 = 8m$

$m^2 - 8m + 16 = 0$

$(m)^2 - 2(4m) + (4)^2 = 0$

$(m - 4)^2 = 0$

$m - 4 = 0$

$m = 4$

The solution is 4.

12.
$$2k^2 - k = 10$$
$$2k^2 - k - 10 = 0$$
$$(2k - 5)(k + 2) = 0$$
$$2k - 5 = 0 \quad \text{or} \quad k + 2 = 0$$
$$k = \frac{5}{2} \quad \text{or} \quad k = -2$$

The solutions are $\frac{5}{2}$ and -2.

13.
$$6x^2 - 5x = 4$$
$$6x^2 - 5x - 4 = 0$$
$$(3x - 4)(2x + 1) = 0$$
$$3x - 4 = 0 \quad \text{or} \quad 2x + 1 = 0$$
$$3x = 4 \qquad\qquad 2x = -1$$
$$x = \frac{4}{3} \quad \text{or} \quad x = -\frac{1}{2}$$

The solutions are $\frac{4}{3}$ and $-\frac{1}{2}$.

14.
$$m(m - 7) = -10$$
$$m^2 - 7m + 10 = 0$$
$$(m - 5)(m - 2) = 0$$
$$m - 5 = 0 \quad \text{or} \quad m - 2 = 0$$
$$m = 5 \quad \text{or} \quad m = 2$$

The solutions are 5 and 2.

15.
$$9x^2 - 16 = 0$$
$$(3x)^2 - (4)^2 = 0$$
$$(3x + 4)(3x - 4) = 0$$
$$3x + 4 = 0 \quad \text{or} \quad 3x - 4 = 0$$
$$3x = -4 \qquad\qquad 3x = 4$$
$$x = -\frac{4}{3} \quad \text{or} \quad x = \frac{4}{3}$$

The solutions are $-\frac{4}{3}$ and $\frac{4}{3}$.

16.
$$z(2z + 7) = 4$$
$$2z^2 + 7z - 4 = 0$$
$$(2z - 1)(z + 4) = 0$$
$$2z - 1 = 0 \quad \text{or} \quad z + 4 = 0$$
$$z = \frac{1}{2} \quad \text{or} \quad z = -4$$

The solutions are $\frac{1}{2}$ and -4.

17.
$$12y^2 - 48y = 0$$
$$12y(y) - 12y(4) = 0$$
$$12y(y - 4) = 0$$
$$12y = 0 \quad \text{or} \quad y - 4 = 0$$
$$y = 0 \quad \text{or} \quad y = 4$$

The solutions are 0 and 4.

18. $3x^2 - 5x + 1 = 0$

Use the quadratic formula.
$$x = \frac{-(-5) \pm \sqrt{(-5)^2 - 4(3)(1)}}{2(3)}$$
$$= \frac{5 \pm \sqrt{25 - 12}}{6}$$
$$x = \frac{5 + \sqrt{13}}{6} \quad \text{or} \quad x = \frac{5 - \sqrt{13}}{6}$$
$$\approx 1.434 \qquad\qquad \approx .232$$

The solutions are $\frac{5+\sqrt{13}}{6} \approx 1.434$ and $\frac{5-\sqrt{13}}{6} \approx .232$.

19.
$$2m^2 = m + 4$$
$$2m^2 - m - 4 = 0$$

Use the quadratic formula.
$$x = \frac{-(-1) \pm \sqrt{(-1)^2 - 4(2)(-4)}}{2(2)}$$
$$x = \frac{1 \pm \sqrt{1 + 32}}{4}$$
$$x = \frac{1 \pm \sqrt{33}}{4}$$

The solutions are $\frac{1+\sqrt{33}}{4} \approx 1.686$ and $\frac{1-\sqrt{33}}{4} \approx -1.186$.

20. $p^2 + p - 1 = 0$
$$p = \frac{-1 \pm \sqrt{1^2 - 4(1)(-1)}}{2(1)}$$
$$= \frac{-1 \pm \sqrt{5}}{2}$$

The solutions are $\frac{-1+\sqrt{5}}{2} \approx .618$ and $\frac{-1-\sqrt{5}}{2} \approx -1.618$.

21.
$$k^2 - 10k - 20$$
$$k^2 - 10k + 20 = 0$$
$$k = \frac{-(-10) \pm \sqrt{(-10)^2 - 4(1)(20)}}{2(1)}$$
$$k = \frac{10 \pm \sqrt{100 - 80}}{2}$$
$$k = \frac{10 \pm \sqrt{20}}{2}$$
$$k = \frac{10 \pm \sqrt{4}\sqrt{5}}{2}$$

$$k = \frac{10 \pm 2\sqrt{5}}{2}$$

$$k = \frac{2(5 \pm 2\sqrt{5})}{2}$$

$$k = 5 \pm \sqrt{5}$$

The solutions are $5 + \sqrt{5} \approx 7.236$ and $5 - \sqrt{5} \approx 2.764$.

22. $2x^2 + 12x + 5 = 0$

$$x = \frac{-12 \pm \sqrt{(12)^2 - 4(2)(5)}}{2(2)}$$

$$= \frac{-12 \pm \sqrt{104}}{4} = \frac{-12 \pm \sqrt{4 \cdot 26}}{4}$$

$$= \frac{-12 \pm \sqrt{4}\sqrt{26}}{4} = \frac{-12 \pm 2\sqrt{26}}{4}$$

$$= \frac{2(-6 \pm \sqrt{26})}{2 \cdot 2} = \frac{-6 \pm \sqrt{26}}{2}$$

The solutions are $\frac{-6+\sqrt{26}}{2} \approx -.450$ and $\frac{-6-\sqrt{26}}{2} \approx -5.550$.

23. $2r^2 - 7r + 5 = 0$

$$(2r - 5)(r - 1) = 0$$

$$2r - 5 = 0 \quad \text{or} \quad r - 1 = 0$$

$$2r = 5$$

$$r = \frac{5}{2} \quad \text{or} \quad r = 1$$

The solutions are $\frac{5}{2}$ and 1.

24. $2x^2 - 7x + 30 = 0$

$$x = \frac{-(-7) \pm \sqrt{(-7)^2 - 4(2)(30)}}{2(2)}$$

$$x = \frac{7 \pm \sqrt{49 - 240}}{4}$$

$$x = \frac{7 \pm \sqrt{-191}}{4}$$

Since there is a negative number under the radical sign, $\sqrt{-191}$ is not a real number. Thus, there are no real-number solutions.

25. $3k^2 + k = 6$

$$3k^2 + k - 6 = 0$$

$$k = \frac{-1 \pm \sqrt{1 - 4(3)(-6)}}{2(3)}$$

$$= \frac{-1 \pm \sqrt{73}}{6}$$

The solutions are $\frac{-1+\sqrt{73}}{6} \approx 1.257$ and $\frac{-1-\sqrt{73}}{6} \approx -1.591$.

26. $5m^2 + 5m = 0$

$$5m(m + 1) = 0$$

$$5m = 0 \quad \text{or} \quad m + 1 = 0$$

$$m = 0 \quad \text{or} \quad m = -1$$

The solutions are 0 and -1.

27. $\dfrac{3x - 2}{7} = \dfrac{x + 2}{5}$

$$35\left(\frac{3x - 2}{7}\right) = 35\left(\frac{x + 2}{5}\right)$$

$$5(3x - 2) = 7(x + 2)$$

$$15x - 10 = 7x + 14$$

$$8x = 24$$

$$x = 3$$

28. $\dfrac{x}{3} - 7 = 6 - \dfrac{3x}{4}$

Multiply both sides by 12, the least common denominator of 3 and 4.

$$12\left(\frac{x}{3} - 7\right) = 12\left(6 - \frac{3x}{4}\right)$$

$$12\left(\frac{x}{3}\right) - (12)(7) = (12)(6) - (12)\left(\frac{3x}{4}\right)$$

$$4x - 84 = 72 - 9x$$

$$13x - 84 = 72$$

$$13x = 156$$

$$x = 12$$

The solution is 12.

29. $\dfrac{4}{x - 3} - \dfrac{8}{2x + 5} + \dfrac{3}{x - 3} = 0$

$$\frac{4}{x - 3} + \frac{3}{x - 3} - \frac{8}{2x + 5} = 0$$

$$\frac{7}{x - 3} - \frac{8}{2x + 5} = 0$$

Multiply both sides by $(x - 3)(2x + 5)$. Note that $x \neq 3$ and $x \neq -\frac{5}{2}$.

$$(x-3)(2x+5)\left(\frac{7}{x-3} - \frac{8}{2x+5}\right) = (x-3)(2x+5)(0)$$

$$7(2x + 5) - 8(x - 3) = 0$$

$$14x + 35 - 8x + 24 = 0$$

$$6x + 59 = 0$$

$$6x = -59$$

$$x = -\frac{59}{6}$$

Note: It is especially important to check solutions of equations that involve rational expressions. Here, a check shows that $-\frac{59}{6}$ is a solution.

30. $\dfrac{5}{2p+3} - \dfrac{3}{p-2} = \dfrac{4}{2p+3}$

Multiply both sides by $(2p+3)(p-2)$. Note that $p \neq -\frac{3}{2}$ and $p \neq 2$.

$$(2p+3)(p-2)\left(\dfrac{5}{2p+3} - \dfrac{3}{p-2}\right)$$

$$= (2p+3)(p-2)\left(\dfrac{4}{2p+3}\right)$$

$$(2p+3)(p-2)\left(\dfrac{5}{2p+3}\right) - (2p+3)(p-2)\left(\dfrac{3}{p-2}\right)$$

$$= (2p+3)(p-2)\left(\dfrac{4}{2p+3}\right)$$

$$(p-2)(5) - (2p+3)(3) = (p-2)(4)$$
$$5p - 10 - 6p - 9 = 4p - 8$$
$$-p - 19 = 4p - 8$$
$$-5p - 19 = -8$$
$$-5p = 11$$
$$p = -\dfrac{11}{5}$$

The solutions is $-\frac{11}{5}$.

31. $\dfrac{2}{m} + \dfrac{m}{m+3} = \dfrac{3m}{m^2+3m}$

$$\dfrac{2}{m} + \dfrac{m}{m+3} = \dfrac{3m}{m(m+3)}$$

Multiply both sides by $m(m+3)$.
Note that $m \neq 0$, and $m \neq -3$.

$$m(m+3)\left(\dfrac{2}{m} + \dfrac{m}{m+3}\right) = m(m+3)\left(\dfrac{3m}{m(m+3)}\right)$$

$$2(m+3) + m(m) = 3m$$
$$2m + 6 + m^2 = 3m$$
$$m^2 - m + 6 = 0$$

$$m = \dfrac{-(-1) \pm \sqrt{(-1)^2 - 4(1)(6)}}{2(1)}$$

$$= \dfrac{1 \pm \sqrt{1-24}}{2}$$

$$= \dfrac{1 \pm \sqrt{-23}}{2}$$

There are no real number solutions.

32. $\dfrac{2y}{y-1} = \dfrac{5}{y} + \dfrac{10-8y}{y^2-y}$

$$\dfrac{2y}{y-1} = \dfrac{5}{y} + \dfrac{10-8y}{y(y-1)}$$

Multiply both sides by $y(y-1)$.
Note that $y \neq 0$ and $y \neq 1$.

$$y(y-1)\left(\dfrac{2y}{y-1}\right) = y(y-1)\left[\dfrac{5}{y} + \dfrac{10-8y}{y(y-1)}\right]$$

$$y(y-1)\left(\dfrac{2y}{y-1}\right) = y(y-1)\left(\dfrac{5}{y}\right)$$
$$+ y(y-1)\left[\dfrac{10-8y}{y(y-1)}\right]$$

$$y(2y) = (y-1)(5) + (10-8y)$$
$$2y^2 = 5y - 5 + 10 - 8y$$
$$2y^2 = 5 - 3y$$
$$2y^2 + 3y - 5 = 0$$
$$(2y+5)(y-1) = 0$$
$$2y + 5 = 0 \quad \text{or} \quad y - 1 = 0$$
$$y = -\dfrac{5}{2} \quad \text{or} \quad y = 1$$

Since $y \neq 1$, 1 is not a solution.
The solution is $-\frac{5}{2}$.

33. $\dfrac{1}{x-2} - \dfrac{3x}{x-1} = \dfrac{2x+1}{x^2-3x+2}$

$$\dfrac{1}{x-2} - \dfrac{3x}{x-1} = \dfrac{2x+1}{(x-2)(x-1)}$$

Multiply both sides by $(x-2)(x-1)$.
Note that $x \neq 2$ and $x \neq 1$.

$$(x-2)(x-1)\left(\dfrac{1}{x-2} - \dfrac{3x}{x-1}\right) = (x-2)(x-1)$$
$$\cdot \left[\dfrac{2x+1}{(x-2)(x-1)}\right]$$

$$(x-2)(x-1)\left(\dfrac{1}{x-2}\right)$$
$$- (x-2)(x-1)\left(\dfrac{3x}{x-1}\right) = \dfrac{(x-2)(x-2)(2x+1)}{(x-2)(x-1)}$$

$$(x-1) - (x-2)(3x) = 2x+1$$
$$x - 1 - 3x^2 + 6x = 2x + 1$$
$$-3x^2 + 7x - 1 = 2x + 1$$
$$-3x^2 + 5x - 2 = 0$$
$$3x^2 - 5x + 2 = 0$$
$$(3x-2)(x-1) = 0$$
$$3x - 2 = 0 \quad \text{or} \quad x - 1 = 0$$
$$x = \dfrac{2}{3} \quad \text{or} \quad x = 1$$

1 is not a solution since $x \neq 1$.
The solution is $\frac{2}{3}$.

34. $\dfrac{5}{a} + \dfrac{-7}{a+1} = \dfrac{a^2-2a+4}{a^2+a}$

$$a(a+1)\left(\dfrac{5}{a} + \dfrac{-7}{a+1}\right) = a(a+1)\left(\dfrac{a^2-2a+4}{a^2+a}\right)$$

Note that $a \neq 0$ and $a \neq -1$.

$$5(a+1) + (-7)(a) = a^2 - 2a + 4$$
$$5a + 5 - 7a = a^2 - 2a + 4$$
$$5 - 2a = a^2 - 2a + 4$$
$$5 = a^2 + 4$$
$$0 = a^2 - 1$$
$$0 = (a+1)(a-1)$$
$$a + 1 = 0 \quad \text{or} \quad a - 1 = 0$$
$$a = -1 \quad \text{or} \quad a = 1$$

Since -1 would make two denominators zero, 1 is the only solution.

35. $\dfrac{2b^2 + 5b - 8}{b^2 + 2b} + \dfrac{5}{b+2} = -\dfrac{3}{b}$

$$\frac{2b^2 + 5b - 8}{b(b+2)} + \frac{5}{b+2} = \frac{-3}{b}$$

Multiply both sides by $b(b+2)$.
Note that $b \neq 0$ and $b \neq -2$.

$$b(b+2)\left(\frac{2b^2 + 5b - 8}{b^2 + 2b}\right)$$
$$+ b(b+2)\left(\frac{5}{b+2}\right) = b(b+2)\left(-\frac{3}{b}\right)$$
$$2b^2 + 5b - 8 + 5b = (b+2)(-3)$$
$$2b^2 + 10b - 8 = -3b - 6$$
$$2b^2 + 13b - 2 = 0$$

$$b = \frac{-(-13) \pm \sqrt{(13)^2 - 4(2)(-2)}}{2(2)}$$

$$= \frac{-13 \pm \sqrt{169 + 16}}{4}$$

$$b = \frac{-13 \pm \sqrt{185}}{4}$$

The solutions are $\frac{-13 + \sqrt{185}}{4} \approx .150$ and

$\frac{-13 - \sqrt{185}}{4} \approx -6.650$.

36. $\dfrac{2}{x^2 - 2x - 3} + \dfrac{5}{x^2 - x - 6} = \dfrac{1}{x^2 + 3x + 2}$

$$\frac{2}{(x-3)(x+1)} + \frac{5}{(x-3)(x+2)} = \frac{1}{(x+2)(x+1)}$$

Multiply both sides by $(x-3)(x+1)(x+2)$.
Note that $x \neq 3$, $x \neq -1$, and $x \neq -2$.

$$(x-3)(x+1)(x+2)\left(\frac{2}{(x-3)(x+1)}\right)$$
$$+ (x-3)(x+1)(x+2)\left(\frac{5}{(x-3)(x+2)}\right)$$
$$= (x-3)(x+1)(x+2)\left(\frac{1}{(x+2)(x+1)}\right)$$

$$2(x+2) + 5(x+1) = x - 3$$
$$2x + 4 + 5x + 5 = x - 3$$
$$7x + 9 = x - 3$$
$$6x + 9 = -3$$
$$6x = -12$$
$$x = -2$$

However, $x \neq -2$. Therefore there is no solution.

37. $\dfrac{2}{y^2 + 7y + 12} - \dfrac{1}{y^2 + 5y + 6} = \dfrac{5}{y^2 + 6y + 8}$

$$\frac{2}{(y+4)(y+3)} - \frac{1}{(y+3)(y+2)} = \frac{5}{(y+4)(y+2)}$$

Multiply both sides by $(y+4)(y+3)(y+2)$.
Note that $y \neq -4$, $y \neq -3$, and $y \neq -2$.

$$(y+4)(y+3)(y+2)\left(\frac{2}{(y+4)(y+3)}\right)$$
$$- (y+4)(y+3)(y+2)\left(\frac{1}{(y+3)(y+2)}\right)$$
$$= (y+4)(y+3)(y+2)\left(\frac{5}{(y+4)(y+2)}\right)$$

$$2(y+2) - (y+4) = 5(y+3)$$
$$2y + 4 - y - 4 = 5y + 15$$
$$y = 5y + 15$$
$$-4y = 15$$
$$y = -\frac{15}{4}$$

The solution is $-\frac{15}{4}$.

R.5 Inequalities

For Exercises 1-6, see the answer graphs in the back of the textbook.

1. $x < 0$

Because the inequality symbol means "less than," the endpoint at 0 is not included. This inequality is written in interval notation as $(-\infty, 0]$. To graph this interval on a number line, place a right bracket at 0 and draw a heavy arrow pointing to the left.

2. $x \geq -3$

Because the inequality sign means "greater than or equal to," the endpoint at -3 is included. This inequality is written in interval notation as $[-3, \infty)$.

To graph this interval on a number line, place a left bracket at -3 and draw a heavy arrow pointing to the right.

3. $-1 \leq x < 2$

The endpoint at -1 is included, but the endpoint at 2 is not. This inequality is written in interval notation as $[-1, 2)$. To graph this interval, place a left bracket at -1 and a right parenthesis at 2; then draw a heavy line segment between them.

4. $-5 < x \leq -4$

The endpoint at -4 is included, but the endpoint at -5 is not. This inequality is written in interval notation as $(-5, 4]$. To graph this interval, place a left parenthesis at -5 and a right bracket at 4; then draw a heavy line segment between them.

5. $-9 > x$

This inequality may be rewritten as $x < -9$, and is written in interval notation as $(-\infty, -9)$. Note that the endpoint at -9 is not included. To graph this interval, place a right parenthesis at -9 and draw a heavy arrow pointing to the left.

6. $6 \leq x$

This inequality may be written as $x \geq 6$, and is written in interval notation as $[6, \infty)$. Note that the endpoint at 6 is included. To graph this interval, place a left bracket at 6 and draw a heavy arrow pointing to the right.

7. $(-4, 3)$

This represents all the numbers between -4 and 3, not including the endpoints. This interval can be written as the inequality $-4 < x < 3$.

8. $[2, 7)$

This represents all the numbers between 2 and 7, including 2 but not including 7. This interval can be written as the inequality $2 \leq x < 7$.

9. $(-\infty, -1]$

This represents all the numbers to the left of -1 on the number line and includes the endpoint. This interval can be written as the inequality $x \leq -1$.

10. $(3, \infty)$

This represents all the numbers to the right of 3, and does not include the endpoint. This interval can be written as the inequality $x > 3$.

11. Notice that the endpoint -2 is included, but 6 is not. The interval show in the graph can be written as the inequality $-2 \leq x < 6$.

12. Notice that neither endpoint is included. The interval shown in the graph can be written as the inequality $0 < x < 8$.

13. Notice that both endpoints are included. The interval shown in the graph can be written as $x \leq -4$ or $x \geq 4$.

14. Notice that the endpoint 0 is not included, but 3 is included. The interval shown in the graph can be written as $x < 0$ or $x \geq 3$.

For Exercises 15-38, see the answer graphs in the back of the textbook.

15.
$$-3p - 2 \geq 1$$
$$-3p \geq 3$$
$$\left(-\frac{1}{3}\right)(-3p) \leq \left(-\frac{1}{3}\right)(3)$$
$$p \leq -1$$

The solution in interval notation is $(-\infty, -1]$.

16.
$$6k - 4 < 3k - 1$$
$$6k < 3k + 3$$
$$3k < 3$$
$$k < 1$$

The solution in interval notation is $(-\infty, 1)$.

17.
$$m - (4 + 2m) + 3 < 2m + 2$$
$$m - 4 - 2m + 3 < 2m + 2$$
$$-m - 1 < 2m + 2$$
$$-3m - 1 < 2$$
$$-3m < 3$$
$$-\frac{1}{3}(-3m) > -\frac{1}{3}(3)$$
$$m > -1$$

The solution is $(-1, \infty)$.

18.
$$-2(3y - 8) \geq 5(4y - 2)$$
$$-6y + 16 \geq 20y - 10$$
$$-6y + 16 + (-16) \geq 20y - 10 + (-16)$$
$$-6y \geq 20y - 26$$
$$-6y + (-20y) \geq 20y - 26$$
$$-26y \geq -26$$
$$-\frac{1}{26}(-26)y \leq -\frac{1}{26}(-26)$$
$$y \leq 1$$

The solution is $(-\infty, 1]$.

19. $3p - 1 < 6p + 2(p - 1)$
$3p - 1 < 6p + 2p - 2$
$3p - 1 < 8p - 2$
$-5p - 1 < -2$
$-5p < -1$
$-\frac{1}{5}(-5p) > -\frac{1}{5}(-1)$
$p > \frac{1}{5}$

The solution is $\left(\frac{1}{5}, \infty\right)$.

20. $x + 5(x + 1) > 4(2 - x) + x$
$x + 5x + 5 > 8 - 4x + x$
$6x + 5 > 8 - 3x$
$6x > 3 - 3x$
$9x > 3$
$x > \frac{1}{3}$

The solution is $\left(\frac{1}{3}, \infty\right)$.

21. $-7 < y - 2 < 4$
$-7 + 2 < y - 2 + 2 < 4 + 2$
$-5 < y < 6$

The solution is $(-5, 6)$.

22. $8 \le 3r + 1 \le 13$
$8 + (-1) \le 3r + 1 + (-1) \le 13 + (-1)$
$7 \le 3r \le 12$
$\frac{1}{3}(7) \le \frac{1}{3}(3r) \le \frac{1}{3}(12)$
$\frac{7}{3} \le r \le 4$

The solution is $\left[\frac{7}{3}, 4\right]$.

23. $-4 \le \dfrac{2k - 1}{3} \le 2$
$3(-4) \le 3\left(\dfrac{2k - 1}{3}\right) \le 3(2)$
$-12 \le 2k - 1 \le 6$
$-11 \le 2k \le 7$
$-\frac{11}{2} \le k \le \frac{7}{2}$

The solution is $\left[-\frac{11}{2}, \frac{7}{2}\right]$.

24. $-1 \le \dfrac{5y + 2}{3} \le 4$
$3(-1) \le 3\left(\dfrac{5y + 2}{3}\right) \le 3(4)$
$-3 \le 5y + 2 \le 12$
$-5 \le 5y \le 10$
$-1 \le y \le 2$

The solution is $[-1, 2]$.

25. $\frac{3}{5}(2p + 3) \ge \frac{1}{10}(5p + 1)$
$10\left(\frac{3}{5}\right)(2p + 3) \ge 10\left(\frac{1}{10}\right)(5p + 1)$
$6(2p + 3) \ge 5p + 1$
$12p + 18 \ge 5p + 1$
$7p \ge -17$
$p \ge -\frac{17}{7}$

The solution is $\left[-\frac{17}{7}, \infty\right)$.

26. $\frac{8}{3}(z - 4) \le \frac{2}{9}(3z + 2)$
$(9)\frac{8}{3}(z - 4) \le (9)\frac{2}{9}(3z + 2)$
$24(z - 4) \le 2(3z + 2)$
$24z - 96 \le 6z + 4$
$24z \le 6z + 100$
$18z \le 100$
$z \le \frac{100}{18}$
$z \le \frac{50}{9}$

The solution is $\left(-\infty, \frac{50}{9}\right]$.

27. $(m + 2)(m - 4) < 0$

Solve $(m + 2)(m - 4) = 0$.

$$m = -2 \quad \text{or} \quad m = 4$$

Intervals: $(-\infty, -2), (-2, 4), (4, \infty)$

For $(-\infty, -2)$, choose -3 to test for m.

$$(-3 + 2)(-3 - 4) = -1(-7) = 8 \not< 0$$

For $(-2, 4)$, choose 0.

$$(0 + 2)(0 - 4) = 2(-4) = -8 < 0$$

For $(4, \infty)$, choose 5.

$$(5 + 2)(5 - 4) = 7(1) = 7 \not< 0$$

The solution is $(-2, 4)$.

28. $(t + 6)(t - 1) \ge 0$

Solve $(t + 6)(t - 1) = 0$.

$$(t + 6)(t - 1) = 0$$
$$t = -6 \quad \text{or} \quad t = 1$$

Intervals: $(-\infty, -6)$, $(-6, 1)$, $(1, \infty)$

For $(-\infty, -6)$, choose -7 to test for t.

$$(-7+6)(-7-1) = (-1)(-8) = 8 \geq 0$$

For $(-6, 1)$, choose 0.

$$(0+6)(0-1) = (6)(-1) = -6 \not\geq 0$$

For $(1, \infty)$, choose 2.

$$(2+6)(2-1) = (8)(1) = 8 \geq 0$$

Because the symbol $\geq$ is used, the endpoints -6 and 1 are included in the solution, $(-\infty, -6] \cup [1, \infty)$.

29. $y^2 - 3y + 2 < 0$
$(y-2)(y-1) < 0$

Solve $(y-2)(y-1) = 0$.

$$y = 2 \quad \text{or} \quad y = 1$$

Intervals: $(-\infty, 1)$, $(1, 2)$, $(2, \infty)$

For $(-\infty, 1)$, choose $y = 0$.

$$0^2 - 3(0) + 2 = 2 \not< 0$$

For $(1, 2)$, choose $y = \frac{3}{2}$.

$$\left(\frac{3}{2}\right)^2 - 3\left(\frac{3}{2}\right) + 2 = \frac{9}{4} - \frac{9}{2} + 2$$
$$= \frac{9 - 18 + 8}{4}$$
$$= -\frac{1}{4} < 0$$

For $(2, \infty)$, choose 3.

$$3^2 - 3(3) + 2 = 2 \not< 0$$

The solution is $(1, 2)$.

30. $2k^2 + 7k - 4 > 0$

Solve $2k^2 + 7k - 4 = 0$.

$$2k^2 + 7k - 4 = 0$$
$$(2k-1)(k+4) = 0$$
$$k = \frac{1}{2} \quad \text{or} \quad k = -4$$

Intervals: $(-\infty, -4)$, $\left(-4, \frac{1}{2}\right)$ $\left(\frac{1}{2}, \infty\right)$

For $(-\infty, -4)$, choose -5.

$$2(-5)^2 + 7(-5) - 4 = 11 > 0$$

For $\left(-4, \frac{1}{2}\right)$, choose 0.

$$2(0)^2 + 7(0) - 4 = -4 \not> 0$$

For $\left(\frac{1}{2}, \infty\right)$, choose 1.

$$2(1)^2 + 7(1) - 4 = 5 > 0$$

The solution is $(-\infty, -4) \cup \left(\frac{1}{2}, \infty\right)$.

31. $q^2 - 7q + 6 \leq 0$

Solve $q^2 - 7q + 6 = 0$.

$$(q-1)(q-6) = 0$$
$$q = 1 \quad \text{or} \quad q = 6$$

These solutions are also solutions of the given inequality because the symbol $\leq$ indicates that the endpoints are included.

Intervals $(-\infty, 1)$, $(1, 6)$, $(6, \infty)$

For $(-\infty, 1)$, choose 0.

$$0^2 - 7(0) + 6 = 6 \not\leq 0$$

For $(1, 6)$, choose 2.

$$2^2 - 7(2) + 6 = -4 \leq 0$$

For $(6, \infty)$, choose 7.

$$7^2 - 7(7) + 6 = 6 \not\leq 0$$

The solution is $[1, 6]$.

32. $2k^2 - 7k - 15 \leq 0$

Solve $2k^2 - 7k - 15 = 0$.

$$2k^2 - 7k - 15 = 0$$
$$(2k+3)(k-5) = 0$$
$$k = -\frac{3}{2} \quad \text{or} \quad k = 5$$

Intervals: $\left(-\infty, -\frac{3}{2}\right)$, $\left(-\frac{3}{2}, 5\right)$, $(5, \infty)$

For $\left(-\infty, -\frac{3}{2}\right)$, choose -2.

$$2(-2)^2 - 7(-2) - 15 = 7 \not\leq 0$$

For $\left(-\frac{3}{2}, 5\right)$, choose 0.

$$2(0)^2 - 7(0) - 15 = -15 \leq 0$$

For $(5, \infty)$, choose 6.

$$2(6)^2 - 7(6) - 15 \not\leq 0$$

The solution is $\left[-\frac{3}{2}, 5\right]$.

33. $6m^2 + m > 1$

Solve $6m^2 + m = 1$.

$$6m^2 + m = 0$$
$$(2m + 1)(3m - 1) = 0$$
$$m = -\frac{1}{2} \quad \text{or} \quad m = \frac{1}{3}$$

Intervals: $\left(-\infty, -\frac{1}{2}\right), \left(-\frac{1}{2}, \frac{1}{3}\right), \left(\frac{1}{3}, \infty\right)$

For $\left(-\infty, -\frac{1}{2}\right)$, choose -1.

$$6(-1)^2 + (-1) = 5 > 1$$

For $\left(-\frac{1}{2}, \frac{1}{3}\right)$, choose 0.

$$6(0)^2 + 0 = 0 \not> 1$$

For $\left(\frac{1}{3}, \infty\right)$ choose 1.

$$6(1)^2 + 1 = 7 > 1$$

The solution is $\left(-\infty, -\frac{1}{2}\right) \cup \left(\frac{1}{3}, \infty\right)$.

34. $10r^2 + r \le 2$

Solve $10r^2 + r = 2$.

$$10r^2 + r = 2$$
$$10r^2 + r - 2 = 0$$
$$(5r - 2)(2r + 1) = 0$$
$$r = \frac{2}{5} \quad \text{or} \quad r = -\frac{1}{2}$$

Intervals: $\left(-\infty, -\frac{1}{2}\right), \left(-\frac{1}{2}, \frac{2}{5}\right), \left(\frac{2}{5}, \infty\right)$

For $\left(-\infty, -\frac{1}{2}\right)$, choose -1.

$$10(-1)^2 + (-1) = 9 \not\le 2$$

For $\left(-\frac{1}{2}, \frac{2}{5}\right)$, choose 0.

$$10(0)^2 + 0 = 0 \le 2$$

For $\left(\frac{2}{5}, \infty\right)$, choose 1.

$$10(1)^2 + 1 = 11 \not\le 2$$

The solution is $\left[-\frac{1}{2}, \frac{2}{5}\right]$.

35. $2y^2 + 5y \le 3$

Solve $2y^2 + 5y = 3$.

$$2y^2 + 5y - 3 = 0$$
$$(y + 3)(2y - 1) = 0$$
$$y = -3 \quad \text{or} \quad y = \frac{1}{2}$$

Intervals: $\left(-\infty, -3\right), \left(-3, \frac{1}{2}\right), \left(\frac{1}{2}, \infty\right)$

For $\left(-\infty, -3\right)$, choose -4.

$$2(-4)^2 + 5(-4) = 12 \not\le 3$$

For $\left(-3, \frac{1}{2}\right)$, choose 0.

$$2(0)^2 + 5(0) = 0 \le 3$$

For $\left(\frac{1}{2}, \infty\right)$, choose 1.

$$2(1)^2 + 5(1) = 7 \not\le 3$$

The solution is $\left[-3, \frac{1}{2}\right]$.

36. $3a^2 + a > 10$

Solve $3a^2 + a = 10$.

$$3a^2 + a = 10$$
$$3a^2 + a - 10 = 0$$
$$(3a - 5)(a + 2) = 0$$
$$a = \frac{5}{3} \quad \text{or} \quad a = -2$$

Intervals: $\left(-\infty, -2\right), \left(-2, \frac{5}{3}\right), \left(\frac{5}{3}, \infty\right)$

For $\left(-\infty, -2\right)$, choose -3.

$$3(-3)^2 + (-3) = 24 > 10$$

For $\left(-2, \frac{5}{3}\right)$, choose 0.

$$3(0)^2 + 0 = 0 \not> 10$$

For $\left(\frac{5}{3}, \infty\right)$, choose 2.

$$3(2)^2 + 2 = 14 > 10$$

The solution is $\left(-\infty, -2\right) \cup \left(\frac{5}{3}, \infty\right)$.

37. $x^2 \le 25$

Solve $x^2 = 25$.

$$x = -5 \quad \text{or} \quad x = 5$$

Intervals: $(-\infty, -5), (-5, 5), (5, \infty)$

For $(-\infty, -5)$, choose -6.

$$(-6)^2 = 36 \not\le 25$$

For $(-5, 5)$, choose 0.

$$0^2 = 0 \le 25$$

For $(5, \infty)$, choose 6.

$$6^2 = 36 \not\le 25$$

The solution is $[-5, 5]$.

38. $p^2 - 16p > 0$

Solve $p^2 - 16p = 0$.

$$p^2 - 16p = 0$$
$$p(p - 16) = 0$$
$$p = 0 \quad \text{or} \quad p = 16$$

Intervals: $(-\infty, 0), (0, 16), (16, \infty)$

For $(-\infty, 0)$, choose -1.

$$(-1)^2 - 16(-1) = 17 > 0$$

For $(0, 16)$, choose 1.

$$(1)^2 - 16(1) = -15 \ngtr 0$$

For $(16, \infty)$, choose 17.

$$(17)^2 - 16(17) = 17 > 0$$

The solution is $(-\infty, 0) \cup (16, \infty)$.

39. $\dfrac{m-3}{m+5} \le 0$

Solve $\dfrac{m-3}{m+5} = 0$.

$$(m+5)\frac{m-3}{m+5} = (m+5)(0)$$
$$m - 3 = 0$$
$$m = 3$$

Set the denominator equal to 0 and solve.

$$m + 5 = 0$$
$$m = -5$$

Intervals: $(-\infty, -5), (-5, 3), (3, \infty)$

For $(-\infty, -5)$, choose -6.

$$\frac{-6-3}{-6+5} = 9 \nleq 0$$

For $(-5, 3)$, choose 0.

$$\frac{0-3}{0+5} = -\frac{3}{5} \le 0$$

For $(3, \infty)$, choose 4.

$$\frac{4-3}{4+5} = \frac{1}{9} \nleq 0$$

Although the $\le$ symbol is used, including -5 in the solution would cause the denominator to be zero.
The solution is $(-5, 3]$.

40. $\dfrac{r+1}{r-1} > 0$

Solve the equation $\dfrac{r+1}{r-1} = 0$.

$$\frac{r+1}{r-1} = 0$$
$$(r-1)\frac{r+1}{r-1} = (r-1)(0)$$
$$r + 1 = 0$$
$$r = -1$$

Find the value for which the denominator equals zero.

$$r - 1 = 0$$
$$r = 1$$

Intervals: $(-\infty, -1), (-1, 1), (1, \infty)$

For $(-\infty, -1)$, choose -2.

$$\frac{-2+1}{-2-1} = \frac{-1}{-3} = \frac{1}{3} > 0$$

For $(-1, 1)$, choose 0.

$$\frac{0+1}{0-1} = \frac{1}{-1} = -1 \ngtr 0$$

For $(1, \infty)$, choose 2.

$$\frac{2+1}{2-1} = \frac{3}{1} = 3 > 0$$

The solution is $(-\infty, -1) \cup (1, \infty)$.

41. $\dfrac{k-1}{k+2} > 1$

Solve $\dfrac{k-1}{k+2} = 1$.

$$k - 1 = k + 2$$
$$-1 \ne 2$$

The equation has no solution.
Solve $k + 2 = 0$.

$$k = -2$$

Intervals: $(-\infty, -2), (-2, \infty)$

For $(-\infty, -2)$, choose -3.

$$\frac{-3-1}{-3+2} = 4 > 1$$

For $(-2, \infty)$, choose 0.

$$\frac{0-1}{0+2} = -\frac{1}{2} \ngtr 1$$

The solution is $(-\infty, -2)$.

42. $\dfrac{a-5}{a+2} < -1$

Solve the equation $\dfrac{a-5}{a+2} = -1$.

$$\dfrac{a-5}{a+2} = -1$$
$$a - 5 = -1(a+2)$$
$$a - 5 = -a - 2$$
$$2a = 3$$
$$a = \dfrac{3}{2}$$

Set the denominator equal to zero and solve for a.

$$a + 2 = 0$$
$$a = -2$$

Intervals: $(-\infty, -2), \left(-2, \frac{3}{2}\right), \left(\frac{3}{2}, \infty\right)$

For $(-\infty, -2)$, choose -3.

$$\dfrac{-3-5}{-3+2} = \dfrac{-8}{-1} = 8 \nless -1$$

For $\left(-2, \frac{3}{2}\right)$, choose 0.

$$\dfrac{0-5}{0+2} = \dfrac{-5}{2} = -\dfrac{5}{2} < -1$$

For $\left(\frac{3}{2}, \infty\right)$, choose 2.

$$\dfrac{2-5}{2+2} = \dfrac{-3}{4} = -\dfrac{3}{4} \nless -1$$

The solution is $\left(-2, \frac{3}{2}\right)$.

43. $\dfrac{2y+3}{y-5} \leq 1$

Solve $\dfrac{2y+3}{y-5} = 1$.

$$2y + 3 = y - 5$$
$$y = -8$$

Solve $y - 5 = 0$.
$$y = 5$$

Intervals: $(-\infty, -8), (-8, 5), (5, \infty)$

For $(-\infty,, -8)$, choose $y = -10$.

$$\dfrac{2(-10)+3}{-10-5} = \dfrac{17}{15} \nleq 1$$

For $(-8, 5)$, choose $y = 0$.

$$\dfrac{2(0)+3}{0-5} = -\dfrac{3}{5} \leq 1$$

For $(5, \infty)$, choose $y = 6$.

$$\dfrac{2(6)+3}{6-5} = \dfrac{15}{1} \nleq 1$$

The solution is $[-8, 5)$.

44. $\dfrac{a+2}{3+2a} \leq 5$

For the equation $\dfrac{a+2}{3+2a} = 5$.

$$\dfrac{a+2}{3+2a} = 5$$
$$a + 2 = 5(3+2a)$$
$$a + 2 = 15 + 10a$$
$$-9a = 13$$
$$a = -\dfrac{13}{9}$$

Set the denominator equal to zero and solve for a.

$$3 + 2a = 0$$
$$2a = -3$$
$$a = -\dfrac{3}{2}$$

Intervals: $\left(-\infty, -\frac{3}{2}\right), \left(-\frac{3}{2}, -\frac{13}{9}\right), \left(-\frac{13}{9}, \infty\right)$

For $\left(-\infty, -\frac{3}{2}\right)$, choose -2.

$$\dfrac{-2+2}{3+2(-2)} = \dfrac{0}{-1} = 0 \leq 5$$

For $\left(-\frac{3}{2}, -\frac{13}{9}\right)$, choose -1.46.

$$\dfrac{-1.46+2}{3+2(-1.46)} = \dfrac{.54}{.08} = 6.75 \nleq 5$$

For $\left(-\frac{13}{9}, \infty\right)$, choose 0.

$$\dfrac{0+2}{3+2(0)} = \dfrac{2}{3} \leq 5$$

The value $-\frac{3}{2}$ cannot be included in the solution since it would make the denominator zero. The solution is $\left(-\infty, -\frac{3}{2}\right) \cup \left[-\frac{13}{9}, \infty\right)$.

45. $\dfrac{7}{k+2} \geq \dfrac{1}{k+2}$

Solve $\dfrac{7}{k+2} = \dfrac{1}{k+2}$.

$$\dfrac{7}{k+2} - \dfrac{1}{k+2} = 0$$
$$\dfrac{6}{k+2} = 0$$

The equation has no solution.

Solve $k + 2 = 0$.
$$k = -2$$

Intervals: $(-\infty, -2), (-2, \infty)$

For $(-\infty, -2)$, choose $k = -3$.
$$\frac{6}{-3+2} = -6 \not\geq 0$$

For $(-2, \infty)$, choose $k = 0$.
$$\frac{6}{0+2} = 3 \geq 0$$

The solution is $(-2, \infty)$.

46. $\dfrac{5}{p+1} > \dfrac{12}{p+1}$

Solve the equation $\dfrac{5}{p+1} = \dfrac{12}{p+1}$.
$$\frac{5}{p+1} = \frac{12}{p+1}$$
$$5 = 12$$

The equation has no solution.
Set the denominator equal to zero and solve for p.
$$p + 1 = 0$$
$$p = -1$$

Intervals: $(-\infty, -1), (-1, \infty)$
For $(-\infty, -1)$, choose -2.
$$\frac{5}{-2+1} = -5 \text{ and } \frac{12}{-2+1} = -12, \text{ so}$$
$$\frac{5}{-2+1} > \frac{12}{-2+1}.$$

For $(-1, \infty)$, choose 0.
$$\frac{5}{0+1} = 5 \text{ and } \frac{12}{0+1} = 12, \text{ so}$$
$$\frac{5}{0+1} \not> \frac{12}{0+1}.$$

The solution is $(-\infty, -1)$.

47. $\dfrac{3x}{x^2 - 1} < 2$

Solve
$$\frac{3x}{x^2 - 1} = 2.$$
$$3x = 2x^2 - 2$$
$$-2x^2 + 3x + 2 = 0$$
$$(2x + 1)(-x + 2) = 0$$

$$x = -\frac{1}{2} \quad \text{or} \quad x = 2$$

Set $x^2 - 1 = 0$.

$$x = 1 \quad \text{or} \quad x = -1$$

Intervals: $(-\infty, -1), \left(-1, -\frac{1}{2}\right), \left(-\frac{1}{2}, 1\right),$
$(1, 2), (2, \infty)$

For $(-\infty, -1)$, choose $x = -2$.

$$\frac{3(-2)}{(-2)^2 - 1} = -\frac{6}{3} = -2 < 2$$

For $\left(-1, -\frac{1}{2}\right)$, choose $x = -\frac{3}{4}$.

$$\frac{3\left(-\frac{3}{4}\right)}{\left(-\frac{3}{4}\right)^2 - 1} = \frac{-\frac{9}{4}}{\frac{9}{16} - 1} = \frac{36}{7} \not< 2$$

For $\left(-\frac{1}{2}, 1\right)$, choose $x = 0$.

$$\frac{3(0)}{0^2 - 1} = 0 < 2$$

For $(1, 2)$, choose $x = \frac{3}{2}$.

$$\frac{3\left(\frac{3}{2}\right)}{\left(\frac{3}{2}\right)^2 - 1} = \frac{\frac{9}{2}}{\frac{5}{4}} = \frac{18}{5} \not< 2$$

For $(2, \infty)$, choose $x = 3$.

$$\frac{3(3)}{3^2 - 1} = \frac{9}{8} < 2$$

The solution is

$$(-\infty, -1) \cup \left(-\frac{1}{2}, 1\right) \cup (2, \infty).$$

48. $\dfrac{8}{p^2 + 2p} > 1$

Solve the equation $\dfrac{8}{p^2 + 2p} = 1$.

$$\frac{8}{p^2 + 2p} = 1$$
$$8 = p^2 + 2p$$
$$0 = p^2 + 2p - 8$$
$$0 = p + 4)(p - 2)$$
$$p + 4 = 0 \quad \text{or} \quad p - 2 = 0$$
$$p = -4 \quad \text{or} \quad p = 2$$

Set the denominator equal to zero and solve for p.

$$p^2 + 2p = 0$$
$$p(p + 2) = 0$$
$$p = 0 \quad \text{or} \quad p + 2 = 0$$
$$p = -2$$

Intervals: $(-\infty, -4), (-4, -2), (-2, 0),$
$(0, 2), (2, \infty)$

For $(-\infty, -4)$, choose -5.

$$\frac{8}{(-5)^2 + 2(-5)} = \frac{8}{15} \not> 1$$

For $(-4, -2)$, choose -3.

$$\frac{8}{(-3)^2 + 2(-3)} = \frac{8}{9 - 6} = \frac{8}{3} > 1$$

For $(-2, 0)$, choose -1.

$$\frac{8}{(-1)^2 + 2(-1)} = \frac{8}{-1} = -8 \not> 1$$

For $(0, 2)$, choose 1.

$$\frac{8}{(1)^2 + 2(1)} = \frac{8}{3} > 1$$

For $(2, \infty)$, choose 3.

$$\frac{8}{(3)^2 + (2)(3)} = \frac{8}{15} \not> 1$$

The solution is $(-4, -2) \cup (0, 2)$.

49. $\dfrac{z^2 + z}{z^2 - 1} \geq 3$

Solve

$$\frac{z^2 + z}{z^2 - 1} = 3.$$

$$z^2 + z = 3z^2 - 3$$
$$-2z^2 + z + 3 = 0$$
$$-1(2z^2 - z - 3) = 0$$
$$-1(z + 1)(2z - 3) = 0$$
$$z = -1 \quad \text{or} \quad z = \frac{3}{2}$$

Set $z^2 - 1 = 0$.

$$z^2 = 1$$
$$z = -1 \quad \text{or} \quad z = 1$$

Intervals: $(-\infty, -1), (-1, 1), \left(1, \frac{3}{2}\right), \left(\frac{3}{2}, \infty\right)$

For $(-\infty, -1)$, choose $x = -2$.

$$\frac{(-2)^2 + 3}{(-2)^2 - 1} = \frac{7}{3} \not\geq 3$$

For $(-1, 1)$, choose $x = 0$.

$$\frac{0^2 + 3}{0^2 - 1} = -3 \not\geq 3$$

For $\left(1, \frac{3}{2}\right)$, choose $x = \frac{3}{2}$.

$$\frac{\left(\frac{3}{2}\right)^2 + 3}{\left(\frac{3}{2}\right)^2 - 1} = \frac{21}{5} \geq 3$$

For $\left(\frac{3}{2}, \infty\right)$, choose $x = 2$.

$$\frac{2^2 + 3}{2^2 - 1} = \frac{7}{3} \not\geq 3$$

The solution is $\left(1, \frac{3}{2}\right]$.

50. $\dfrac{a^2 + 2a}{a^2 - 4} \leq 2$

Solve the equation $\dfrac{a^2 + 2a}{a^2 - 4} = 2$.

$$\frac{a^2 + 2a}{a^2 - 4} = 2$$
$$a^2 + 2a = 2(a^2 - 4)$$
$$a^2 + 2a = 2a^2 - 8$$
$$0 = a^2 - 2a - 8$$
$$0 = (a - 4)(a + 2)$$
$$a - 4 = 0 \quad \text{or} \quad a + 2 = 0$$
$$a = 4 \quad \text{or} \quad a = -2$$

But -2 is not a possible solution.
Set the denominator equal to zero and solve for a.

$$a^2 - 4 = 0$$
$$(a + 2)(a - 2) = 0$$
$$a + 2 = 0 \quad \text{or} \quad a - 2 = 0$$
$$a = -2 \quad \text{or} \quad a = 2$$

Intervals: $(-\infty, -2), (-2, 2),$
$(2, 4), (4, \infty)$

For $(-\infty, -2)$, choose -3.

$$\frac{(-3)^2 + 2(-3)}{(-3)^2 - 4} = \frac{9 - 6}{9 - 4} = \frac{3}{5} \leq 2$$

For $(-2, 2)$, choose 0.

$$\frac{(0)^2 + 2(0)}{0 - 4} = \frac{0}{-4} = 0 \leq 2$$

For $(2, 4)$, choose 3.

$$\frac{(3)^2 + 2(3)}{(3)^2 - 4} = \frac{9 + 6}{9 - 5} = \frac{15}{4} \not\leq 2$$

For $(4, \infty)$, choose 5.

$$\frac{(5)^2 + 2(5)}{(5)^2 - 4} = \frac{25 + 10}{25 - 4} = \frac{35}{21} \leq 2$$

The value 4 will satisfy the original inequality, but the values -2 and 2 will not since they make the denominator zero. The solution is $(-\infty, -2) \cup (-2, 2) \cup [4, \infty)$.

R.6   Exponents

1. $8^{-2} = \dfrac{1}{8^2} = \dfrac{1}{64}$

2. $3^{-4} = \dfrac{1}{3^4} = \dfrac{1}{81}$

3. $5^0 = 1$, by definition.

4. $(-12)^0 = 1$, by definition.

5. $-(-3)^{-2} = -\dfrac{1}{(-3)^2} = -\dfrac{1}{9}$

6. $-(-3^{-2}) = -\left(-\dfrac{1}{3^2}\right) = -\left(-\dfrac{1}{9}\right) = \dfrac{1}{9}$

7. $\left(\dfrac{2}{7}\right)^{-2} = \dfrac{1}{\left(\frac{2}{7}\right)^2} = \dfrac{1}{\frac{4}{49}} = \dfrac{49}{4}$

8. $\left(\dfrac{4}{3}\right)^{-3} = \dfrac{1}{\left(\frac{4}{3}\right)^3} = \dfrac{1}{\frac{64}{27}} = \dfrac{27}{64}$

9. $\dfrac{3^{-4}}{3^2} = 3^{(-4)-2} = 3^{-4-2} = 3^{-6} = \dfrac{1}{3^6}$

10. $\dfrac{8^9 \cdot 8^{-7}}{8^{-3}} = 8^{9+(-7)-(-3)} = 8^{9-7+3} = 8^5$

11. $\dfrac{10^8 \cdot 10^{-10}}{10^4 \cdot 10^2}$

$\qquad = \dfrac{10^{8+(-10)}}{10^{4+2}} = \dfrac{10^{-2}}{10^6}$

$\qquad = 10^{-2-6} = 10^{-8}$

$\qquad = \dfrac{1}{10^8}$

12. $\left(\dfrac{5^{-6} \cdot 5^3}{5^{-2}}\right)^{-1} = (5^{-6+3-(-2)})^{-1}$

$\qquad = (5^{-6+3+2})^{-1} = (5^{-1})^{-1}$

$\qquad = 5^{(-1)(-1)} = 5^1 = 5$

13. $\dfrac{x^4 \cdot x^3}{x^5} = \dfrac{x^{4+3}}{x^5} = \dfrac{x^7}{x^5} = x^{7-5} = x^2$

14. $\dfrac{y^9 y^7}{y^{13}} = y^{9+7-13} = y^3$

15. $\dfrac{(4k^{-1})^2}{2k^{-5}} = \dfrac{4^2 k^{-2}}{2k^{-5}} = \dfrac{16k^{-2-(-5)}}{2}$

$\qquad = 8k^{-2+5} = 8k^3$

$\qquad = 2^3 k^3$

16. $\dfrac{(3z^2)^{-1}}{z^5} = \dfrac{3^{-1}(z^2)^{-1}}{z^5} = \dfrac{3^{-1}z^{2(-1)}}{z^5}$

$\qquad = \dfrac{3^{-1}z^{-2}}{z^5} = 3^{-1}a^{-2-5}$

$\qquad = 3^{-1}z^{-7} = \dfrac{1}{3} \cdot \dfrac{1}{z^7} = \dfrac{1}{3z^7}$

17. $\dfrac{2^{-1}x^3 y^{-3}}{xy^{-2}} = 2^{-1}x^{3-1}y^{-3-(-2)}$

$\qquad = 2^{-1}x^2 y^{-3+2} = 2^{-1}x^2 y^{-1}$

$\qquad = \dfrac{1}{2}x^2 \cdot \dfrac{1}{y} = \dfrac{x^2}{2y}$

18. $\dfrac{5^{-2}m^2 y^{-2}}{5^2 m^{-1}y^{-2}} = \dfrac{5^{-2}}{5^2} \cdot \dfrac{m^2}{m^{-1}} \cdot \dfrac{y^{-2}}{y^{-2}}$

$\qquad = 5^{-2-2}m^{2-(-1)}y^{-2-(-2)}$

$\qquad = 5^{-2-2}m^{2+1}y^{-2+2}$

$\qquad = 5^{-4}m^3 y^0 = \dfrac{1}{5^4} \cdot m^3 \cdot 1$

$\qquad = \dfrac{m^3}{5^4}$

19. $\left(\dfrac{a^{-1}}{b^2}\right)^{-3} = \dfrac{(a^{-1})^{-3}}{(b^2)^{-3}} = \dfrac{a^{(-1)(-3)}}{b^{2(-3)}}$

$\qquad = \dfrac{a^3}{b^{-6}} = a^3 b^6$

20. $\left(\dfrac{2c^2}{d^3}\right)^{-2} = \dfrac{2^{-2}(c^2)^{-2}}{(d^3)^{-2}}$

$\qquad = \dfrac{2^{-2}c^{(2)(-2)}}{d^{(3)(-2)}} = \dfrac{2^{-2}c^{-4}}{d^{-6}}$

$\qquad = \dfrac{d^6}{2^2 c^4}$

21. $\left(\dfrac{x^6 y^{-3}}{x^{-2}y^5}\right)^{1/2} = (x^{6-(-2)}y^{-3-5})^{1/2}$

$\qquad = (x^8 y^{-8})^{1/2}$

$\qquad = (x^8)^{1/2}(y^{-8})^{1/2}$

$\qquad = x^4 y^{-4}$

$\qquad = \dfrac{x^4}{y^4}$

22. $\left(\dfrac{a^{-7}b^{-1}}{b^{-4}a^2}\right)^{1/3} = \left(a^{-7-2}b^{-1-(-4)}\right)^{1/3}$

$= \left(a^{-9}b^3\right)^{1/3}$

$= \left(a^{-9}\right)^{1/3}\left(b^3\right)^{1/3}$

$= a^{-3}b^1$

$= \dfrac{b}{a^3}$

23. $a^{-1} + b^{-1} = \dfrac{1}{a} + \dfrac{1}{b}$

$= \left(\dfrac{b}{b}\right)\left(\dfrac{1}{a}\right) + \left(\dfrac{a}{a}\right)\left(\dfrac{1}{b}\right)$

$= \dfrac{b}{ab} + \dfrac{a}{ab}$

$= \dfrac{b+a}{ab}$

$= \dfrac{a+b}{ab}$

24. $b^{-2} - a = \dfrac{1}{b^2} - a$

$= \dfrac{1}{b^2} - a\left(\dfrac{b^2}{b^2}\right)$

$= \dfrac{1}{b^2} - \dfrac{ab^2}{b^2}$

$= \dfrac{1 - ab^2}{b^2}$

25. $\dfrac{2n^{-1} - 2m^{-1}}{m + n^2} = \dfrac{\frac{2}{n} - \frac{2}{m}}{m + n^2}$

$= \dfrac{\frac{2}{n}\cdot\frac{m}{m} - \frac{2}{m}\cdot\frac{n}{n}}{mn(m + n^2)}$

$= \dfrac{2m - 2n}{mn(m + n^2)}$ or $\dfrac{2(m - n)}{mn(m + n^2)}$

26. $\left(\dfrac{m}{3}\right)^{-1} + \left(\dfrac{n}{2}\right)^{-2} = \left(\dfrac{3}{m}\right)^1 + \left(\dfrac{2}{n}\right)^2$

$= \dfrac{3}{m} + \dfrac{4}{n^2}$

$= \left(\dfrac{3}{m}\right)\left(\dfrac{n^2}{n^2}\right) + \left(\dfrac{4}{n^2}\right)\left(\dfrac{m}{m}\right)$

$= \dfrac{3n^2}{mn^2} + \dfrac{4m}{mn^2}$

$= \dfrac{3n^2 + 4m}{mn^2}$

27. $\left(x^{-1} - y^{-1}\right)^{-1} = \dfrac{1}{\frac{1}{x} - \frac{1}{y}}$

$= \dfrac{1}{\frac{1}{x}\cdot\frac{y}{y} - \frac{1}{y}\cdot\frac{x}{x}}$

$= \dfrac{1}{\frac{y}{xy} - \frac{x}{xy}}$

$= \dfrac{1}{\frac{y-x}{xy}}$

$= \dfrac{xy}{y - x}$

28. $\left(x^{-2} + y^{-2}\right)^{-2} = \left(\dfrac{1}{x^2} + \dfrac{1}{y^2}\right)^{-2}$

$= \left[\left(\dfrac{1}{x^2}\right)\left(\dfrac{y^2}{y^2}\right) + \left(\dfrac{x^2}{x^2}\right)\left(\dfrac{1}{y^2}\right)\right]^{-2}$

$= \left(\dfrac{y^2}{x^2y^2} + \dfrac{x^2}{x^2y^2}\right)^{-2}$

$= \left(\dfrac{y^2 + x^2}{x^2y^2}\right)^{-2} = \left(\dfrac{x^2y^2}{y^2 + x^2}\right)^2$

$= \dfrac{(x^2)^2(y^2)^2}{(x^2 + y^2)^2} = \dfrac{x^4y^4}{(x^2 + y^2)^2}$

29. $81^{1/2} = (9^2)^{1/2} = 9^{2(1/2)} = 9^1 = 9$

30. $27^{1/3} = \sqrt[3]{27} = 3$

31. $32^{2/5} = (32^{1/5})^2 = 2^2 = 4$

32. $-125^{2/3} = (125^{1/3})^2 = -5^2 = -25$

33. $\left(\dfrac{4}{9}\right)^{1/2} = \dfrac{4^{1/2}}{9^{1/2}} = \dfrac{2}{3}$

34. $\left(\dfrac{64}{27}\right)^{1/3} = \dfrac{64^{1/3}}{27^{1/3}} = \dfrac{4}{3}$

35. $16^{-5/4} = (16^{1/4})^{-5} = 2^{-5}$

$= \dfrac{1}{2^5}$ or $\dfrac{1}{32}$

36. $625^{-1/4} = \dfrac{1}{625^{1/4}} = \dfrac{1}{5}$

37. $\left(\dfrac{27}{64}\right)^{-1/3} = \dfrac{27^{-1/3}}{64^{-1/3}} = \dfrac{64^{1/3}}{27^{1/3}}$

$= \dfrac{4}{3}$

38. $\left(\dfrac{121}{100}\right)^{-3/2} = \dfrac{1}{\left(\frac{121}{100}\right)^{3/2}} = \dfrac{1}{\left[\left(\frac{121}{100}\right)^{1/2}\right]^3}$

$= \dfrac{1}{\left(\frac{11}{10}\right)^3} = \dfrac{1}{\frac{1331}{1000}} = \dfrac{1000}{1331}$

39. $2^{1/2} \cdot 2^{3/2} = 2^{1/2+3/2} = 2^{4/2}$
$$= 2^2$$

40. $27^{2/3} \cdot 27^{-1/3} = 27^{(2/3)+(-1/3)}$
$$= 27^{2/3-1/3}$$
$$= 27^{1/3}$$

41. $\dfrac{4^{2/3} \cdot 4^{5/3}}{4^{1/3}} = \dfrac{4^{2/3+5/3}}{4^{1/3}}$
$$= 4^{7/3-1/3} = 4^{6/3}$$
$$= 4^2$$

42. $\dfrac{3^{-5/2} \cdot 3^{3/2}}{3^{7/2} \cdot 3^{-9/2}}$
$$= 3^{(-5/2)+(3/2)-(7/2)-(-9/2)}$$
$$= 3^{-5/2+3/2-7/2+9/2}$$
$$= 3^0 = 1$$

43. $\dfrac{7^{-1/3} \cdot 7r^{-3}}{7^{2/3} \cdot (r^{-2})^2}$
$$= \dfrac{7^{-1/3+1}r^{-3}}{7^{2/3} \cdot r^{-4}}$$
$$= 7^{-1/3+3/3-2/3}r^{-3-(-4)}$$
$$= 7^0 r^{-3+4} = 1 \cdot r^1 = r$$

44. $\dfrac{12^{3/4} \cdot 12^{5/4} \cdot y^{-2}}{12^{-1} \cdot (y^{-3})^{-2}}$
$$= \dfrac{12^{3/4+5/4} \cdot y^{-2}}{12^{-1} \cdot y^{(-3)(-2)}} = \dfrac{12^{8/4} \cdot y^{-2}}{12^{-1} \cdot y^6}$$
$$= \dfrac{12^2 \cdot y^{-2}}{12^{-1}y^6}$$
$$= 12^{2-(-1)} \cdot y^{-2-(-6)} = 12^3 y^{-8}$$
$$= \dfrac{12^3}{y^8}$$

45. $\dfrac{6k^{-4} \cdot (3k^{-1})^{-2}}{2^3 \cdot k^{1/2}}$
$$= \dfrac{2 \cdot 3k^{-4}(3^{-2})(k^2)}{2^3 k^{1/2}}$$
$$= 2^{1-3}3^{1+(-2)}k^{-4+2-1/2}$$
$$= 2^{-2}3^{-1}k^{-5/2}$$
$$= \dfrac{1}{2^2} \cdot \dfrac{1}{3} \cdot \dfrac{1}{k^{5/2}} = \dfrac{1}{2^2 3 k^{5/2}}$$
$$\text{or} \quad \dfrac{1}{12k^{5/2}}$$

46. $\dfrac{8p^{-3}(4p^2)^{-2}}{p^{-5}} = \dfrac{8p^{-3} \cdot 4^{-2}p^{(2)(-2)}}{p^{-5}}$
$$= \dfrac{8p^{-3}4^{-2}p^{-4}}{p^{-5}}$$
$$= 8 \cdot 4^{-2}p^{(-3)+(-4)-(-5)}$$
$$= 8 \cdot 4^{-2}p^{-3-4+5}$$
$$= 8 \cdot 4^{-2}p^{-2}$$
$$= 8 \cdot \dfrac{1}{4^2} \cdot \dfrac{1}{p^2}$$
$$= 8 \cdot \dfrac{1}{16} \cdot \dfrac{1}{p^2}$$
$$= \dfrac{8}{16p^2} = \dfrac{1}{2p^2}$$

47. $\dfrac{a^{4/3}}{a^{2/3}} \cdot \dfrac{b^{1/2}}{b^{-3/2}} = a^{4/3-2/3}b^{1/2-(-3/2)}$
$$= a^{2/3}b^2$$

48. $\dfrac{x^{1/3} \cdot y^{2/3} \cdot z^{1/4}}{x^{5/3} \cdot y^{-1/3} \cdot z^{3/4}}$
$$= x^{1/3-(5/3)}y^{(2/3)-(-1/3)}z^{1/4-(3/4)}$$
$$= x^{1/3-5/3}y^{2/3+1/3}z^{1/4-3/4}$$
$$= x^{-4/3}y^{3/3}z^{-2/4}$$
$$= \dfrac{y}{x^{4/3}z^{2/4}}$$
$$= \dfrac{y}{x^{4/3}z^{1/2}}$$

49. $\dfrac{k^{-3/5} \cdot h^{-1/3} \cdot t^{2/5}}{k^{-1/5} \cdot h^{-2/3} \cdot t^{1/5}}$
$$= k^{-3/5-(-1/5)}h^{-1/3-(-2/3)}t^{2/5-1/5}$$
$$= k^{-3/5+1/5}h^{-1/3+2/3}t^{2/5-1/5}$$
$$= k^{-2/5}h^{1/3}t^{1/5}$$
$$= \dfrac{h^{1/3}t^{1/5}}{k^{2/5}}$$

50. $\dfrac{m^{7/3} \cdot n^{-2/5} \cdot p^{3/8}}{m^{-2/3} \cdot n^{3/5} \cdot p^{-5/8}}$
$$= m^{7/3+2/3}n^{-2/5-(3/5)}p^{3/8-(-5/8)}$$
$$= m^{7/3-(-2/3)}n^{-2/5-3/5}p^{3/8+5/8}$$
$$= m^{9/3}n^{-5/5}p^{8/8}$$
$$= m^3 n^{-1}p^1$$
$$= \dfrac{m^3 p}{n}$$

51. $12x^2(x^2+2)^2 - 4x(4x^3+1)(x^2+2)$
$= (4x)(3x)(x^2+2)(x^2+2)$
$\quad - 4x(4x^3+1)(x^2+2)$
$= 4x(x^2+2)[3x(x^2+2) - (4x^3+1)]$
$= 4x(x^2+2)(3x^3+6x-4x^3-1)$
$= 4x(x^2+2)(-x^3+6x-1)$

52. $6x(x^3+7)^2 - 6x^2(3x^2+5)(x^3+7)$
$= 6x(x^3+7)(x^3+7) - 6x(x)(3x^2+5)(x^3+7)$
$= 6x(x^3+7)[(x^3+7) - x(3x^2+5)]$
$= 6x(x^3+7)(x^3+7-3x^3-5x)$
$= 6x(x^3+7)(-2x^3-5x+7)$

53. $(x^2+2)(x^2-1)^{-1/2}(x) + (x^2-1)^{1/2}(2x)$
$= (x^2+2)(x^2-1)^{-1/2}(x)$
$\quad + (x^2-1)^1(x^2-1)^{-1/2}(2x)$
$= x(x^2-1)^{-1/2}[(x^2+2) + (x^2-1)(2)]$
$= x(x^2-1)^{-1/2}(x^2+2+2x^2-2)$
$= x(x^2-1)^{-1/2}(3x^2)$
$= 3x^3(x^2-1)^{-1/2}$

54. $9(6x+2)^{1/2} + 3(9x-1)(6x+2)^{-1/2}$
$= 3 \cdot 3(6x+2)^{-1/2}(6x+2)^1$
$\quad + 3(9x-1)(6x+2)^{-1/2}$
$= 3(6x+2)^{-1/2}[3(6x+2) + (9x-1)]$
$= 3(6x+2)^{-1/2}(18x+6+9x-1)$
$= 3(6x+2)^{-1/2}(27x+5)$

55. $x(2x+5)^2(x^2-4)^{-1/2} + 2(x^2-4)^{1/2}(2x+5)$
$= (2x+5)^2(x^2-4)^{-1/2}(x)$
$\quad + (x^2-4)^1(x^2-4)^{-1/2}(2)(2x+5)$
$= (2x+5)(x^2-4)^{-1/2}$
$\quad \cdot [(2x+5)(x) + (x^2-4)(2)]$
$= (2x+5)(x^2-4)^{-1/2}$
$\quad \cdot (2x^2+5x+2x^2-8)$
$= (2x+5)(x^2-4)^{-1/2}(4x^2+5x-8)$

56. $(4x^2+1)^2(2x-1)^{-1/2} + 16x(4x^2+1)(2x-1)^{1/2}$
$= (4x^2+1)(4x^2+1)(2x-1)^{-1/2}$
$\quad + 16x(4x^2+1)(2x-1)^{-1/2}(2x-1)$
$= (4x^2+1)(2x-1)^{-1/2}$
$\quad \cdot [(4x^2+1) + 16x(2x-1)]$
$= (4x^2+1)(2x-1)^{-1/2}(4x^2+1+32x^2-16x)$
$= (4x^2+1)(2x-1)^{-1/2}(36x^2-16x+1)$

R.7   Radicals

1. $\sqrt[3]{125} = 5$ because $5^3 = 125$.

2. $\sqrt[4]{1296} = \sqrt[4]{6^4} = 6$

3. $\sqrt[5]{-3125} = -5$ because $(-5)^5 = -3125$.

4. $\sqrt{50} = \sqrt{25 \cdot 2} = \sqrt{25}\sqrt{2} = 5\sqrt{2}$

5. $\sqrt{2000} = \sqrt{4 \cdot 100 \cdot 5} = 2 \cdot 10\sqrt{5}$
$\quad = 20\sqrt{5}$

6. $\sqrt{32y^5} = \sqrt{(16y^4)(2y)} = \sqrt{16y^4}\sqrt{2y}$
$\quad = 4y^2\sqrt{2y}$

7. $7\sqrt{2} - 8\sqrt{18} + 4\sqrt{72}$
$= 7\sqrt{2} - 8\sqrt{9 \cdot 2} + 4\sqrt{36 \cdot 2}$
$= 7\sqrt{2} - 8(3)\sqrt{2} + 4(6)\sqrt{2}$
$= 7\sqrt{2} - 24\sqrt{2} + 24\sqrt{2}$
$= 7\sqrt{2}$

8. $4\sqrt{3} - 5\sqrt{12} + 3\sqrt{75}$
$= 4\sqrt{3} - 5(\sqrt{4}\sqrt{3}) + 3(\sqrt{25}\sqrt{3})$
$= 4\sqrt{3} - 5(2\sqrt{3}) + 3(5\sqrt{3})$
$= 4\sqrt{3} - 10\sqrt{3} + 15\sqrt{3}$
$= (4 - 10 + 15)\sqrt{3} = 9\sqrt{3}$

9. $2\sqrt{5} - 3\sqrt{20} + 2\sqrt{45}$
$= 2\sqrt{5} - 3\sqrt{4 \cdot 5} + 2\sqrt{9 \cdot 5}$
$= 2\sqrt{5} - 3(2)\sqrt{5} + 2(3)\sqrt{5}$
$= 2\sqrt{5} - 6\sqrt{5} + 6\sqrt{5}$
$= 2\sqrt{5}$

10. $3\sqrt{28} - 4\sqrt{63} + \sqrt{112}$
$= 3(\sqrt{4}\sqrt{7}) - 4(\sqrt{9}\sqrt{7}) + (\sqrt{16}\sqrt{7})$
$= 3(2\sqrt{7}) - 4(3\sqrt{7}) + (4\sqrt{7})$
$= 6\sqrt{7} - 12\sqrt{7} + 4\sqrt{7}$
$= (6 - 12 + 4)\sqrt{7}$
$= -2\sqrt{7}$

11. $\sqrt[3]{2} - \sqrt[3]{16} + 2\sqrt[3]{54}$
$= \sqrt[3]{2} - (\sqrt[3]{8 \cdot 2}) + 2(\sqrt[3]{27 \cdot 2})$
$= \sqrt[3]{2} - \sqrt[3]{8}\sqrt[3]{2} + 2(\sqrt[3]{27}\sqrt[3]{2})$
$= \sqrt[3]{2} - 2\sqrt[3]{2} + 2(3\sqrt[3]{2})$
$= \sqrt[3]{2} - 2\sqrt[3]{2} + 6\sqrt[3]{2}$
$= 5\sqrt[3]{2}$

12. $2\sqrt[3]{3} + 4\sqrt[3]{24} - \sqrt[3]{81}$
$= 2\sqrt[3]{3} + 4\sqrt[3]{8 \cdot 3} - \sqrt[3]{27 \cdot 3}$
$= 2\sqrt[3]{3} + 4(2)\sqrt[3]{3} - 3\sqrt[3]{3}$
$= 2\sqrt[3]{3} + 8\sqrt[3]{3} - 3\sqrt[3]{3}$
$= 7\sqrt[3]{3}$

13. $\sqrt[3]{32} - 5\sqrt[3]{4} + 2\sqrt[3]{108}$
$= \sqrt[3]{8 \cdot 4} - 5\sqrt[3]{4} + 2\sqrt[3]{27 \cdot 4}$
$= \sqrt[3]{8}\sqrt[3]{4} - 5\sqrt[3]{4} + 2\sqrt[3]{27}\sqrt[3]{4}$
$= 2\sqrt[3]{4} - 5\sqrt[3]{4} + 2(3\sqrt[3]{4})$
$= 2\sqrt[3]{4} - 5\sqrt[3]{4} + 6\sqrt[3]{4}$
$= 3\sqrt[3]{4}$

14. $\sqrt{2x^3y^2z^4} = \sqrt{x^2y^2z^4 \cdot 2x}$
$= xyz^2\sqrt{2x}$

15. $\sqrt{98r^3s^4t^{10}}$
$= \sqrt{(49 \cdot 2)(r^2 \cdot r)(s^4)(t^{10})}$
$= \sqrt{(49r^2s^4t^{10})(2r)}$
$= \sqrt{49r^2s^4t^{10}}\sqrt{2r}$
$= 7rs^2t^5\sqrt{2r}$

16. $\sqrt[3]{16z^5x^8y^4} = \sqrt[3]{8z^3x^6y^3 \cdot 2z^2z^2y}$
$= 2zx^2y\sqrt[3]{2z^2x^2y}$

17. $\sqrt[4]{x^8y^7z^{11}} = \sqrt[4]{(x^8)(y^4 \cdot y^3)(z^8z^3)}$
$= \sqrt[4]{(x^8y^4z^8)(y^3z^3)}$
$= \sqrt[4]{x^8y^4z^8}\sqrt[4]{y^3z^3}$
$= x^2yz^2\sqrt[4]{y^3z^3}$

18. $\sqrt{a^3b^5} - 2\sqrt{a^7b^3} + \sqrt{a^3b^9}$
$= \sqrt{a^2b^4ab} - 2\sqrt{a^6b^2ab} + \sqrt{a^2b^8ab}$
$= ab^2\sqrt{ab} - 2a^3b\sqrt{ab} + ab^4\sqrt{ab}$
$= (ab^2 - 2a^3b + ab^4)\sqrt{ab}$
$= ab\sqrt{ab}(b - 2a^2 + b^3)$

19. $\sqrt{p^7q^3} - \sqrt{p^5q^9} + \sqrt{p^9q}$
$= \sqrt{(p^6p)(q^2q)} - \sqrt{(p^4p)(q^8q)}$
$\quad + \sqrt{(p^8p)q}$
$= \sqrt{(p^6q^2)(pq)} - \sqrt{(p^4q^8)(pq)}$
$\quad + \sqrt{(p^8)(pq)}$
$= \sqrt{p^6q^2}\sqrt{pq} - \sqrt{p^4q^8}\sqrt{pq} + \sqrt{p^8}\sqrt{pq}$
$= p^3q\sqrt{pq} - p^2q^4\sqrt{pq} + p^4\sqrt{pq}$
$= p^2pq\sqrt{pq} - p^2q^4\sqrt{pq} + p^2p^2\sqrt{pq}$
$= p^2\sqrt{pq}(pq - q^4 + p^2)$

20. $\dfrac{5}{\sqrt{7}} = \dfrac{5}{\sqrt{7}} \cdot \dfrac{\sqrt{7}}{\sqrt{7}} = \dfrac{5\sqrt{7}}{7}$

21. $\dfrac{-2}{\sqrt{3}} = \dfrac{-2}{\sqrt{3}} \cdot \dfrac{\sqrt{3}}{\sqrt{3}} = \dfrac{-2\sqrt{3}}{\sqrt{9}} = -\dfrac{2\sqrt{3}}{3}$

22. $\dfrac{-3}{\sqrt{12}} = \dfrac{-3}{\sqrt{4 \cdot 3}} = \dfrac{-3}{2\sqrt{3}} \cdot \dfrac{\sqrt{3}}{\sqrt{3}}$
$= \dfrac{-3\sqrt{3}}{6}$
$= -\dfrac{\sqrt{3}}{2}$

23. $\dfrac{4}{\sqrt{8}} = \dfrac{4}{\sqrt{8}} \cdot \dfrac{\sqrt{2}}{\sqrt{2}} = \dfrac{4\sqrt{2}}{\sqrt{16}} = \dfrac{4\sqrt{2}}{4} = \sqrt{2}$

24. $\dfrac{3}{1 - \sqrt{5}} = \dfrac{3}{1 - \sqrt{5}} \cdot \dfrac{1 + \sqrt{5}}{1 + \sqrt{5}}$
$= \dfrac{3(1 + \sqrt{5})}{1 - 5}$
$= \dfrac{-3(1 + \sqrt{5})}{4}$

25. $\dfrac{5}{2 - \sqrt{6}} = \dfrac{5}{2 - \sqrt{6}} \cdot \dfrac{2 + \sqrt{6}}{2 + \sqrt{6}}$
$= \dfrac{5(2 + \sqrt{6})}{4 + 2\sqrt{6} - 2\sqrt{6} - \sqrt{36}}$
$= \dfrac{5(2 + \sqrt{6})}{4 - \sqrt{36}} = \dfrac{5(2 + \sqrt{6})}{4 - 6}$
$= \dfrac{5(2 + \sqrt{6})}{-2}$
$= -\dfrac{5(2 + \sqrt{6})}{2}$

26. $\dfrac{-2}{\sqrt{3} - \sqrt{2}}$
$= \dfrac{-2}{\sqrt{3} - \sqrt{2}} \cdot \dfrac{\sqrt{3} + \sqrt{2}}{\sqrt{3} + \sqrt{2}}$
$= \dfrac{-2(\sqrt{3} + \sqrt{2})}{3 - 2} = \dfrac{-2(\sqrt{3} + \sqrt{2})}{1}$
$= -2(\sqrt{3} + \sqrt{2})$

27. $\dfrac{1}{\sqrt{10} + \sqrt{3}} = \dfrac{1}{\sqrt{10} + \sqrt{3}} \cdot \dfrac{\sqrt{10} - \sqrt{3}}{\sqrt{10} - \sqrt{3}}$
$= \dfrac{\sqrt{10} - \sqrt{3}}{\sqrt{100} - \sqrt{30} + \sqrt{30} - \sqrt{9}}$
$= \dfrac{\sqrt{10} - \sqrt{3}}{\sqrt{100} - \sqrt{9}} = \dfrac{\sqrt{10} - \sqrt{3}}{10 - 3}$
$= \dfrac{\sqrt{10} - \sqrt{3}}{7}$

28. $\dfrac{1}{\sqrt{r} - \sqrt{3}}$
$= \dfrac{1}{\sqrt{r} - \sqrt{3}} \cdot \dfrac{\sqrt{r} + \sqrt{3}}{\sqrt{r} + \sqrt{3}}$
$= \dfrac{\sqrt{r} + \sqrt{3}}{r - 3}$

29. $\dfrac{5}{\sqrt{m} - \sqrt{5}} = \dfrac{5}{\sqrt{m} - \sqrt{5}} \cdot \dfrac{\sqrt{m} + \sqrt{5}}{\sqrt{m} + \sqrt{5}}$
$= \dfrac{5(\sqrt{m} + \sqrt{5})}{\sqrt{m^2} + \sqrt{5m} - \sqrt{5m} - \sqrt{25}}$
$= \dfrac{5(\sqrt{m} + \sqrt{5})}{\sqrt{m^2} - \sqrt{25}} = \dfrac{5(\sqrt{m} + \sqrt{5})}{m - 5}$

30. $\dfrac{y-5}{\sqrt{y}-\sqrt{5}}$

$= \dfrac{y-5}{\sqrt{y}-\sqrt{5}} \cdot \dfrac{\sqrt{y}+\sqrt{5}}{\sqrt{y}+\sqrt{5}}$

$= \dfrac{(y-5)(\sqrt{y}+\sqrt{5})}{y-5}$

$= \sqrt{y}+\sqrt{5}$

31. $\dfrac{z-11}{\sqrt{z}-\sqrt{11}} = \dfrac{z-11}{\sqrt{z}-\sqrt{11}} \cdot \dfrac{\sqrt{z}+\sqrt{11}}{\sqrt{z}+\sqrt{11}}$

$= \dfrac{(z-11)(\sqrt{z}+\sqrt{11})}{\sqrt{z^2}+\sqrt{11z}-\sqrt{11z}-\sqrt{121}}$

$= \dfrac{(z-11)(\sqrt{z}+\sqrt{11})}{\sqrt{z^2}-\sqrt{121}}$

$= \dfrac{(z-11)(\sqrt{z}+\sqrt{11})}{(z-11)}$

$= \sqrt{z}+\sqrt{11}$

32. $\dfrac{\sqrt{x}+\sqrt{x+1}}{\sqrt{x}-\sqrt{x+1}}$

$= \dfrac{\sqrt{x}+\sqrt{x+1}}{\sqrt{x}-\sqrt{x+1}} \cdot \dfrac{\sqrt{x}+\sqrt{x+1}}{\sqrt{x}+\sqrt{x+1}}$

$= \dfrac{x+2\sqrt{x(x+1)}+(x+1)}{x-(x+1)}$

$= \dfrac{2x+2\sqrt{x(x+1)}+1}{-1}$

$= -2x-2\sqrt{x(x+1)}-1$

33. $\dfrac{\sqrt{p}+\sqrt{p^2-1}}{\sqrt{p}-\sqrt{p^2-1}}$

$= \dfrac{\sqrt{p}+\sqrt{p^2-1}}{\sqrt{p}-\sqrt{p^2-1}} \cdot \dfrac{\sqrt{p}+\sqrt{p^2-1}}{\sqrt{p}+\sqrt{p^2-1}}$

$= \dfrac{(\sqrt{p})^2+2\sqrt{p}\sqrt{p^2-1}+(\sqrt{p^2-1})^2}{\sqrt{p^2}+\sqrt{p}\sqrt{p^2-1}-\sqrt{p}\sqrt{p^2-1}-(\sqrt{p^2-1})^2}$

$= \dfrac{p+2\sqrt{p}\sqrt{p^2-1}+(p^2-1)}{p-(p^2-1)}$

$= \dfrac{p^2+p+2\sqrt{p(p^2-1)}-1}{-p^2+p+1}$

34. $\dfrac{1+\sqrt{2}}{2} = \dfrac{(1+\sqrt{2})(1-\sqrt{2})}{2(1-\sqrt{2})}$

$= \dfrac{1-2}{2(1-\sqrt{2})}$

$= -\dfrac{1}{2(1-\sqrt{2})}$

35. $\dfrac{1-\sqrt{3}}{3} = \dfrac{1-\sqrt{3}}{3} \cdot \dfrac{1+\sqrt{3}}{1+\sqrt{3}}$

$= \dfrac{1^2-(\sqrt{3})^2}{3(1+\sqrt{3})}$

$= \dfrac{1-3}{3(1+\sqrt{3})}$

$= -\dfrac{2}{3(1+\sqrt{3})}$

36. $\dfrac{\sqrt{x}+\sqrt{x+1}}{\sqrt{x}-\sqrt{x+1}}$

$= \dfrac{\sqrt{x}+\sqrt{x+1}}{\sqrt{x}-\sqrt{x+1}} \cdot \dfrac{\sqrt{x}-\sqrt{x+1}}{\sqrt{x}-\sqrt{x+1}}$

$= \dfrac{x-(x+1)}{x-2\sqrt{x}\cdot\sqrt{x+1}+(x+1)}$

$= \dfrac{-1}{2x-2\sqrt{x(x+1)}+1}$

37. $\dfrac{\sqrt{p}+\sqrt{p^2-1}}{\sqrt{p}-\sqrt{p^2-1}}$

$= \dfrac{\sqrt{p}+\sqrt{p^2-1}}{\sqrt{p}-\sqrt{p^2-1}} \cdot \dfrac{\sqrt{p}-\sqrt{p^2-1}}{\sqrt{p}-\sqrt{p^2-1}}$

$= \dfrac{(\sqrt{p})^2-(\sqrt{p^2-1})^2}{(\sqrt{p})^2-2\sqrt{p}\sqrt{p^2-1}+(\sqrt{p^2-1})^2}$

$= \dfrac{p-(p^2-1)}{p-2\sqrt{p(p^2-1)}+p^2-1}$

$= \dfrac{p-p^2+1}{p-2\sqrt{p(p^2-1)}+p^2-1}$

$= \dfrac{-p^2+p+1}{p^2+p-2\sqrt{p(p^2-1)}-1}$

38. $\sqrt{16-8x+x^2}$
$= \sqrt{(4-x)^2}$
$= |4-x|$

Since $\sqrt{}$ denotes the nonnegative root, we must have $4-x \geq 0$.

39. $\sqrt{4y^2+4y+1}$
$= \sqrt{(2y+1)^2}$
$= |2y+1|$

Since $\sqrt{}$ denotes the nonnegative root, we must have $2y+1 \geq 0$.

40. $\sqrt{4-25z^2} = \sqrt{(2+5z)(2-5z)}$

This factorization does not produce a perfect square, so the expression $\sqrt{4-25z^2}$ cannot be simplified.

41. $\sqrt{9k^2 + h^2}$

The expression $9k^2 + h^2$ is the sum of two squares and cannot be factored. Therefore, $\sqrt{9k^2 + h^2}$ cannot be simplified.

Chapter 1

LINEAR FUNCTIONS

1.1 Slope and Equations of a Line

1. Find the slope of the line through $(4, 5)$ and $(-1, 2)$.

$$m = \frac{5 - 2}{4 - (-1)}$$

$$= \frac{3}{5}$$

3. Find the slope of the line through $(8, 4)$ and $(8, -7)$.

$$m = \frac{4 - (-7)}{8 - 8}$$

$$= \frac{11}{0}$$

The slope is undefined; the line is vertical.

5. $y = 2x$

Using the slope-intercept form, $y = mx + b$, we see that the slope is 2.

7. $5x - 9y = 11$

Rewrite the equation in slope-intercept form.

$$9y = 5x - 11$$

$$y = \frac{5}{9}x - \frac{11}{9}$$

The slope is $\frac{5}{9}$.

9. $x = -6$

This is vertical line, the slope is undefined.

11. $y = 8$

This is a horizontal line, which has a slope of 0.

13. $y + 2 = 0$

Putting this equation into the slope-intercept form, $y = 0x - 2$, we can see that the slope is 0.

15. $x = -8$

This is a vertical line, the slope is undefined.

17. Find the slope of a line parallel to $2y - 4x = 7$. Rewrite the equation in slope-intercept form.

$$2y = 4x + 7$$

$$y = 2x + \frac{7}{2}$$

This slope is 2, so a parallel line will also have slope 2.

19. Find the slope of the line through $(-1.978, 4.806)$ and $(3.759, 8.125)$. (Note that there are four digits in each number.)

$$m = \frac{8.125 - 4.806}{3.759 - (-1.978)}$$

$$= \frac{3.319}{5.737}$$

$$= .5785$$

(We give the answer correct to four significant digits.)

21. The line goes through $(1, 3)$, with slope $m = -2$. Use point-slope form.

$$y - 3 = -2(x - 1)$$

$$y = -2x + 2 + 3$$

$$2x + y = 5$$

23. The line goes through $(6, 1)$, with slope $m = 0$. Use point-slope form.

$$y - 1 = 0(x - 6)$$

$$y - 1 = 0$$

$$y = 1$$

25. The line goes through $(4, 2)$ and $(1, 3)$. Find the slope, then use point-slope form with either of the two given points.

$$m = \frac{3 - 2}{1 - 4}$$

$$= -\frac{1}{3}$$

$$y - 3 = -\frac{1}{3}(x - 1)$$

$$y = -\frac{1}{3}x + \frac{1}{3} + 3$$

$$y = -\frac{1}{3}x + \frac{10}{3}$$

$$3y = -x + 10$$

$$x + 3y = 10$$

27. The line goes through $\left(\frac{1}{2}, \frac{5}{3}\right)$ and $\left(3, \frac{1}{6}\right)$.

$$m = \frac{\frac{1}{6} - \frac{5}{3}}{3 - \frac{1}{2}} = \frac{\frac{1}{6} - \frac{10}{6}}{\frac{6}{2} - \frac{1}{2}}$$

$$= \frac{-\frac{9}{6}}{\frac{5}{2}} = -\frac{18}{30} = -\frac{3}{5}$$

$$y - \frac{5}{3} = -\frac{3}{5}\left(x - \frac{1}{2}\right)$$

$$y - \frac{5}{3} = -\frac{3}{5}x + \frac{3}{10}$$

$$30\left(y - \frac{5}{3}\right) = 30\left(-\frac{3}{5}x + \frac{3}{10}\right)$$

$$30y - 50 = -18x + 9$$

$$18x + 30y = 59$$

29. The line goes through $(-8, 4)$ and $(-8, 6)$.

$$m = \frac{4 - 6}{-8 - (-8)} = \frac{-2}{0};$$

which is undefined.
This is a vertical line; the value of x is always -8.
The equation of this line is $x = -8$.

31. The line goes through $(-1, 3)$ and $(0, 3)$.

$$m = \frac{3 - 3}{-1 - 0} = \frac{0}{-1} = 0$$

This is a horizontal line; the value of y is always 3. The equation of this line is $y = 3$.

33. The line has x-intercept 3 and y-intercept -2.
Two points on the line are $(3, 0)$ and $(0, -2)$. Find the slope; then use slope-intercept form.

$$m = \frac{0 - (-2)}{3 - 0}$$

$$= \frac{2}{3}$$

$$b = -2$$

$$y = \frac{2}{3}x - 2$$

$$3y = 2x - 6$$

$$2x - 3y = 6$$

35. The line has an x-intercept of -3 and a y-intercept of 2.

Two points on the line are $(-3, 0)$ and $(0, 2)$. Find the slope; then use slope-intercept form.

$$m = \frac{2 - 0}{0 - (-3)} = \frac{2}{3}$$

$$b = 2$$

$$y = \frac{2}{3}x + 2$$

$$3y = 3\left(\frac{2}{3}x + 2\right)$$

$$3y = 2x + 6$$

$$2x - 3y = -6$$

37. The vertical line through $(-6, 5)$ goes through the point $(-6, 0)$, so the equation is $x = -6$.

39. The line goes through $(-1.76, 4.25)$ and has slope -5.081.
Use point-slope form.

$$y - 4.25 = -5.081[x - (-1.76)]$$

$$y = -5.081x - 8.94256 + 4.25$$

$$y = -5.081x - 4.69256$$

$$5.081x + y = -4.69$$

41. Write an equation of the line through $(-1, 4)$, parallel to $x + 3y = 5$.

Rewrite the equation of the given line in slope-intercept form.

$$x + 3y = 5$$

$$3y = -x + 5$$

$$y = -\frac{1}{3}x + \frac{5}{3}$$

The slope is $-\frac{1}{3}$.

Use $m = -\frac{1}{3}$ and the point $(-1, 4)$ in the point-slope form.

$$y - 4 = -\frac{1}{3}[x - (-1)]$$

$$y = -\frac{1}{3}(x + 1) + 4$$

$$y = -\frac{1}{3}x - \frac{1}{3} + 4$$

$$= -\frac{1}{3}x + \frac{11}{3}$$

$$x + 3y = 11$$

43. Write an equation of the line through $(3, -4)$, perpendicular to $x + y = 4$.

Rewrite the equation of the given line as

$$y = -x + 4.$$

The slope of this line is -1. To find the slope of a perpendicular line, solve

$$-1m = -1.$$
$$m = 1$$

Use $m = 1$ and $(3, -4)$ in the point-slope form.

$$y - (-4) = 1(x - 3)$$
$$y = x - 3 - 4$$
$$y = x - 7$$
$$x - y = 7$$

45. Write an equation of the line with x-intercept -2, parallel to $y = 2x$.

The given line has slope 2. A parallel line will also have $m = 2$. Since the x-intercept is -2, the point $(-2, 0)$ is on the required line. Using point-slope form, we have

$$y - 0 = 2[x - (-2)]$$
$$y = 2x + 4$$
$$-2x + y = 4.$$

47. Write an equation of the line with y-intercept 2, perpendicular to $3x + 2y = 6$.

Find the slope of the given line.

$$3x + 2y = 6$$
$$2y = -3x + 6$$
$$y = -\frac{3}{2}x + 3$$

The slope is $-\frac{3}{2}$, so the slope of the perpendicular line will be $\frac{2}{3}$. If the y-intercept is 2, then the point $(0, 2)$ lies on the line. Using point-slope form, we have

$$y - 2 = \frac{2}{3}(x - 0)$$
$$y = \frac{2}{3}x + 2$$
$$2x - 3y = -6.$$

49. Do $(4, 3)$, $(2, 0)$, and $(-18, -12)$ lie on the same line?

We will find the equation of the line through $(4, 3)$ and $(2, 0)$, and then determine whether $(-18, -12)$ lies on this line.

First, find the slope.

$$m = \frac{0 - 3}{2 - 4} = \frac{-3}{-2} = \frac{3}{2}.$$

Now use the point-slope form with $m = \frac{3}{2}$ and $(x_1, y_1) = (2, 0)$.

$$y - y_1 = m(x - x_1)$$
$$y - 0 = \frac{3}{2}(x - 2)$$
$$y = \frac{3}{2}(x - 2)$$
$$2y = 3(x - 2)$$
$$2y = 3x - 6$$
$$6 = 3x - 2y$$
$$\text{or} \quad 3x - 2y = 6$$

To determine whether the third point, $(-18, -12)$, lies on this line, substitute -18 for x and -12 for y in the equation obtained using the first two points.

$$3x - 2y = 6$$
$$3(-18) - 2(-12) = 6 \quad ? \quad Let \quad x = -18,$$
$$ y = -12$$
$$-54 + 24 = 6 \quad ?$$
$$-30 = 6 \qquad False$$

Thus, the three given points do not lie on the same line.

51. A parallelogram has 4 sides, with opposite sides parallel. The slope of the line through $(1, 3)$ and $(2, 1)$ is

$$m = \frac{3 - 1}{1 - 2} = \frac{2}{-1} = -2.$$

The slope of the line through $\left(-\frac{5}{2}, 2\right)$ and $\left(-\frac{7}{2}, 4\right)$ is

$$m = \frac{2 - 4}{-\frac{5}{2} - \left(-\frac{7}{2}\right)} = \frac{-2}{1} = -2.$$

Since these slopes are equal, these two sides are parallel.

The slope of the line through $\left(-\frac{7}{2}, 4\right)$ and $(1, 3)$ is

$$m = \frac{4 - 3}{-\frac{7}{2} - 1} = \frac{1}{-\frac{9}{2}} = -\frac{2}{9}.$$

Slope of the line through $\left(-\frac{5}{2}, 2\right)$ and $(2, 1)$ is

$$m = \frac{2 - 1}{-\frac{5}{2} - 2} = \frac{1}{-\frac{9}{2}} = -\frac{2}{9}.$$

Since these slopes are equal, these two sides are parallel.

Since both pairs of opposite sides are parallel, the quadrilateral is a parallelogram.

53. The line goes through $(0, 2)$ and $(-2, 0)$

$$m = \frac{2 - 0}{0 - (-2)} = \frac{2}{2} = 1$$

The correct choice is (a).

55. The line appears to go through $(0, 0)$ and $(-1, 4)$.

$$m = \frac{4 - 0}{-1 - 0} = \frac{4}{-1} = -4$$

57. (a) See the figure in the textbook.
Segment MN is drawn perpendicular to segment PQ. Recall that MQ is the length of segment MQ.

$$m_1 = \frac{\triangle y}{\triangle x} = \frac{MQ}{PQ}$$

From the diagram, we know that $PQ = 1$. Thus, $m_1 = \frac{MQ}{1}$, so MQ has length m_1.

(b) $\quad m_2 = \dfrac{\triangle y}{\triangle x} = \dfrac{-QN}{PQ} = \dfrac{-QN}{1}$

$$QN = -m_2$$

(c) Triangles MPQ, PNQ, and MNP are right triangles by construction. In triangles MPQ and MNP,

$$\text{angle } M = \text{angle } M,$$

and in the right triangles PNQ and MNP,

$$\text{angle } N = \text{angle } N.$$

Since all right angles are equal, and since triangles with two equal angles are similar, triangle MPQ is similar to triangle MNP and triangle PNQ is similar to triangle MNP.

Therefore, triangles MPQ and PNQ are similar to each other.

(d) Since corresponding sides in similar triangles are proportional,

$$MQ = k \cdot PQ \quad \text{and} \quad PQ = k \cdot QN.$$

$$\frac{MQ}{PQ} = \frac{k \cdot PQ}{k \cdot QN}$$

$$\frac{MQ}{PQ} = \frac{PQ}{QN}$$

From the diagram, we know that $PQ = 1$.

$$MQ = \frac{1}{QN}$$

From (a) and (b), $m_1 = MQ$ and $-m_2 = QN$.

Substituting, we get

$$m_1 = \frac{1}{-m_2}.$$

Multiplying both sides by m_2, we have

$$m_1 m_2 = -1.$$

59. (a) The line goes through $(20, 13{,}900)$ and $(10, 7500)$.

$$m = \frac{13{,}900 - 7500}{20 - 10} = \frac{6400}{10} = 640$$

$$y - 7500 = 640(x - 10)$$
$$y - 7500 = 640x - 6400$$
$$y = 640x + 1100$$

(b) Let $y = 23{,}500$; find x.

$$23{,}500 = 640x + 1100$$
$$22{,}400 = 640x$$
$$35 = x$$

(c) Let $y = 1000$; find x.

$$1000 = 640x + 1100$$
$$-100 = 640x$$
$$-\frac{100}{640} = x$$

Since x is negative, this answer makes no sense. No solar heaters can be made for $1000.

61. (a) The line goes through $(4, .17)$ and $(7, 33)$.

$$m = \frac{.33 - .17}{7 - 4} = \frac{.16}{3} \approx .053$$

$$y - .33 = \frac{.16}{3}(x - 7)$$

$$y - .33 = .053x - .373$$
$$y \approx .053x - .043$$

(b) Let $y = .5$; solve for x.

$$.5 = .053x - .043$$
$$.543 = .053x$$
$$10.2 = x$$

In about 10.2 years, half of these patients will have AIDS.

63. (a) We have the points $(45, 32.5)$ and $(60, 70)$.

$$m = \frac{70 - 32.5}{60 - 45} = \frac{37.5}{15} = 2.5$$

Using point-slope form with $m = 2.5$ and $(x_1, y_1) = (60, 70)$, we have

$$y - 70 = 2.5(x - 60)$$
$$y - 70 = 2.5x - 150$$
$$y = 2.5x - 80.$$

(b) To make the prediction, subsititute $x = 50$ into the formula.

$$y = 2.5(50) - 80$$
$$= 125 - 80 = 45$$

If the Republicans get 50% of the vote, we predict that they will win 45% of the seats.

(c) If $y = 60$,

$$60 = 2.5x - 80$$
$$120 = 5x - 160$$
$$280 = 5x$$
$$x = 56.$$

The Republicans would need to win 56% of the vote to capture 60% of the seats.

65. (a) The line goes through $(0, 152,000)$ and $(39, 330,870)$.

$$m = \frac{330,870 - 152,000}{39 - 0}$$
$$= \frac{178,870}{39}$$
$$\approx 4586.4$$

Since $b = 152,000$, the equation is

$$y = 4586.4x + 152,000.$$

(b) The year 2000 is equivalent to $x = 57$.

$$y = 4586.4(57) + 152,000$$
$$y = 413,424.8$$

There would be about 413,420 illnesses and injuries.

67. (a) Plot the points $(15, 1600)$, $(200, 15,000)$, $(290, 24,000)$, and $(520, 40,000)$.

The points lie approximately on a line, so there appears to be a linear relationship between distance and time.

(b) The graph of any equation of the form $y = mx$ goes through the origin, so the line goes through $(520, 40,000)$ and $(0, 0)$.

$$m = \frac{40,000 - 0}{520 - 0} \approx 76.9$$
$$b = 0$$
$$y = 76.9x + 0$$
$$y = 76.9x$$

(c) Let $y = 60,000$; solve for x.

$$60,000 = 76.9x$$
$$780.23 \approx x$$

Hydra is about 780 megaparsecs from earth.

(d) $A = \dfrac{9.5 \times 10^{11}}{m}, m = 76.9$

$$A = \frac{9.5 \times 10^{11}}{76.9}$$
$$= 12.4 \text{ billion years}$$

1.2   Linear Functions

1. This statement is true.
   When we solve $y = f(x) = 0$, we are finding the value of x when $y = 0$, which is the x-intercept. When we evaluate $f(0)$, we are finding the value of y when $x = 0$, which is the y-intercept.

3. This statement is true.
   Only a vertical line has an undefined slope, but a vertical line is not the graph of a function. Therefore, the slope of a linear function cannot be defined.

5. $f(x) = -2x - 4$
   (a) $f(4) = -2(4) - 4 = -8 - 4 = -12$
   (b) $f(-3) = -2(-3) - 4 = 6 - 4 = 2$
   (c) $f(0) = -2(0) - 4 = 0 - 4 = -4$
   (d) $f(a) = -2(a) - 4 = -2a - 4$

7. $f(x) = 6$
   (a) $f(4) = 6$
   (b) $f(-3) = 6$
   (c) $f(0) = 6$
   (d) $f(a) = 6$

For Exercises 9-23, see the answer graphs in the back of the textbook.

9. $y = x - 1$

Three ordered pairs that satisfy this equation are $(0, -1)$, $(1, 0)$, and $(4, 3)$. Plot these points and draw a line through them.

11. $y = -4x + 9$

Three ordered pairs that satisfy this equation are $(0, 9)$, $(1, 5)$, and $(2, 1)$. Plot these points and draw a line through them.

13. $y = 2x + 1$

Three ordered pairs that satisfy this equation are $(0, 1)$, $(2, 5)$, and $(-2, -3)$. Plot these points and draw a line through them.

15. $3y + 4x = 12$

Find the intercepts.
If $y = 0$, then

$$3(0) + 4x = 12$$
$$4x = 12$$
$$x = 3,$$

so the x-intercept is 3.
If $x = 0$, then

$$3y + 4(0) = 12$$
$$3y = 12$$
$$y = 4,$$

so the y-intercept is 4.
Plot the ordered pairs $(3, 0)$ and $(0, 4)$ and draw a line through these points. (A third point may be used as a check.)

17. $y = -2$

The equation $y = -2$, or, equivalently, $y = 0x - 2$, always gives the same y-value, -2, for any value of x. The graph of this equation is the horizontal line with y-intercept -2.

19. $x + 5 = 0$

This equation may be rewritten as $x = -5$. For any value of y, the x-value is -5. Because all ordered pairs that satisfy this equation have the same first number, this equation does not represent a function. The graph is the vertical line with x-intercept -5.

21. $y = 2x$

Three ordered pairs that satisfy this equation are $(0, 0)$, $(-2, -4)$, and $(2, 4)$. Use these points to draw the graph.

23. $x + 4y = 0$

If $y = 0$, then $x = 0$, so the x-intercept is 0. If $x = 0$, then $y = 0$, so the y-intercept is 0. Both intercepts give the same ordered pair, $(0, 0)$.
To get a second point, choose some other value of x (or y). For example if $x = 4$, then

$$x + 4y = 0$$
$$4 + 4y = 0$$
$$4y = -4$$
$$y = -1,$$

giving the ordered pair $(4, -1)$. Graph the line through $(0, 0)$ and $(4, -1)$.

27. (a) Using the points $(.7, 1.4)$ and $(5.3, 10.9)$, we obtain

$$m = \frac{10.9 - 1.4}{5.3 - .7} = \frac{9.5}{4.6}$$
$$\approx 2.065.$$

To avoid round-off error, keep all digits for the value of m in your calculator; then round the decimals in the final step.
Use the point-slope form.

$$y - 1.4 = \frac{9.5}{4.6}(x - .7)$$
$$y - 1.4 = \frac{9.5}{4.6}x - \frac{9.5}{4.6}(.7)$$
$$y = \frac{9.5}{4.6}x - \frac{9.5}{4.6}(.7) + 1.4$$
$$y = 2.065x - .0456$$

(b) The slope of 2.065 indicates that the number of passengers at these airports is predicted to approximately double between 1992 and 2005.

(c) $f(4.9) = 2.065(4.9) - .0456 \approx 10.1$ million passengers; this agrees favorably with the FAA predication of 10.3 million.

29. $D(q) = 16 - \frac{5}{4}q$

(a) $D(0) = 16 - \frac{5}{4}(0) = 16 - 0 = 16$

When 0 can openers are demanded, the price is $16.

(b) $D(4) = 16 - \dfrac{5}{4}(4) = 16 - 5 = 11$

When 400 can openers are demanded, the price is $11.

(c) $D(8) = 16 - \dfrac{5}{4}(8) = 16 - 10 = 6$

When 800 can openers are demanded, the price is $6.

(d) Let $D(q) = 6$. Find q.

$$6 = 16 - \frac{5}{4}q$$

$$\frac{5}{4}q = 10$$

$$q = 8$$

When the price is $6, 800 can openers are demanded.

(e) Let $D(q) = 11$. Find q.

$$11 = 16 - \frac{5}{4}q$$

$$\frac{5}{4}q = 5$$

$$q = 4$$

When the price is $16, 400 can openers are demanded.

(f) Let $D(q) = 16$. Find q.

$$16 = 16 - \frac{5}{4}q$$

$$0 = -\frac{5}{4}q$$

$$0 = q$$

(g) See the answer graph in the back of the textbook.

(h) $S(q) = \dfrac{3}{4}q$

Let $S(q) = 0$. Find q.

$$0 = \frac{3}{4}q$$

$$0 = q$$

When the price is $0, 0 can openers are supplied.

(i) Let $S(q) = 10$. Find q.

$$10 = \frac{3}{4}q$$

$$\frac{40}{3} = q$$

$$q = 13.\overline{3}$$

When the price is $10, about 1333 can openers are supplied.

(j) Let $S(q) = 20$. Find q.

$$20 = \frac{3}{4}q$$

$$\frac{80}{3} = q$$

$$q = 26.\overline{6}$$

When the price is $20, about 2667 can openers are demanded.

(k) See the answer graph for part (g) in the back of the textbook.

(l) $D(q) = S(q)$

$$16 - \frac{5}{4}q = \frac{3}{4}q$$

$$16 = 2q$$

$$8 = q$$

$$S(8) = \frac{3}{4}(8) = 6$$

The equilibrium quantity is 800, and the equilibrium price is $6.

31. $p = S(q); \ p = D(q) = 100 - \dfrac{2}{5}q$

(a) See the answer graph in the back of the textbook.

(b) $S(q) = D(q)$

$$\frac{2}{5}q = 100 - \frac{2}{5}q$$

$$\frac{4}{5}q = 100$$

$$q = 125$$

$$S(125) = \frac{2}{5}(125) = 50$$

The equilibrium quantity is 125, the equilibrium price is $50.

33. $y = .07x + 135$

(a) $x = 0$

$$y = .07(0) + 135 = \$135$$

(b) $x = 1000$

$$y = .07(1000) + 135$$
$$= 70 + 135 = \$205$$

(c) $x = 2000$

$$y = .07(2000) + 135$$
$$= 140 + 135 = \$275$$

(d) $x = 3000$

$$y = .07(3000) + 135$$
$$= 210 + 135 = \$345$$

(e) See the answer graph in the back of the textbook.

35. (a) Using the points $(2, 9)$ and $(22, 17)$,

$$m = \frac{17 - 9}{22 - 2} = \frac{8}{20} = .4.$$

$$p - 9 = .4(t - 2)$$
$$p - 9 = .4t - .8$$
$$p = .4t + 8.2$$

(b) Let $t = 40$; find p.

$$p = .4(40) + 8.2$$
$$= 16 + 8.2$$
$$= 24.2 \quad \text{or about } 24\%$$

(c) Let $p = 31$; find t.

$$31 = .4t + 8.2$$
$$22.8 = .4t$$
$$57 = t$$

The percentage will reach 31% in the year $1970 + 57 = 2027$.

37. (a) Using the points $(1995, 157)$ and $(2000, 247)$,

$$m = \frac{247 - 157}{2000 - 1995} = \frac{90}{5} = 18.$$

$$y - 157 = 18(x - 1995)$$
$$y - 157 = 18x - 35,910$$
$$y = 18x - 35,753$$

(b) From (a), we see that the slope is 18. This indicates that costs are increasing at a rate of 18 billion dollars per year for the years 1995 to 2000.

1.3   Linear Mathematical Models

1. $12 is the fixed cost and $1 is the cost per hour.

Let $x =$ number of hours;
$C(x) =$ cost of renting a saw for x hours.

Thus,

$$C(x) = \text{fixed cost} + (\text{cost per hour})$$
$$\cdot (\text{number of hours})$$
$$C(x) = 12 + 1x$$
$$12 + x.$$

3. 50¢ is the fixed cost and 35¢ is the cost per half-hour.

Let $x =$ the number of half-hours;
$C(x) =$ the cost of parking a car for x half-hours.

Thus,

$$C(x) = 50 + 35x$$
$$= 35x + 50.$$

5. Fixed cost, \$100; 50 items cost \$1600 to produce.

Let $C(x) =$ cost of producing x items.
$C(x) = mx + b$, where b is the fixed cost.

$$C(x) = mx + 100$$

Now,

$C(x) = 1600$ when $x = 50$, so

$$1600 = m(50) + 100$$
$$1500 = 50m$$
$$30 = m.$$

Thus, $C(x) = 30x + 100$.

7. Marginal cost, \$90; 150 items cost \$16,000 to produce.

$$C(x) = 90x + b$$

Now, $C(x) = 16,000$ when $x = 150$.

$$16,000 = 90(150) + b$$
$$16,000 = 13,500 + b$$
$$2500 = b$$

Thus, $C(x) = 90x + 2500$.

9. For data that can be modeled with a linear function, the average rate of change will be the same as the slope of the line. In cases where the data is not linear, the average rate of change is the slope of the secant line connecting the beginning and ending points.

11. (a) $C(x) = mx + b$
$C(1000) = 2675$; $b = 525$

Find m.

$$2675 = m(1000) + 525$$
$$2150 = 1000m$$
$$2.15 = m$$
$$C(x) = 2.15x + 525$$

(b) $R(x) = 4.95x$

$C(x) = R(x)$

$$2.15x + 525 = 4.95x$$
$$525 = 2.80x$$
$$187.5 = x$$

In order to break even, he must produce and sell 188 books.

(c) $P(x) = R(x) - C(x); \ P(x) = 1000$

$$1000 = 4.95x - (2.15x + 525)$$
$$1000 = 4.95x - 2.15x - 525$$
$$1000 = 2.80x - 525$$
$$1525 = 2.80x$$
$$544.6 = x$$

In order to make a profit of \$1000, he must produce and sell 545 books.

13. $C(10,000) = 547,500; \ C(50,000) = 737,500$

(a) $C(x) = mx + b$

$$m = \frac{737,500 - 547,500}{50,000 - 10,000}$$

$$= \frac{190,000}{40,000} = 4.75$$

$$y - 547,500 = 4.75(x - 10,000)$$
$$y - 547,500 = 4.75x - 47,500$$
$$y = 4.75x + 500,000$$
$$C(x) = 4.75x + 500,000$$

(b) $C(100,000) = 4.75(100,000) + 500,000$

$$= 475,000 + 500,000$$
$$= 975,000$$

The total cost to produce 100,000 items is \$975,000.

(c) The marginal cost is given by the slope of the linear function $C(x)$, namely \$4.75.

15. (a) $R(4) = 45$ (in millions)

$R(x) = mx + b; \ m = .5$

Find b.

$$45 = .5(4) + b$$
$$45 = 2 + b$$
$$43 = b$$
$$R(x) = .5x + 43 \quad \text{(in millions)}$$

(b) Let $R(x) = 50$. Solve for x.

$$50 = .5x + 43$$
$$7 = .5x$$
$$14 = x$$

Revenue will reach \$50 million in $1990 + 14 = 2004$.

17. $\overline{C}(x) = \dfrac{C(x)}{x} = \dfrac{500,000 + 4.75x}{x}$

$$= \frac{500,000}{x} + \frac{4.75x}{x}$$

$$C(x) = \frac{500,000}{x} + 4.75$$

(a) $\overline{C}(1000) = \dfrac{500,000}{1000} + 4.75$

$$= 500 + 4.75$$
$$\overline{C}(1000) = \$504.75$$

(b) $\overline{C}(5000) = \dfrac{500,000}{5000} + 4.75$

$$= 100 + 4.75$$
$$\overline{C}(5000) = \$104.75$$

(c) $\overline{C}(10,000) = \dfrac{500,000}{10,000} + 4.75$

$$= 100 + 4.75$$
$$\overline{C}(10,000) = \$54.75$$

21. $C(x) = 5x + 20; \ R(x) = 15x$

(a) $C(x) = R(x)$

$$5x + 20 = 15x$$
$$20 = 10x$$
$$2 = x$$

The break-even quantity is 2 units.

(b) $P(x) = R(x) - C(x)$

$$P(x) = 15x - (5x + 20)$$
$$P(100) = 15(100) - (5 \cdot 100 + 20)$$
$$= 1500 - 520$$
$$= 980$$

The profit from 100 units is \$980.

(c) $P(x) = 500$

$$15x - (5x + 20) = 500$$
$$10x - 20 = 500$$
$$10x = 520$$
$$x = 52$$

For a profit of \$500, 52 units must be produced.

23. $C(x) = 85x + 900$

$R(x) = 105x$

Set $C(x) = R(x)$ to find the break-even quantity.

$$85x + 900 = 105x$$
$$900 = 20x$$
$$45 = x$$

The break-even quantity is 45 units. You should decide not to produce since no more than 38 units can be sold.

25. $C(x) = 70x + 500$
$R(x) = 60x$

$$70x + 500 = 60x$$
$$10x = -500$$
$$x = -50$$

This represents a break-even quantity of -50 units. It is impossible to make a profit when the break-even quantity is negative. Cost will always be greater than revenue.

27. $C(x) = 85x + 900;\ R(x) = 105x$
$P(x) = R(x) - C(x)$
$P(x) = 105x - (85x + 900)$
$\quad = 105x - 85x - 900$
$\quad = 20x - 900$

29. $C(x) = 70x + 500;\ R(x) = 60x$
$P(x) = R(x) - C(x)$
$P(x) = 60x - (70x + 500)$
$\quad = 60x - 70x - 500$
$\quad = -10x - 500$ (always a loss)

31. Using the points $(1995, 85.635)$ and $(1990, 99.235)$,

$$\frac{\triangle y}{\triangle x} = \frac{99.235 - 85.635}{1990 - 1995}$$
$$= \frac{13.6}{-5}$$
$$= -2.72 \quad \text{(in thousands)}.$$

This means that the number of new car dealers selling used cars is decreasing at the rate of 2720 per year.

33. Let x represent the force and y represent the speed. The linear function contains the points $(.75, 2)$ and $(.93, 3)$.

$$m = \frac{3 - 2}{.93 - .75} = \frac{1}{.18} = \frac{1}{\frac{18}{100}}$$
$$= \frac{100}{18} = \frac{50}{9}$$

Use point-slope form to write the equation.

$$y - 2 = \frac{50}{9}(x - .75)$$
$$y - 2 = \frac{50}{9}x - \frac{50}{9}(.75)$$
$$y = \frac{50}{9}x - \frac{75}{18} + 2$$
$$= \frac{50}{9}x - \frac{39}{18}$$

Now determine y, the speed, when x, the force, is 1.16.

$$y = \frac{50}{9}(1.16) - \frac{39}{18}$$
$$= \frac{58}{9} - \frac{39}{18}$$
$$= \frac{77}{18} \approx 4.3$$

The pony switches from a trot to a gallop at approximately 4.3 meters per second.

35. (a) Using the points $(0, 3022)$ and $(30, 13{,}359)$,

$$m = \frac{13{,}359 - 3022}{30 - 0}$$
$$= \frac{10{,}337}{30} \approx 344.6.$$

$$b = 3022$$
$$y = f(x) = 344.6x + 3022$$

(b) The average rate of change of the cutoff is the slope of the line, \$344.60 per year, which indicates that the cutoff has increased annually by that amount.

37. (a) Using the points $(0, 504)$ and $(14, 374)$ (using 1980 as $x = 0$),

$$m = \frac{504 - 374}{0 - 14} = \frac{130}{-14} \approx -9.3.$$

$$b = 504$$
$$y = f(x) = -9.3x + 504$$

(b) 1987 corresponds to $x = 7$.

$$f(7) = -9.3(7) + 504$$
$$= -65.1 + 504$$
$$= 438.9 \approx 439$$

The average payment in 1987 was about \$439.

(c) Since the slope of the linear function is -9.3, the average payment is decreasing by about \$9.30 per month.

39. Use the formula derived in Example 1 in this section of the textbook.

$$F = \frac{9}{5}C + 32$$

$$C = \frac{5}{9}(F - 32)$$

(a) $C = 37$; find F.

$$F = \frac{9}{5}(37) + 32$$

$$F = \frac{333}{5} + 32$$

$$F = 98.6$$

The Fahrenheit equivalent of $37°C$ is $98.6°F$.

(b) $C = 36.5$; find F.

$$F = \frac{9}{5}(36.5) + 32$$

$$F = 65.7 + 32$$
$$F = 97.7$$

$C = 37.5$; find F.

$$F = \frac{9}{5}(37.5) + 32$$

$$= 67.5 + 32 = 99.5$$

The range is between $97.7°F$ and $99.5°F$.

1.4 Constructing A Mathematical Model

3.

x	y	xy	x^2	y^2
6.8	.8	5.44	46.24	.64
7.0	1.2	8.4	49.0	1.44
7.1	.9	6.39	50.41	.81
7.2	.9	6.48	51.84	.81
7.4	1.5	11.1	54.76	2.25
35.5	5.3	37.81	252.25	5.95

$$r = \frac{5(37.81) - (35.5)(5.3)}{\sqrt{5(252.25) - (35.5)^2} \cdot \sqrt{5(5.95) - (5.3)^2}}$$

$$\approx .6985$$
$$r^2 = (.6985)^2 \approx .5$$

The answer is choice (c).

5. (a)
$$6b + 555m = 754$$
$$555b + 51{,}355m = 69{,}869$$

$$6b = 754 - 555m$$

$$b = \frac{754 - 555m}{6}$$

$$555\left(\frac{754 - 555m}{6}\right) + 51{,}355m = 69{,}869$$

$$555(754 - 555m) + 308{,}130m = 419{,}214$$
$$418{,}470 - 308{,}025m + 308{,}130m = 419{,}214$$
$$105m = 744$$
$$m \approx 7.09$$

$$b = \frac{754 - 55(7.09)}{6} \approx -530$$

$$Y = 7.09x - 530$$

(b) Let $x = 97$; find Y.

$$Y = 7.09(97) - 530 \approx 157.73$$

There will be about 158,000 jobs in 1997.

(c) Let $Y = 170$, find x.

$$170 = 7.09x - 530$$
$$700 = 7.09x$$
$$x = 98.73$$

The number of new jobs will reach 170,000 in 1999.

(d) $$r = \frac{6(69{,}869) - (555)(754)}{\sqrt{6(51{,}355) - 555^2} \cdot \sqrt{6(95{,}740) - 754^2}}$$
$$= .943$$

This indicates that the points on the line give good approximations of the data points.

7. (a)

x	y	xy	x^2	y^2
1979	159	314,661	3,916,441	25,281
1715	148	253,820	2,941,225	21,904
802	110	88,220	643,204	12,100
1552	162	251,424	2,408,704	26,244
1771	179	317,009	3,136,441	32,041
1608	155	249,240	2,585,664	24,025
2548	169	430,612	6,492,304	28,561
2786	169	470,834	7,761,796	28,561
892	104	92,,760	795,664	10,816
1448	139	201,272	2,096,704	19,321
2411	169	407,459	5,812,921	28,561
489	84	41,076	239,121	7056
1792	142	254,464	3,211,264	20,164
2782	169	470,150	7,739,524	28,561
2934	179	525,186	8,608,356	32,041
2815	169	475,735	7.924,225	28,561
1344	147	197,568	1,806,336	21,609
31,668	2553	5,041,490	68,119,894	395,407

$$r = \frac{17(5{,}041{,}490) - (31{,}668)(2553)}{\sqrt{17(68{,}119{,}894) - (31{,}688)^2} \cdot \sqrt{17(395{,}407) - 2553^2}}$$
$$= .863$$

Yes, there is positive linear correlation between the cost of the ticket and the distance flown.

(b)

$$17b + 31{,}668m = 2553$$
$$31{,}668b + 68{,}119{,}894m = 5{,}041{,}490$$

$$17b = 2553 - 31{,}668m$$

$$b = \frac{2553 - 31{,}668m}{17}$$

$$31{,}668 \left(\frac{2553 - 31{,}668m}{17} \right) + 68{,}119{,}894m = 5{,}041{,}490$$

$$31{,}668(2553 - 31{,}668m) + 1{,}158{,}038{,}198m = 85{,}705{,}330$$
$$80{,}848{,}404 - 1{,}002{,}862{,}224m = 85{,}705{,}330$$
$$+ 1{,}158{,}038{,}198m$$
$$155{,}175{,}974m = 4{,}856{,}926$$
$$m \approx .0313$$

$$b = \frac{2553 - 31{,}668(.0313)}{17} \approx 91.868$$

$$Y = .0313x + 91.868$$

The marginal cost per mile to fly is 3.13¢ per mile.

9. $Y = 49.2755 - 1.1924x_1 + .1631x_2$

(a) $x_1 = 42;\ x_2 = 170.0$

$Y = 49.2755 - 1.1924(42) + .1631(170.0)$
  $= 26.9217$ (in thousands) or approximately
  $26,920$

$x_1 = 45;\ x_2 = 170.0$

$Y = 49.2755 - 1.1924(45) + .1631(170.0)$
  $= 23.3445$ (in thousands) or approximately
  $23,340$

$x_1 = 48;\ x_2 = 170.0$

$Y = 49.2755 - 1.1924(48) + .1631(170.0)$
  $= 19.7673$ (in thousands) or approximately
  $19,770$

(b) $x_1 = 42;\ x_2 = 185.0$

$Y = 49.2755 - 1.1924(42) + .1631(185.0)$
  $= 29.3682$ (in thousands) or approximately
  $29,370$

$x_1 = 45;\ x_2 = 185.0$

$Y = 49.2755 - 1.1924(45) + .1631(185.0)$
  $= 25.791$ (in thousands) or approximately
  $25,790$

$x_1 = 48;\ x_2 = 185.0$

$Y = 49.2755 - 1.1924(48) + .1631(185.0)$
  $= 22.2138$ (in thousands) or approximately
  $22,210$

(c) Total estimated sales are greatest for a price level of $42.

11. (a) See the answer graph in the back of the textbook.

If all points are included, the pattern is not linear.

(b)

x	y	xy	x^2	y^2
88	27	2376	7744	729
89	27	2403	7921	729
90	29	2610	8100	841
91	33	3003	8281	1089
92	37	3404	8464	1369
93	43	3999	8649	1849
94	45	4230	8836	2025
637	241	22,025	57,995	8631

$$7b + 636m = 241$$
$$637b + 57{,}995m = 22{,}025$$

$$7b = 241 - 637m$$

$$b = \frac{241 - 637m}{7}$$

$$637 \left(\frac{241 - 637m}{7} \right) + 57{,}995m = 22{,}025$$

$$637(241 - 637m) + 405{,}965m = 154{,}175$$
$$153{,}517 - 405{,}769m + 405{,}965m = 154{,}175$$
$$196m = 658$$
$$m \approx 3.36$$

$$b = \frac{241 - 637(3.36)}{7} \approx -271$$

$$Y = 3.36x - 271.$$

This line fits the data fairly well.
See the answer graph in the back of the textbook.

(c) $r = \dfrac{7(22,025) - (637)(241)}{\sqrt{7(57,995) - (637)^2} \cdot \sqrt{7(8631) - 241^2}}$

$$= .972$$

The value of r does agree with the estimate of fit in part (b).

(d) Even though the points are not linear, the least square line fits nicely between the points with none of the points being far off the line.

13. (a)

x	y	xy	x^2	y^2
88.6	20.0	1772	7849.96	400.0
71.6	16.0	1145.6	5126.56	256.0
93.3	19.8	1847.34	8704.89	392.04
84.3	18.4	1551.12	7106.49	338.56
80.6	17.1	1378.26	6496.36	292.41
75.2	15.5	1165.6	5655.04	240.25
69.7	14.7	1024.59	4858.09	216.09
82.0	17.1	1402.2	6724	292.41
69.4	15.4	1068.76	4816.36	237.16
83.3	16.2	1349.46	6938.89	262.44
79.6	15.0	1194	6336.16	225
82.6	17.2	1420.72	6822.76	295.84
80.6	16.0	1289.6	6496.36	256.0
83.5	17.0	1419.5	6972.25	289.0
76.3	14.4	1098.72	5821.69	207.36
1200.6	249.8	20,127.47	96,725.86	4200.56

$$15b + 1200.6m = 249.8$$
$$1200.6b + 96,725.86m = 20,127.47$$

$$15b = 249.8 - 1200.6m$$
$$b = \frac{249.8 - 1200.6m}{15}$$

$$1200.6\left(\frac{249.8 - 1200.6m}{15}\right) + 96,725.86m = 20,127.47$$
$$1200.6(249.8 - 1200.6m) + 1,450,887.9m = 301,912.05$$
$$299,909.88 - 1,441,440.36m = 301,912.05$$
$$+ 1,450,887.9m$$
$$9447.54m = 2002.17$$
$$m \approx .212$$

$$b = \frac{249.8 - 1200.6(.212)}{15} = -.309$$

$$Y = .212x - .309$$

(b) Let $x = 73$; find Y.
$$Y = .212(73) - .309$$
$$\approx 15.2$$
If the temperature were 73°F, you would expect to hear 15.2 chirps per second.

(c) Let $Y = 18$; find x.
$$18 = .212x - .309$$
$$18.309 = .212x$$
$$86.4 \approx x$$
When the crickets are chirping 18 times per second, the temperature is 86.4°F.

(d)

$$r = \frac{15(20,127) - (1200.6)(249.8)}{\sqrt{15(96,725.86) - (1200.6)^2} \cdot \sqrt{15(4200.56) - 249.8)^2}}$$

$$= .835$$

15. (a)

x	y	xy	x^2	y^2
150	5000	750,000	22,500	25,000,000
175	5500	962,500	30,625	30,250,000
215	6000	1,290,000	46,225	36,000,000
250	6500	1,625,000	62,500	42,250,000
280	7000	1,960,000	78,400	49,000,000
310	7500	2,325,000	96,100	56,250,000
350	8000	2,800,000	122,500	64,000,000
370	8500	3,145,000	136,900	72,250,000
420	9000	3,780,000	176,400	81,000,000
450	9500	4,275,000	202,500	90,250,000
2970	72,500	22,912,500	974,650	546,250,000

$$10b + 2970m = 72,500$$
$$2970b + 974,650m = 22,912,500$$

$$10b = 72,500 - 2970m$$
$$b = 7250 - 297m$$

$$2970(7250 - 297m) + 974,650m = 22,912,500$$
$$21,532,500 - 882,090m + 974,650m = 22,912,500$$
$$92,560m = 1,380,000$$
$$m = 14.9$$

$$b = 7250 - 297(14.9) \approx 2820$$
$$Y = 14.9x + 2820$$

(b) Let $x = 150$; find Y.

$$Y = 14.9(150) + 2820$$
$$Y \approx 5060, \text{ compared to actual } 5000$$

Let $x = 280$; find Y.

$$Y = 14.9(280) + 2820$$
$$\approx 6990, \text{ compared to actual 7000}$$

Let $x = 420$; find Y.

$$Y = 14.9(420) + 2820$$
$$\approx 9080, \text{ compared to actual 9000}$$

(c) Let $x = 230$; find Y.

$$Y = 14.9(230) + 2820$$
$$\approx 6250$$

Adam would need to buy a 6500 BTU air conditioner.

(d) Since cold air is heavier than hot air, it is not necessary to fill the entire room with cold air. Therefore, ft^2 is used (representing area) rather that the volume, ft^3.

$$r = \frac{38(46,209,266) - (175,878)(9989)}{\sqrt{38(872,066,218) - (175,878)^2} \cdot \sqrt{38(2,629,701) - (9989)^2}}$$
$$= -.049$$

No, there does not appear to be a trend.

19. (a)

x	y	xy	x^2	y^2
5	113.4	567	25	12,859.56
15	111.9	1678.5	225	12,521.61
25	111.9	2797.5	625	12,521.61
35	109.7	3839.5	1225	12,034.09
45	106.6	4797	2025	11,363.56
55	105.7	5813.5	3025	11,172.49
65	104.3	6779.5	4225	10,878.49
75	103.7	7777.5	5625	10,753.69
85	101.73	8647.05	7225	10,348.9929
405	968.93	42,697.05	24,225	104,454.0929

$$9b + 405m = 968.93$$
$$405b + 24,225m = 42,697.05$$

$$9b + 405m = 968.93$$
$$9b = 968.93 - 405m$$
$$b = \frac{968.93 - 405m}{9}$$

$$405\left(\frac{968.93 - 405m}{9}\right) + 24,225m = 42,697.05$$

$$405(968.93 - 405m) + 218,025m = 384,273.45$$
$$392,416.65 - 164,025m + 218,025m = 384,273.45$$
$$54,000m = -8143.2$$
$$m = -.1508$$

$$b = \frac{968.93 - 405(-.1508)}{9}$$
$$\approx 114.44$$
$$Y = -.1508x + 114.44$$

(b)

x	y	xy	x^2	y^2
25	144.0	3600.0	625	20,736
35	135.6	4746	1225	10,387.36
45	132.0	5940.0	2025	17,424
55	125.0	6875.0	3025	15,625
65	118.0	7670.0	4225	13,924
75	117.48	8811	5625	13,801.5504
85	113.28	9628.8	7225	12,832.3584
385	885.36	47,270.8	23,975	112,730.2688

$$7b + 385m = 885.36$$
$$385b + 23,975m = 47,270.8$$
$$7b = 885.36 - 385m$$
$$b = 126.48 - 55m$$

$$385(126.48 - 55m) + 23,975m = 47,270.8$$
$$48,694.8 - 21,175m + 23,975m = 47,270.8$$
$$2800m = -1424$$
$$m = -.5086$$

$$b = 126.48 - 55(-.5086)$$
$$\approx 154.45$$
$$Y = -.5086x + 154.45$$

(c) $-.1508x + 114.44 = -.5086x + 154.45$
$$.3578x = 40.01$$
$$x = 111.82 \approx 112$$

The women will catch up with the men in the year $1900 + 112 = 2012$.

(d)

$$r_{men} = \frac{9(42,697.05) - (405)(968.93)}{\sqrt{9(24,225) - 405^2} \cdot \sqrt{9(104,454.0929) - (968.93)^2}}$$
$$= -.9842$$

$$r_{women} = \frac{7(42,270.8) - (385)(885.36)}{\sqrt{7(23,975) - 385^2} \cdot \sqrt{7(112,730.2688) - (885.36)^2}}$$
$$= -.9826$$

21. (a)

x	y	xy	x^2
1	33	33	1
2	34	68	4
3	36	108	9
4	35	140	16
5	40	200	25
6	44	264	36
7	48	336	49
8	45	360	64
9	46	414	81
10	48	480	100
11	49	539	121
12	49	588	144
13	48	624	169
14	54	756	196
15	57	855	225
120	666	5765	1240

$$15b + 120m = 666$$
$$120b + 1240m = 5765$$
$$15b = 666 - 120m$$
$$b = \frac{666 - 120m}{15}$$

$$120\left(\frac{666 - 120m}{15}\right) + 1240m = 5765$$

$$8(666 - 120m) + 1240m = 5765$$
$$5328 - 960m + 1240m = 5765$$
$$280m = 437$$
$$m \approx 1.5607$$

$$b = \frac{666 - 120(1.5607)}{15} \approx 31.914$$

$$Y = 1.5607x + 31.914$$

(b) Let $Y = 75$ (1 hour and 15 minutes beyond 2 hours); find x.

$$75 = 1.5607x + 31.914$$
$$43.086 = 1.5607x$$
$$27.61 \approx x$$

If the trend continues, the average completion time will be 3 hours and 15 minutes in the year $1980 + 27.61 \approx 2008$.

Chapter 1 Review Exercises

1. If the domain is not specified, the domain is assumed to be all real numbers. The range is also assumed to be all real numbers. If the slope is 0, the range is just one number.

3. To complete the coefficient of correlation, you need to compute the following quantities: $\sum x$, $\sum y$, $\sum xy$, $\sum x^2$, $\sum y^2$, and n.

5. Through $(4, -1)$ and $(3, -3)$.

$$m = \frac{-3 - (-1)}{3 - 4}$$
$$= \frac{-3 + 1}{-1}$$
$$= \frac{-2}{-1} = 2$$

7. Through the origin and $(0, 7)$

$$m = \frac{7 - 0}{0 - 0} = \frac{7}{0}$$

The slope of the line is undefined.

9. $4x - y = 7$
$$-y = -4x + 7$$
$$y = 4x - 7$$
$$m = 4$$

11. $3y - 1 = 14$
$$3y = 14 + 1$$
$$3y = 15$$
$$y = 5$$

This is a horizontal line. The slope of a horizontal line is 0.

13. $x = 5y$

$$\frac{1}{5}x = y$$

$$m = \frac{1}{5}$$

15. Through $(8, 0)$, with slope $-\frac{1}{4}$

Use point-slope form.

$$y - 0 = -\frac{1}{4}(x - 8)$$

$$y = -\frac{1}{4}x + 2$$

$$4y = -x + 8$$
$$x + 4y = 8$$

17. Through $(2, -3)$ and $(-3, 4)$

$$m = \frac{4 - (-3)}{-3 - 2} = -\frac{7}{5}$$

Use point-slope form.

$$y - (-3) = -\frac{7}{5}(x - 2)$$

$$y + 3 = -\frac{7}{5}x + \frac{14}{5}$$

$$5y + 15 = -7x + 14$$
$$7x + 5y = -1$$

19. Slope 0, through $(-2, 5)$

Horizontal lines have 0 slope and an equation of the form $y = k$.
The line passes through $(-2, 5)$ so $k = 5$. An equation of the line is $y = 5$.

21. Through $(0, 5)$ and perpendicular to $8x + 5y = 3$
Find the slope of the given line first.

$$8x + 5y = 3$$
$$5y = -8x + 3$$

$$y = \frac{-8}{5}x + \frac{3}{5}$$

$$m = -\frac{8}{5}$$

The perpendicular line has $m = \frac{5}{8}$.
Use point-slope form.

$$y - 5 = \frac{5}{8}(x - 0)$$

$$y = \frac{5}{8}x + 5$$

$$5x - 8y = -40$$

23. Through $(3, -5)$, parallel to $y = 4$

Find the slope of the given line.
$y = 0x + 4$, so $m = 0$, and the required line will also have slope 0.
Use the point-slope form.

$$y - (-5) = 0(x - 3)$$
$$y + 5 = 0$$
$$y = -5$$

For Exercises 25-31, see the answer graphs in the back of the textbook.

25. $y = 4x + 3$

Let $x = 0$.

$$y = 4(0) + 3$$
$$y = 3$$

Let $y = 0$.

$$0 = 4x + 3$$
$$-3 = 4x$$

$$-\frac{3}{4} = x$$

Draw the line through $(0, 3)$ and $\left(-\frac{3}{4}, 0\right)$.

27. $3x - 5y = 15$

$$-5y = -3x + 15$$

$$y = \frac{3}{5}x + 3$$

When $x = 0$, $y = -3$.
When $y = 0$, $x = 5$.
Draw the line through $(0, -3)$ and $(5, 0)$.

29. $x + 2 = 0$

$$x = -2$$

This is the vertical line through $(-2, 0)$.

31. $y = 2x$

When $x = 0$, $y = 0$.
When $x = 1$, $y = 2$.
Draw the line through $(0, 0)$ and $(1, 2)$.

33. (a) $E = 352 + 42x$ (where x is in thousands)

(b) $R = 130x$ (where x is in thousands)

(c)
$$R > E$$
$$130x > 352 + 42x$$
$$88x > 352$$
$$x > 4$$

For a profit to be made, more than 4000 chips must be sold.

35. Using the points $(60, 40)$ and $(100, 60)$,

$$m = \frac{60 - 40}{100 - 60} = \frac{20}{40} = .5.$$

$$p - 40 = .5(q - 60)$$
$$p - 40 = .5q - 30$$
$$p = .5q + 10$$
$$S(q) = .5q + 10$$

37.
$$S(q) = D(q)$$
$$.5q + 10 = -.5q + 72.50$$
$$q = 62.5$$
$$S(62.5) = .5(62.5) + 10$$
$$= 31.25 + 10$$
$$= 41.25$$

The equilibrium price is \$41.25, and the equilibrium quantity is 62.5 diet pills.

39. Fixed cost is \$2000; 36 units cost \$8480.

Two points on the line are $(0, 2000)$ and $(36, 8480)$, so

$$m = \frac{8480 - 2000}{36 - 0} = \frac{6480}{36} = 180.$$

Use point-slope form.

$$y = 180x + 2000$$
$$C(x) = 180x + 2000$$

$$\overline{C}(x) = \frac{C(x)}{x} = \frac{180x + 2000}{x}$$

$$= 180 + \frac{2000}{x}$$

41. Thirty units cost \$1500; 120 units cost \$5640.

Two points on the line are $(30, 1500)$, $(120, 5640)$, so

$$m = \frac{5640 - 1500}{120 - 30}$$

$$= \frac{4140}{90}$$

$$= 46.$$

Use point-slope form.

$$y - 1500 = 46(x - 30)$$
$$y = 46x - 1380 + 1500$$
$$y = 46x + 120$$
$$C(x) = 46x + 120$$

$$\overline{C}(x) = \frac{C(x)}{x} = \frac{46x + 120}{x}$$

$$= 46 + \frac{120}{x}$$

43. (a) $C(x) = 3x + 160; \ R(x) = 7x$

$$C(x) = R(x)$$
$$3x + 160 = 7x$$
$$160 = 4x$$
$$40 = x$$

The break-even quantity is 40 pounds.

(b) $R(40) = 7 \cdot 40 = \$280$

The revenue for 40 pounds is \$280.

45. $Y = 3.9x - 7.9$

Let $x = 6$; find Y.

$$Y = 3.9(6) - 7.9$$
$$= 15.5$$

This number represents 15.5 (\$10,000) = \$155,000, so the company's earnings after 6 years were about \$155,000.

47. Using the points $(88, 8)$ and $(95, 39)$,

$$m = \frac{39 - 8}{95 - 88} = \frac{31}{7} \approx 4.43.$$

$$y - 8 = 4.43(x - 88)$$
$$y - 8 = 4.43x - 389.84$$
$$y = 4.43x - 381.84$$

49. (a)

x	y	xy	x^2	y^2
75	6000	450,000	5625	36,000,000
80	7500	600,000	6400	56,250,000
85	12,000	1,020,000	7225	144,000,000
90	16,000	1,440,000	8100	256,000,000
95	20,400	1,938,000	9025	416,160,000
425	61,900	5,448,000	36,375	908,410,000

$$5b + 425m = 61,900$$
$$425b + 36,375m = 5,448,000$$

$$5b + 425m = 61,900$$
$$5b = 61,900 - 425m$$
$$b = 12,380 - 85m$$

$$425(12,380 - 85m) + 36,375m = 5,448,000$$
$$5,261,500 - 36,125m + 36,375m = 5,448,000$$
$$250m = 186,500$$
$$m = 746$$

$$b = 12,380 - 85(746) = -51,030$$
$$Y = 746x - 51,030$$

(b) Let $x = 100$; find y.

$$y = 746(100) - 51,030$$
$$= 23,570$$

We predict that the average cost of a new car in the year 2000 will be \$23,570.

(c)

$$r = \frac{5(5,448,000) - (425)(61,900)}{\sqrt{5(36,375) - 425^2} \cdot \sqrt{5(908,410,000) - (61,900)^2}}$$

$r = .990$, which means that the line fits the data quite well.

(d) See the answer graph in the back of the textbook.

The scatterplot suggests a curve would fit the data better than a line, even though the coefficient of correlation is high.

51. Using the points $(70, 131.7)$ and $(92, 114.1)$,

$$m = \frac{131.7 - 114.1}{70 - 92} = \frac{17.6}{-22} = -.8.$$

$$y - 131.7 = -.8(x - 70)$$
$$y - 131.7 = -.8x + 56$$
$$y = -.8x + 187.7$$

53. Use a graphing calculator. Follow the steps outlined in Examples 3 and 5 in this section of the textbook.

(a) $Y = 212.9x - 18,227$

(b) $r = .976$

(c) Since $r = .976$, which is close to 1, there is strong positive linear correlation which is well represented by the linear function.

(d) The slope of 212.9 tells us that cellular telephone systems are increasing at the rate of about 213 per year.

(e) $f(100) = 212.9(100) - 18,227$
$$= 3063$$

Extended Application: Using Marginal Cost to Estimate the Cost of Educating Immigrants

1. Since the cost function is linear, the variable cost, m, is the marginal cost, which represents the additional cost per student.

2. The average cost is $\frac{mx+b}{x} = m + \frac{b}{x}$.

3. $\left(m + \frac{b}{x}\right) - m = \frac{b}{x}$, which is the fixed cost per student.

4. As more students enter the system, the fixed cost per student decreases. The variable cost per student, which is usually small, is not a good indicator. For example, if a great deal more students enter the system, new schools have to be built and maintained, which drives the fixed cost up.

Extended Application: Depreciation

1. For straight-line depreciation, the annual depreciation is

$$\frac{\$55,000}{5} = \$11,000.$$

For sum-of-the-years'-digits depreciation,

$$5 + 4 + 3 + 2 + 1 = 15$$

since the machine tool has a 5-yr life. The depreciation is

Year 1: $\frac{5}{15}(\$55,000) = \$18,333.33;$

Year 2: $\frac{4}{15}(\$55,000) = \$14,666.67;$

Year 3: $\frac{3}{15}(\$55,000) = \$11,000.00;$

Year 4: $\frac{2}{15}(\$15,000) = \$7333.33;$

Year 5: $\frac{1}{15}(\$15,000) = \$3666.67.$

For double declining balance depreciation in year one, the depreciation is

$$\frac{2}{5}(\$55,000) = \$22,000.$$

For year two, the depreciation is

$$\$22,000\left(1 - \frac{2}{5}\right) = \$13,200.$$

For year three, the depreciation is

$$\$13,200\left(1 - \frac{2}{5}\right) = \$7920.$$

The total depreciation through year three is

$$\$22,000 + 13,200 + 7920 = \$43,120.$$

The undepreciated balance is

$$\$55,000 - 43,120 = \$11,880.$$

Since the straight-line method is to be used in the last two years, the depreciation for each of these years is

$$\frac{\$11,880}{2} = \$5940.$$

A summary is given below.

Year	Straight-line	Sum-of-the-Years'-Digits	Double Declining Balance
1	$11,000	$18,333.33	$22,000
2	$11,000	$14,666.67	$13,200
3	$11,000	$11,000.00	$7920
4	$11,000	$7333.33	$5940
5	$11,000	$3666.67	$5940

2. The yearly depreciation using the straight-line method is

$$\frac{\$600,000}{10} = \$60,000.$$

Using the sum-of-the-years'-digits method with $n = 10$ and $j = 1$, for year one the depreciation is

$$D(x) = \frac{2(n+1-j)}{n(n+1)}x$$

$$= \frac{2(10+1-1)}{10(10+1)}(\$600,000)$$

$$= \frac{2}{11}(\$600,000)$$

$$= \$109,090.91.$$

For year four the depreciation is

$$D(x) = \frac{2(10+1-4)}{10(10+1)}(\$600,000)$$

$$= \frac{7}{55}(\$600,000)$$

$$= \$76,363.64.$$

Using the double declining balance method, for year one the depreciation is

$$D(x) = \frac{2}{n}x$$

$$= \frac{2}{10}(\$600,000)$$

$$= \$120,000.$$

For year four the depreciation is

$$D(x) = \frac{2}{n} \cdot x \cdot \left(1 - \frac{2}{n}\right)^{j-1}$$

$$= \frac{2}{10}(\$600,000)\left(1 - \frac{2}{10}\right)^{3}$$

$$= \$120,000\left(\frac{8}{10}\right)^{3}$$

$$= \$61,440.$$

A summary is given below.

Year	Straight-line	Sum-of-the-Years'-Digits	Double Declining Balance
1	$60,000	$109,090.91	$120,000
4	$60,000	$76,363.64	$61,440

3. Based on our work in Exercises 1 and 2, both the sum-of-the-years'-digits method and the double declining balance method give the largest deductions in the early years.

Chapter 1 Test

[1.1]

Find the slope of each line that has a slope.

1. Through $(2, -5)$ and $(-1, 7)$

2. Through $(9, 5)$ and $(9, 2)$

3. $3x - 7y = 9$

4. Perpendicular to $2x + 5y = 7$

Find an equation in the form $ax + by = c$ for each line.

5. Through $(-1, 6)$ and $(5, -3)$

6. x–intercept 6, y–intercept -5

7. Through $(0, 3)$, parallel to $2x - 4y = 1$

8. Through $(1, -4)$, perpendicular to $3x + y = 1$

9. Through $(2, 5)$, perpendicular to $x = 5$

[1.2]

Graph each of the following.

10. $3x + 5y = 15$

11. $2x + y = 0$

12. $x - 3 = 0$

13. Let the supply and demand functions for a certain product be given by the following equations.

$$\text{Supply: } p = .20q - 5 \qquad \text{Demand: } p = 100 - .15q,$$

where p represents the price (in dollars) at a supply or demand, respectively, of q units.

   (a) Graph these equations on the same axes.

   (b) Find the equilibrium price.

   (c) Find the equilibrium quantity.

[1.3]

14. For a given product, eight units cost $450, while forty units cost $770.

   (a) Find the appropriate linear cost function.

   (b) What is the fixed cost?

   (c) What is the marginal cost per item?

   (d) Find the average cost function.

15. For a given product, the variable cost is $100, while 150 items cost $16,000 to produce.

   (a) Find the appropriate linear cost function.

   (b) What is the fixed cost?

   (c) What is the marginal cost per item?

   (d) Find the average cost function.

16. Producing x hundred units of widgets costs $C(x) = 4x + 16$; revenue is $12x$, where $R(x)$ and $C(x)$ are in thousands of dollars.

    (a) What is the break-even quantity?

    (b) What is the profit from 300 units?

    (c) How many units will produce a profit of $40,000?

[1.4]

17. An electronics firm was planning to expand its product line and wanted to get an idea of the salary picture for technicians it would hire in this field. The following data was collected.

$$
\begin{aligned}
n &= 12 & \sum x^2 &= 3162 \\
\sum x &= 176 & \sum y^2 &= 10{,}870 \\
\sum y &= 356 \\
\sum xy &= 5629
\end{aligned}
$$

    (a) Find an equation for the least squares line.

    (b) Find the coefficient of correlation.

18. An economist was interested in the production costs for companies supplying chemicals for use in fertilizers. The data below represents the relationship between the number of tons produced during a given year (x) and the production cost per ton (y) for seven companies.

Number of Tons (in thousands)	Cost per Ton (in dollars)
3.0	40
4.0	50
2.4	50
5.0	35
2.6	55
4.0	35
5.5	30

    (a) Find the equation for the least squares line.

    (b) Find the coefficient of correlation.

Chapter 1 Test Answers

1. -4

2. Undefined

3. $\frac{3}{7}$

4. $\frac{5}{2}$

5. $3x + 2y = 9$

6. $5x - 6y = 30$

7. $x - 2y = -6$

8. $x - 3y = 13$

9. $y = 5$

10.

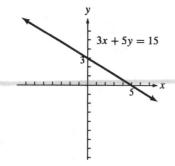

11.

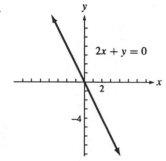

12.

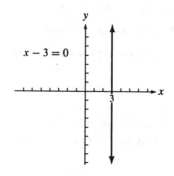

13. (a)

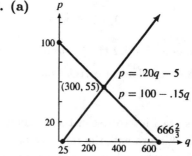

(b) \$55   **(c)** 300 units

14. (a) $y = 10x + 370$ **(b)** \$370
(c) \$10   **(d)** $\overline{C}(x) = 10 + \frac{370}{x}$

15. (a) $y = 100x + 1000$ **(b)** \$1000
(c) \$100   **(d)** $\overline{C}(x) = 100 + \frac{1000}{x}$

16. (a) 200   **(b)** \$8000   **(c)** 700

17. (a) $Y = .70x + 19.37$ **(b)** $r = .96$

18. (a) $Y = -6.37x + 66.24$ **(b)** $r = -.79$

SYSTEMS OF LINEAR EQUATIONS AND MATRICES

2.1 Solution of Linear Systems by the Echelon Method

In Exercises 1-15 and 19-25, check each solution by substitution in the original equations of the system.

1. $x + y = 9$ (1)
$2x - y = 0$ (2)

Multiply equation (1) by -2 and add the result to equation (2). The new system is

$$x + y = 9 \quad (1)$$
$$-2R_1 + R_2 \rightarrow R_2 \qquad -3y = -18. \quad (3)$$

Multiply equation (3) by $-\frac{1}{3}$ to get

$$x + y = 9 \quad (1)$$
$$-\frac{1}{3}R_2 \rightarrow R_2 \qquad y = 6. \quad (4)$$

Substitute 6 for y in equation (1) to get $x = 3$. The solution is $(3, 6)$.

3. $5x + 3y = 7$ (1)
$7x - 3y = -19$ (2)

To eliminate x in equation (2), multiply equation (1) by 7 and equation (2) by -5. Add the results. The new system is

$$5x + 3y = 7 \quad (1)$$
$$7R_1 + (-5)R_2 \rightarrow R_2 \qquad 36y = 144. \quad (3)$$

Now make the leading coefficient in each row equal 1. To accomplish this, multiply equation (1) by $\frac{1}{5}$ and equation (3) by $\frac{1}{36}$.

$$\tfrac{1}{5}R_1 \rightarrow R_1 \quad x + \frac{3}{5}y = \frac{7}{5} \quad (4)$$

$$\tfrac{1}{36}R_2 \rightarrow R_2 \qquad y = 4 \quad (5)$$

Back-substitution of 4 for y in equation (4) gives

$$x + \frac{3}{5}(4) = \frac{7}{5}$$
$$x + \frac{12}{5} = \frac{7}{5}$$
$$x = -1.$$

The solution is $(-1, 4)$.

5. $3x + 2y = -6$ (1)
$5x - 2y = -10$ (2)

Eliminate x in equation (2) to get the system

$$3x + 2y = -6 \quad (1)$$
$$5R_1 + (-3)R_2 \rightarrow R_2 \qquad 16y = 0. \quad (3)$$

Make the leading coefficient in each equation equal 1.

$$\tfrac{1}{3}R_1 \rightarrow R_1 \quad x + \frac{2}{3}y = -2 \quad (4)$$

$$\tfrac{1}{16}R_2 \rightarrow R_2 \qquad y = 0 \quad (5)$$

Substitute 0 for y in equation (4) to get $x = -2$. The solution is $(-2, 0)$.

7. $2x - 3y = -7$ (1)
$5x + 4y = 17$ (2)

Eliminate x in equation (2).

$$2x - 3y = -7 \quad (1)$$
$$5R_1 + (-2)R_2 \rightarrow R_2 \qquad -23y = -69 \quad (3)$$

Make the leading coefficient in each equation equal 1.

$$\tfrac{1}{2}R_1 \rightarrow R_1 \quad x - \frac{3}{2}y = -\frac{7}{2} \quad (4)$$

$$-\tfrac{1}{23}R_2 \rightarrow R_2 \qquad y = 3 \quad (5)$$

Substitute 3 for y in equation (4) to get $x = 1$. The solution is $(1, 3)$.

9. $5p + 7q = 6$ (1)
$10p - 3q = 46$ (2)

Eliminate p in equation (2).

$$5p + 7q = 6 \quad (1)$$
$$(-2)R_1 + R_2 \rightarrow R_2 \qquad -17q = 34 \quad (3)$$

Make the leading coefficient in each equation equal 1.

$$\tfrac{1}{5}R_1 \rightarrow R_1 \quad p + \frac{7}{5}q = \frac{6}{5} \quad (4)$$

$$-\tfrac{1}{17}R_2 \rightarrow R_2 \qquad q = -2 \quad (5)$$

Substitute -2 for q in equation (4) to get $p = 4$. The solution is $(4, -2)$.

11. $6x + 7y = -2$ (1)
     $7x - 6y = 26$ (2)

Eliminate x in equation (2).

$$6x + 7y = -2 \quad (1)$$
$$7R_1 + (-6)R_2 \to R_2 \qquad 85y = -170 \quad (3)$$

Make the leading coefficient in each equation equal 1.

$$\tfrac{1}{6}R_1 \to R_1 \quad x + \tfrac{7}{6}y = -\tfrac{1}{3} \quad (4)$$
$$\tfrac{1}{85}R_2 \to R_2 \qquad y = -2 \quad (5)$$

Substitute -2 for y in equation (4) to get $x = 2$.
The solution is $(2, -2)$.

13. $3x + 2y = 5$ (1)
     $6x + 4y = 8$ (2)

Eliminate x in equation (2).

$$3x + 2y = 5 \quad (1)$$
$$-2R_1 + R_2 \to R_2 \qquad 0 = -2 \quad (3)$$

Equation (3) is a false statement.
The system is inconsistent and has no solution.

15. $4x - y = 9$ (1)
    $-8x + 2y = -18$ (2)

Eliminate x in equation (2).

$$4x - y = 9 \quad (1)$$
$$2R_1 + R_2 \to R_2 \qquad 0 = 0 \quad (3)$$

The true statement in equation (3) indicates that there are an infinite number of solutions for the system. Solve equation (1) for x.

$$4x - y = 9 \qquad (1)$$
$$4x = y + 9$$
$$x = \frac{y + 9}{4} \qquad (4)$$

For each value of y, equation (4) indicates that $x = \frac{y+9}{4}$, and all ordered pairs of the form $\left(\frac{y+9}{4}, y\right)$ are solutions.

17. An inconsistent system has *no* solutions.

19. $\dfrac{x}{2} + \dfrac{y}{3} = 8$ (1)
     $\dfrac{2x}{3} + \dfrac{3y}{2} = 17$ (2)

Rewrite the equations without fractions.

$$6R_1 \to R_1 \quad 3x + 2y = 48 \quad (3)$$
$$6R_2 \to R_2 \quad 4x + 9y = 102 \quad (4)$$

Eliminate x in equation (4).

$$3x + 2y = 48 \quad (3)$$
$$-4R_1 + 3R_2 \to R_2 \qquad 19y = 114 \quad (5)$$

Make each leading coefficient equal 1.

$$\tfrac{1}{3}R_1 \to R_1 \quad x + \tfrac{2}{3}y = 16 \quad (6)$$
$$\tfrac{1}{19}R_2 \to R_2 \qquad y = 6 \quad (7)$$

Substitute 6 for y in equation (6) to get $x = 12$.
The solution is $(12, 6)$.

21. $\dfrac{x}{2} + y = \dfrac{3}{2}$ (1)
     $\dfrac{x}{3} + y = \dfrac{1}{3}$ (2)

Rewrite the equations without fractions.

$$2R_1 \to R_1 \quad x + 2y = 3 \quad (3)$$
$$3R_2 \to R_2 \quad x + 3y = 1 \quad (4)$$

Eliminate x in equation (4).

$$x + 2y = 3 \quad (3)$$
$$-1R_1 + R_2 \to R_2 \qquad y = -2 \quad (5)$$

Substitute -2 for y in equation (3) to get $x = 7$.
The solution is $(7, -2)$.

23. $x + y + z = 2$ (1)
     $2x + y - z = 5$ (2)
     $x - y + z = -2$ (3)

Eliminate x in equations (2) and (3).

$$x + y + z = 2 \quad (1)$$
$$-2R_1 + R_2 \to R_2 \qquad -y - 3z = 1 \quad (4)$$
$$-1R_1 + R_3 \to R_3 \qquad -2y = -4 \quad (5)$$

Eliminate y in equation (5).

$$x + y + z = 2 \quad (1)$$
$$-y - 3z = 1 \quad (4)$$
$$-2R_2 + R_3 \to R_3 \qquad 6z = -6 \quad (6)$$

Make each leading coefficient equal 1.

$$x + y + z = 2 \quad (1)$$
$$-1R_2 \to R_2 \qquad y + 3z = -1 \quad (7)$$
$$\tfrac{1}{6}R_3 \to R_3 \qquad z = -1 \quad (8)$$

Substitute -1 for z in equation (7) to get $y = 2$.
Finally, substitute -1 for z and 2 for y in equation (1) to get $x = 1$. The solution is $(1, 2, -1)$.

25. $x + 3y + 4z = 14$ (1)
$2x - 3y + 2z = 10$ (2)
$3x - y + z = 9$ (3)

Eliminate x in equations (2) and (3).

$$x + 3y + 4z = 14 \quad (1)$$
$$-2R_1 + R_2 \rightarrow R_2 \quad -9y - 6z = -18 \quad (4)$$
$$-3R_1 + R_3 \rightarrow R_3 \quad -10y - 11z = -33 \quad (5)$$

Eliminate y in equation (5).

$$x + 3y + 4z = 14 \quad (1)$$
$$-9y - 6z = -18 \quad (4)$$
$$10R_2 + (-9)R_3 \rightarrow R_3 \quad 39z = 117 \quad (6)$$

Make each leading coefficient equal 1.

$$x + 3y + 4z = 14 \quad (1)$$
$$-\tfrac{1}{9}R_2 \rightarrow R_2 \quad y + \tfrac{2}{3}z = 2 \quad (7)$$
$$\tfrac{1}{39}R_3 \rightarrow R_3 \quad z = 3 \quad (8)$$

Substitute 3 for z in equation (2) to get $y = 0$.
Finally, substitute 3 for z and 0 for y in equation
(1) to get $x = 2$. The solution is $(2, 0, 3)$.

29. $3x + y - z = 0$ (1)
$2x - y + 3z = -7$ (2)

Eliminate x in equation (2).

$$3x + y - z = 0 \quad (1)$$
$$2R_1 + (-3)R_2 \rightarrow R_2 \quad 5y - 11z = 21 \quad (3)$$

Make each leading coefficient equal 1.

$$\tfrac{1}{3}R_1 \rightarrow R_1 \quad x + \tfrac{1}{3}y - \tfrac{1}{3}z = 0 \quad (4)$$
$$\tfrac{1}{5}R_2 \rightarrow R_2 \quad y - \tfrac{11}{5}z = \tfrac{21}{5} \quad (5)$$

Solve equation (5) for y in terms of z.

$$y = \frac{11}{5}z + \frac{21}{5}$$

Substitute this expression for y in equation (4),
and solve the equation for x.

$$x + \frac{1}{3}\left(\frac{11}{5}z + \frac{21}{5}\right) - \frac{1}{3}z = 0$$
$$x + \frac{11}{15}z + \frac{7}{5} - \frac{1}{3}z = 0$$
$$x + \frac{2}{5}z = -\frac{7}{5}$$
$$x = -\frac{2}{5}z - \frac{7}{5}$$

The solution is

$$\left(-\frac{2}{5}z - \frac{7}{5}, \frac{11}{5}z + \frac{21}{5}, z\right) \text{ or}$$
$$\left(\frac{-2z - 7}{5}, \frac{11z + 21}{5}, z\right).$$

31. $-x + y - z = -7$ (1)
$2x + 3y + z = 7$ (2)

Eliminate x in equation (2).

$$-x + y - z = -7 \quad (1)$$
$$2R_1 + R_2 \rightarrow R_2 \quad 5y - z = -7 \quad (3)$$

Make each leading coefficient equal 1.

$$-1R_1 \rightarrow R_1 \quad x - y + z = 7 \quad (4)$$
$$\tfrac{1}{5}R_2 \rightarrow R_2 \quad y - \tfrac{1}{5}z = -\tfrac{7}{5} \quad (5)$$

Solve equation (5) for y in terms of z.

$$y = \frac{1}{5}z - \frac{7}{5}$$

Substitute this expression for y in equation (4),
and solve the equation for x.

$$x - \left(\frac{1}{5}z - \frac{7}{5}\right) + z = 7$$
$$x - \frac{1}{5}z + \frac{7}{5} + z = 7$$
$$x + \frac{4}{5}z = \frac{28}{5}$$
$$x = -\frac{4}{5}z + \frac{28}{5}$$

The solution of the system is

$$\left(-\frac{4}{5}z + \frac{28}{5}, \frac{1}{5}z - \frac{7}{5}, z\right) \text{ or}$$
$$\left(\frac{-4z + 28}{5}, \frac{z - 7}{5}, z\right).$$

35. Let $x =$ the cost per pound of rice
and $y =$ the cost per pound of potatoes.

The system to be solved is

$$20x + 10y = 16.20 \quad (1)$$
$$30x + 12y = 23.04 \quad (2)$$

Multiply equation (1) by $\frac{1}{20}$.

$$\tfrac{1}{20}R_1 \rightarrow R_1 \quad x + .5y = .81 \quad (3)$$
$$30x + 12y = 23.04 \quad (2)$$

Eliminate x in equation (2).

$$x + .5y = .81 \quad (3)$$
$$-30R_1 + R_2 \rightarrow R_2 \qquad -3y = -1.26 \quad (4)$$

Multiply equation (4) by $-\frac{1}{3}$.

$$x + .5y = .81 \quad (3)$$
$$-\frac{1}{3}R_2 \rightarrow R_2 \qquad y = .42 \quad (5)$$

Substitute .42 for y in equation (3).

$$x + .5(.42) = .81$$
$$x + .21 = .81$$
$$x = .60$$

The cost of 10 pounds of rice and 50 pounds of potatoes is

$$10(.60) + 50(.42) = 27,$$

that is, $27.

37. Let $x =$ the number of seats on the main floor and $y =$ the number of seats in the balcony.

The system to be solved is

$$8x + 5y = 4200 \quad (1)$$
$$.25(8x) + .40(5y) = 1200. \quad (2)$$

Make the leading coefficient in equation (1) equal 1.

$$\frac{1}{8}R_1 \rightarrow R_1 \qquad x + \frac{5}{8}y = 525 \quad (3)$$
$$2x + 2y = 1200 \quad (2)$$

Eliminate x in equation (2).

$$x + \frac{5}{8}y = 525 \quad (3)$$
$$-2R_1 + R_2 \rightarrow R_2 \qquad \frac{6}{8}y = 150 \quad (4)$$

Make the leading coefficient in equation (4) equal 1.

$$x + \frac{5}{8}y = 525 \quad (3)$$
$$\frac{8}{6}R_2 \rightarrow R_2 \qquad y = 200 \quad (5)$$

Substitute 200 for y in equation (3).

$$x + \frac{5}{8}(200) = 525$$
$$x + 125 = 525$$
$$x = 400$$

There are 400 main floor seats and 200 balcony seats.

39. Let $x =$ the number of model 201 to make each day

and $y =$ the number of model 301 to make each day.

The system to be solved is

$$2x + 3y = 34 \quad (1)$$
$$18x + 27y = 335. \quad (2)$$

Make the leading coefficient in equation (1) equal 1.

$$\frac{1}{2}R_1 \rightarrow R_1 \qquad x + \frac{3}{2}y = 17 \quad (3)$$
$$18x + 27y = 335 \quad (2)$$

Eliminate x in equation (2).

$$x + \frac{3}{2}y = 17 \quad (3)$$
$$-18R_1 + R_2 \rightarrow R_2 \qquad 0 = 29 \quad (4)$$

Since equation (4) is false, the system is inconsistent. Therefore, this situation is impossible.

41. Let $x =$ the amount invested at 6.5% in mututal funds,

$y =$ the amount invested in government bonds at 6%, and

$z =$ the amount invested in the bank at 5%.

Since she invested $10,000 among these 3 accounts, one equation will be $x + y + z = 10,000$. Since she invested twice as much in bonds as she did in mutual funds, another equation is $y = 2x$. Since her annual return on the investment is $605, the last equation is $.065x + .06y + .05z = 605$.

The system to be solved is

$$x + y + z = 10,000 \quad (1)$$
$$2x - y = 0 \quad (2)$$
$$.065x + .06y + .05z = 605. \quad (3)$$

First eliminate the decimals in equation (3) by multiplying by 1000 to obtain the system

$$x + y + z = 10,000 \quad (1)$$
$$2x - y = 0 \quad (2)$$
$$65x + 60y + 50z = 605,000. \quad (4)$$

Eliminate x in equations (2) and (4).

$$x + y + z = 10,000 \quad (1)$$
$$-2R_1 + R_2 \rightarrow R_2 \qquad -3y - 2z = -20,000 \quad (5)$$
$$-65R_1 + R_3 \rightarrow R_3 \qquad -5y - 15z = -45,000 \quad (6)$$

Eliminate y in equation (6).

$$
\begin{array}{rl}
x + y + z = & 10{,}000 \quad (1) \\
-3y - 2z = & -20{,}000 \quad (5) \\
-35z = & -35{,}000 \quad (7)
\end{array}
$$

$-5R_2 + 3R_3 \to R_3$

Make the leading coefficients equal 1.

$$
\begin{array}{rl}
x + y + z = 10{,}000 & \quad (1)
\end{array}
$$

$-\frac{1}{3}R_2 \to R_2 \qquad y + \dfrac{2}{3}z = \dfrac{20{,}000}{3} \quad (8)$

$-\frac{1}{35}R_3 \to R_3 \qquad\qquad z = 1000 \quad (9)$

Back-substitution of 1000 for z in equation (8) gives

$$
y + \frac{2}{3}(1000) = \frac{20{,}000}{3}
$$

$$
y = \frac{18{,}000}{3} = 6000.
$$

Substitute 6000 for y and 1000 for z in equation (1) to find x.

$$
x + 6000 + 1000 = 10{,}000
$$

$$
x = 3000.
$$

Nui invested \$3000 at 6.5%, \$6000 at 6%, and \$1000 at 5%.

43. Let $x =$ the number of buffets produced each week,

$y =$ the number of chairs produced each week, and

$z =$ the number of tables produced each week.

Make a table.

	Buffet	Chair	Table	Totals
Construction	30	10	10	350
Finishing	10	10	30	150

The system to be solved is

$$
\begin{array}{rl}
30x + 10y + 10z = 350 & \quad (1) \\
10x + 10y + 30z = 150. & \quad (2)
\end{array}
$$

Make the leading coefficient of equation (1) equal 1.

$\frac{1}{30}R_1 \to R_1 \qquad x + \dfrac{1}{3}y + \dfrac{1}{3}z = \dfrac{35}{3} \quad (3)$

$$
10x + 10y + 30z = 150 \quad (2)
$$

Eliminate x from equation (2).

$$
x + \frac{1}{3}y + \frac{1}{3}z = \frac{35}{3} \quad (3)
$$

$-10R_1 + R_2 \to R_2 \qquad \dfrac{20}{3}y + \dfrac{80}{3}z = \dfrac{100}{3} \quad (4)$

Solve equation (4) for y. Multiply by 3.

$$
\begin{array}{rl}
20y + 80z = 100 \\
y + 4z = 5 \\
y = 5 - 4z
\end{array}
$$

Substitute $5 - 4z$ for y in equation (1) and solve for x.

$$
\begin{array}{rl}
30x + 10(5 - 4z) + 10z = 350 \\
30x + 50 - 40z + 10z = 350 \\
30x = 300 + 30z \\
x = 10 + z
\end{array}
$$

The solution is $(10 + z, 5 - 4z, z)$. All variables must be nonnegative integers. Therefore,

$$
\begin{array}{rl}
5 - 4z \geq 0 \\
5 \geq 4z \\
z \leq \dfrac{5}{4},
\end{array}
$$

so $z = 0$ or $z = 1$. (Any larger value of z would cause y to be negative which would make no sense in the problem.) If $z = 0$, then the solution is $(10, 5, 0)$. If $z = 1$, then the solution is $(11, 1, 1)$.

Therefore, the company should make either 10 buffets, 5 chairs, and no tables or 11 buffets, 1 chair, and 1 table.

45. (a) We are given the equation

$$
y = ax^2 + bx.
$$

After substituting the given values for the stopping distances (y) and speeds (x) in mph, the system to be solved is

$$
\begin{array}{rl}
61.7 = a(25)^2 + b(25) & \quad (1) \\
106 = a(35)^2 + b(35). & \quad (2)
\end{array}
$$

These equations can be written as

$$
\begin{array}{rl}
625a + 25b = 61.7 & \quad (1) \\
1225a + 35b = 106. & \quad (2)
\end{array}
$$

Multiply equation (1) by $\frac{1}{625}$; also eliminate the decimal in 61.7 by multiplying numerator and denominator of the fraction by 10.

$\frac{1}{625}R_1 \to R_1 \qquad a + \dfrac{1}{25}b = \dfrac{617}{6250} \quad (3)$

$$
1225a + 35b = 106 \quad (2)
$$

Eliminate a in equation (2).

$$a + \frac{1}{25}b = \frac{617}{6250} \quad (3)$$

$-1225R_1 + R_2 \to R_2 \qquad -14b = -\frac{3733}{250} \quad (4)$

Multiply equation (4) by $-\frac{1}{14}$.

$$a + \frac{1}{25}b = \frac{617}{6250} \quad (3)$$

$-\frac{1}{14}R_2 \to R_2 \qquad b = \frac{3733}{3500} \quad (5)$

Substitute $\frac{3733}{3500}$ for b in equation (3).

$$a + \frac{1}{25}\left(\frac{3733}{3500}\right) = \frac{617}{6250}$$

$$a = \frac{4905}{87,500}$$

Therefore,

$$a = \frac{4905}{87,500} \approx .056057,$$

and $\quad b = \frac{3733}{3500} \approx 1.06657.$

(b) Substitute the values from part (a) for a and b and 55 for x in the equation $y = ax^2 + bx$. Solve for y.

$$y = .056057(55)^2 + 1.06657(55)$$
$$y \approx 228$$

The stopping distance of a car traveling 55 mph is approximately 228 ft.

2.2   Solution of Linear Systems by the Gauss-Jordan Method

1. $2x + 3y = 11$
$\quad x + 2y = 8$

The equations are already in proper form. The augmented matrix obtained from the coefficients and the constants is

$$\begin{bmatrix} 2 & 3 & | & 11 \\ 1 & 2 & | & 8 \end{bmatrix}.$$

3. $2x + y + z = 3$
$\quad 3x - 4y + 2z = -7$
$\quadx + y + z = 2$

leads to the augmented matrix

$$\begin{bmatrix} 2 & 1 & 1 & | & 3 \\ 3 & -4 & 2 & | & -7 \\ 1 & 1 & 1 & | & 2 \end{bmatrix}.$$

5. We are given the augmented matrix

$$\begin{bmatrix} 1 & 0 & | & 2 \\ 0 & 1 & | & 3 \end{bmatrix}.$$

This is equivalent to the system of equations

$$x = 2$$
$$y = 3,$$

or $x = 2$, $y = 3$.

7. $\begin{bmatrix} 1 & 0 & 0 & | & 2 \\ 0 & 1 & 0 & | & 3 \\ 0 & 0 & 1 & | & -2 \end{bmatrix}$

The system associated with this matrix is

$$x = 2$$
$$y = 3$$
$$z = -2,$$

or $x = 2$, $y = 3$, $z = -2$.

9. *Row operations* on a matrix correspond to transformations of a system of equations.

11. $\begin{bmatrix} 2 & 3 & 8 & | & 20 \\ 1 & 4 & 6 & | & 12 \\ 0 & 3 & 5 & | & 10 \end{bmatrix}$

Find $R_1 + (-2)R_2$. In row 2, column 1,

$$2 + (-2)1 = 0.$$

In row 2, column 2,

$$3 + (-2)4 = -5.$$

In row 2, column 3,

$$8 + (-2)6 = -4.$$

Replace R_2 with these values. The new matrix is

$$\begin{bmatrix} 2 & 3 & 8 & | & 20 \\ 0 & -5 & -4 & | & -4 \\ 0 & 3 & 5 & | & 10 \end{bmatrix}.$$

13. $\begin{bmatrix} 1 & 4 & 2 & | & 9 \\ 0 & 1 & 5 & | & 14 \\ 0 & 3 & 8 & | & 16 \end{bmatrix}$

$-4R_2 + R_1 \to R_1$

$$\begin{bmatrix} -4(0)+1 & -4(1)+4 & -4(5)+2 & | & -4(14)+9 \\ 0 & 1 & 5 & | & 14 \\ 0 & 3 & 8 & | & 16 \end{bmatrix}$$

$$= \begin{bmatrix} 1 & 0 & -18 & | & -47 \\ 0 & 1 & 5 & | & 14 \\ 0 & 3 & 8 & | & 16 \end{bmatrix}$$

15. $\begin{bmatrix} 3 & 0 & 0 & | & 18 \\ 0 & 5 & 0 & | & 9 \\ 0 & 0 & 4 & | & 8 \end{bmatrix}$

$\frac{1}{3}R_1 \to R_1$

$\begin{bmatrix} \frac{1}{3}(3) & \frac{1}{3}(0) & \frac{1}{3}(0) & | & \frac{1}{3}(18) \\ 0 & 5 & 0 & | & 9 \\ 0 & 0 & 4 & | & 8 \end{bmatrix} = \begin{bmatrix} 1 & 0 & 0 & | & 6 \\ 0 & 5 & 0 & | & 9 \\ 0 & 0 & 4 & | & 8 \end{bmatrix}$

17. $x + y = 5$
$x - y = -1$

has augmented matrix

$$\begin{bmatrix} 1 & 1 & | & 5 \\ 1 & -1 & | & -1 \end{bmatrix}.$$

Use row operations as follows.

$-1R + R_2 \to R_2 \quad \begin{bmatrix} 1 & 1 & | & 5 \\ 0 & -2 & | & -6 \end{bmatrix}$

$-\frac{1}{2}R_2 \to R_2 \quad \begin{bmatrix} 1 & 1 & | & 5 \\ 0 & 1 & | & 3 \end{bmatrix}$

$-1R_2 + R_1 \to R_1 \quad \begin{bmatrix} 1 & 0 & | & 2 \\ 0 & 1 & | & 3 \end{bmatrix}$

Read the solution from the last column of the matrix. The solution is $(2, 3)$.

19. $2x - 5y = 10$
$4x - 5y = 15$

Write the augmented matrix and use row operations.

$$\begin{bmatrix} 2 & -5 & | & 10 \\ 4 & -5 & | & 15 \end{bmatrix}$$

$-2R_1 + R_2 \to R_2 \quad \begin{bmatrix} 2 & -5 & | & 10 \\ 0 & 5 & | & -5 \end{bmatrix}$

$R_1 + R_2 \to R_1 \quad \begin{bmatrix} 2 & 0 & | & 5 \\ 0 & 5 & | & -5 \end{bmatrix}$

$\frac{1}{2}R_1 \to R_1 \quad \begin{bmatrix} 1 & 0 & | & \frac{5}{2} \\ 0 & 1 & | & -1 \end{bmatrix}$
$\frac{1}{5}R_2 \to R_2$

The solution is $\left(\frac{5}{2}, -1\right)$.

21. $2x - 3y = 2$
$4x - 6y = 1$

Write the augmented matrix and use row operations.

$$\begin{bmatrix} 2 & -3 & | & 2 \\ 4 & -6 & | & 1 \end{bmatrix}$$

$-2R_1 + R_2 \to R_2 \quad \begin{bmatrix} 2 & -3 & | & 2 \\ 0 & 0 & | & -3 \end{bmatrix}$

The system associated with the last matrix is

$$2x - 3y = 2$$
$$0x + 0y = -3.$$

Since the second equation, $0 = -3$, is false, the system is inconsistent and therefore has no solution.

23. $6x - 3y = 1$
$-12x + 6y = -2$

Write the augmented matrix of the system and use row operations.

$$\begin{bmatrix} 6 & -3 & | & 1 \\ -12 & 6 & | & -2 \end{bmatrix}$$

$2R_1 + R_2 \to R_2 \quad \begin{bmatrix} 6 & -3 & | & 1 \\ 0 & 0 & | & 0 \end{bmatrix}$

$\frac{1}{6}R_1 \to R_1 \quad \begin{bmatrix} 1 & -\frac{1}{2} & | & \frac{1}{6} \\ 0 & 0 & | & 0 \end{bmatrix}$

This is as far as we can go with the Gauss-Jordan method. To complete the solution, write the equation that corresponds to the first row of the matrix.

$$x - \frac{1}{2}y = \frac{1}{6} = \frac{3y + 1}{6}$$

Solve this equation for x in terms of y.

$$x = \frac{1}{2}y + \frac{1}{6} = \frac{3y + 1}{6}$$

The solution is $\left(\frac{3y+1}{6}, y\right)$.

25. $y = x - 1$
$y = 6 + z$
$z = -1 - x$

First write the system in proper form.

$$-x + y \quad\quad = -1$$
$$y - z = 6$$
$$x \quad\quad + z = -1$$

Write the augmented matrix and use row operations.

$$\begin{bmatrix} -1 & 1 & 0 & | & -1 \\ 0 & 1 & -1 & | & 6 \\ 1 & 0 & 1 & | & -1 \end{bmatrix}$$

$$R_1 + R_3 \to R_3 \quad \begin{bmatrix} -1 & 1 & 0 & | & -1 \\ 0 & 1 & -1 & | & 6 \\ 0 & 1 & 1 & | & -2 \end{bmatrix}$$

$$\begin{matrix} -1R_2 + R_1 \to R_1 \\ \\ -1R_2 + R_3 \to R_3 \end{matrix} \quad \begin{bmatrix} -1 & 0 & 1 & | & -7 \\ 0 & 1 & -1 & | & 6 \\ 0 & 0 & 2 & | & -8 \end{bmatrix}$$

$$\begin{matrix} R_3 + (-2)R_1 \to R_1 \\ 2R_2 + R_3 \to R_2 \end{matrix} \quad \begin{bmatrix} 2 & 0 & 0 & | & 6 \\ 0 & 2 & 0 & | & 4 \\ 0 & 0 & 2 & | & -8 \end{bmatrix}$$

$$\begin{matrix} \frac{1}{2}R_1 \to R_1 \\ \frac{1}{2}R_2 \to R_2 \\ \frac{1}{2}R_3 \to R_3 \end{matrix} \quad \begin{bmatrix} 1 & 0 & 0 & | & 3 \\ 0 & 1 & 0 & | & 2 \\ 0 & 0 & 1 & | & -4 \end{bmatrix}$$

The solution is $(3, 2, -4)$.

27. $2x - 2y \quad\;\; = -2$
$\qquad\quad\;\; y + z = \;\; 4$
$\;\; x \qquad + z = \;\; 1$

Write the augmented matrix and use row operations.

$$\begin{bmatrix} 2 & -2 & 0 & | & -2 \\ 0 & 1 & 1 & | & 4 \\ 1 & 0 & 1 & | & 1 \end{bmatrix}$$

$$R_1 + (-2)R_3 \to R_3 \quad \begin{bmatrix} 2 & -2 & 0 & | & -2 \\ 0 & 1 & 1 & | & 4 \\ 0 & -2 & -2 & | & -4 \end{bmatrix}$$

$$\begin{matrix} 2R_2 + R_1 \to R_1 \\ \\ 2R_2 + R_3 \to R_3 \end{matrix} \quad \begin{bmatrix} 2 & 0 & 2 & | & 6 \\ 0 & 1 & 1 & | & 4 \\ 0 & 0 & 0 & | & 4 \end{bmatrix}$$

This matrix corresponds to the system

$$2x + 2z = 6$$
$$y + \;\; z = 4$$
$$\qquad 0 = 4.$$

The false statement $0 = 4$ indicates that the system is inconsistent and therefore has no solution.

29. $4x + 4y - 4z = 24$
$\quad\; 2x - \;\; y + \;\; z = -9$
$\quad\;\; x - 2y + 3z = \;\; 1$

Write the augmented matrix and use row operations.

$$\begin{bmatrix} 4 & 4 & -4 & | & 24 \\ 2 & -1 & 1 & | & -9 \\ 1 & -2 & 3 & | & 1 \end{bmatrix}$$

$$\begin{matrix} R_1 + (-2)R_2 \to R_2 \\ R_1 + (-4)R_3 \to R_3 \end{matrix} \quad \begin{bmatrix} 4 & 4 & -4 & | & 24 \\ 0 & 6 & -6 & | & 42 \\ 0 & 12 & -16 & | & 20 \end{bmatrix}$$

$$\begin{matrix} 2R_2 + (-3)R_1 \to R_1 \\ \\ -2R_2 + R_3 \to R_3 \end{matrix} \quad \begin{bmatrix} -12 & 0 & 0 & | & 12 \\ 0 & 6 & -6 & | & 42 \\ 0 & 0 & -4 & | & -64 \end{bmatrix}$$

$$-3R_3 + 2R_2 \to R_2 \quad \begin{bmatrix} -12 & 0 & 0 & | & 12 \\ 0 & 12 & 0 & | & 276 \\ 0 & 0 & -4 & | & -64 \end{bmatrix}$$

$$\begin{matrix} -\frac{1}{12}R_1 \to R_1 \\ -\frac{1}{12}R_2 \to R_2 \\ -\frac{1}{4}R_3 \to R_3 \end{matrix} \quad \begin{bmatrix} 1 & 0 & 0 & | & -1 \\ 0 & 1 & 0 & | & 23 \\ 0 & 0 & 1 & | & 16 \end{bmatrix}$$

The solution is $(-1, 23, 16)$.

31. $\quad 3x + \;\; 5y - \;\; z = 0$
$\qquad 4x - \;\; y + 2z = 1$
$\quad -6x - 10y + 2z = 0$

Write the augmented matrix and use row operations.

$$\begin{bmatrix} 3 & 5 & -1 & | & 0 \\ 4 & -1 & 2 & | & 1 \\ -6 & -10 & 2 & | & 0 \end{bmatrix}$$

$$\begin{matrix} 4R_1 + (-3)R_2 \to R_2 \\ 2R_1 + R_3 \to R_3 \end{matrix} \quad \begin{bmatrix} 3 & 5 & -1 & | & 0 \\ 0 & 23 & -10 & | & -3 \\ 0 & 0 & 0 & | & 0 \end{bmatrix}$$

$$23R_1 + (-5)R_2 \to R_1 \quad \begin{bmatrix} 69 & 0 & 27 & | & 15 \\ 0 & 23 & -10 & | & -3 \\ 0 & 0 & 0 & | & 0 \end{bmatrix}$$

$$\begin{matrix} \frac{1}{69}R_1 \to R_1 \\ \\ \frac{1}{23}R_2 \to R_2 \end{matrix} \quad \begin{bmatrix} 1 & 0 & \frac{9}{23} & | & \frac{5}{23} \\ 0 & 1 & -\frac{10}{23} & | & -\frac{3}{23} \\ 0 & 0 & 0 & | & 0 \end{bmatrix}$$

The row of zeros indicates dependent equations. Solve the first two equations respectively for x and y in terms of z to obtain

$$x = -\frac{9}{23}z + \frac{5}{23} = \frac{-9z+5}{23}$$

and

$$y = \frac{10}{23}z - \frac{3}{23} = \frac{10z-3}{23}.$$

The solution is $\left(\frac{-9z+5}{23}, \frac{10z-3}{23}, z\right)$.

33. $2x + 3y + z = 9$
$4x - y + 3z = -1$
$6x + 2y - 4z = -8$

Write the augmented matrix and use row operations.

$$\begin{bmatrix} 2 & 3 & 1 & 9 \\ 4 & -1 & 3 & -1 \\ 6 & 2 & -4 & -8 \end{bmatrix}$$

$\begin{matrix} \\ -2R_1+R_2 \rightarrow R_2 \\ -3R_1+R_3 \rightarrow R_3 \end{matrix}$ $\begin{bmatrix} 2 & 3 & 1 & 9 \\ 0 & -7 & 1 & -19 \\ 0 & -7 & -7 & -35 \end{bmatrix}$

$\begin{matrix} 3R_2+7R_1 \rightarrow R_1 \\ \\ R_2+(-1)R_3 \rightarrow R_3 \end{matrix}$ $\begin{bmatrix} 14 & 0 & 10 & 6 \\ 0 & -7 & 1 & -19 \\ 0 & 0 & 8 & 16 \end{bmatrix}$

$\begin{matrix} -10R_3+8R_1 \rightarrow R_1 \\ R_3+(-8R_2) \rightarrow R_2 \end{matrix}$ $\begin{bmatrix} 112 & 0 & 0 & -112 \\ 0 & 56 & 0 & 168 \\ 0 & 0 & 8 & 16 \end{bmatrix}$

$\begin{matrix} \frac{1}{112}R_1 \rightarrow R_1 \\ \frac{1}{56}R_2 \rightarrow R_2 \\ \frac{1}{8}R_3 \rightarrow R_3 \end{matrix}$ $\begin{bmatrix} 1 & 0 & 0 & -1 \\ 0 & 1 & 0 & 3 \\ 0 & 0 & 1 & 2 \end{bmatrix}$

The solution is $(-1, 3, 2)$.

35. $5x - 4y + 2z = 4$
$5x + 3y - z = 17$
$15x - 5y + 3z = 25$

Write the augmented matrix and use row operations.

$$\begin{bmatrix} 5 & -4 & 2 & 4 \\ 5 & 3 & -1 & 17 \\ 15 & -5 & 3 & 25 \end{bmatrix}$$

$\begin{matrix} -1R_1+R_2 \rightarrow R_2 \\ -3R_1+R_3 \rightarrow R_3 \end{matrix}$ $\begin{bmatrix} 5 & -4 & 2 & 4 \\ 0 & 7 & -3 & 13 \\ 0 & 7 & -3 & 13 \end{bmatrix}$

$\begin{matrix} 4R_2+7R_1 \rightarrow R_1 \\ \\ -1R_2+R_3 \rightarrow R_3 \end{matrix}$ $\begin{bmatrix} 35 & 0 & 2 & 80 \\ 0 & 7 & -3 & 13 \\ 0 & 0 & 0 & 0 \end{bmatrix}$

$\begin{matrix} \frac{1}{35}R_1 \rightarrow R_1 \\ \frac{1}{7}R_2 \rightarrow R_2 \end{matrix}$ $\begin{bmatrix} 1 & 0 & \frac{2}{35} & \frac{16}{7} \\ 0 & 1 & -\frac{3}{7} & \frac{13}{7} \\ 0 & 0 & 0 & 0 \end{bmatrix}$

The row of zeros indicates dependent equations. Solve the first two equations respectively for x and y in terms of z to obtain

$$x = -\frac{2}{35}z + \frac{16}{7} = \frac{-2z+80}{35}$$

and

$$y = \frac{3}{7}z + \frac{13}{7} = \frac{3z+13}{7}.$$

The solution is $\left(\frac{-2z+80}{35}, \frac{3z+13}{7}, z\right)$.

37. $x + 2y - w = 3$
$2x + 4z + 2w = -6$
$x + 2y - z = 6$
$2x - y + z + w = -3$

Write the augmented matrix and use row operations.

$$\begin{bmatrix} 1 & 2 & 0 & -1 & 3 \\ 2 & 0 & 4 & 2 & -6 \\ 1 & 2 & -1 & 0 & 6 \\ 2 & -1 & 1 & 1 & -3 \end{bmatrix}$$

$\begin{matrix} \\ -2R_1+R_2 \rightarrow R_2 \\ -1R_1+R_3 \rightarrow R_3 \\ -2R_1+R_4 \rightarrow R_4 \end{matrix}$ $\begin{bmatrix} 1 & 2 & 0 & -1 & 3 \\ 0 & -4 & 4 & 4 & -12 \\ 0 & 0 & -1 & 1 & 3 \\ 0 & -5 & 1 & 3 & -9 \end{bmatrix}$

$\begin{matrix} R_2+2R_1 \rightarrow R_1 \\ \\ \\ -5R_2+4R_4 \rightarrow R_4 \end{matrix}$ $\begin{bmatrix} 2 & 0 & 4 & 2 & -6 \\ 0 & -4 & 4 & 4 & -12 \\ 0 & 0 & -1 & 1 & 3 \\ 0 & 0 & -16 & -8 & 24 \end{bmatrix}$

$\begin{matrix} 4R_3+R_1 \rightarrow R_1 \\ 4R_3+R_2 \rightarrow R_2 \\ \\ 16R_3+(-1)R_4 \rightarrow R_4 \end{matrix}$ $\begin{bmatrix} 2 & 0 & 0 & 6 & 6 \\ 0 & -4 & 0 & 8 & 0 \\ 0 & 0 & -1 & 1 & 3 \\ 0 & 0 & 0 & 24 & 24 \end{bmatrix}$

$R_4 + (-4R_1) \rightarrow R_1$
$R_4 + (-3R_2) \rightarrow R_2$
$R_4 + (-24R_3) \rightarrow R_3$

$$\begin{bmatrix} -8 & 0 & 0 & 0 & 0 \\ 0 & 12 & 0 & 0 & 24 \\ 0 & 0 & 24 & 0 & -48 \\ 0 & 0 & 0 & 24 & 24 \end{bmatrix}$$

$-\frac{1}{8}R_1 \rightarrow R_1$
$\frac{1}{12}R_2 \rightarrow R_2$
$\frac{1}{24}R_3 \rightarrow R_3$
$\frac{1}{24}R_4 \rightarrow R_4$

$$\begin{bmatrix} 1 & 0 & 0 & 0 & 0 \\ 0 & 1 & 0 & 0 & 2 \\ 0 & 0 & 1 & 0 & -2 \\ 0 & 0 & 0 & 1 & 1 \end{bmatrix}$$

The solution is $x = 0$, $y = 2$, $z = -2$, $w = 1$, or $(0, 2, -2, 1)$.

39. $x + y - z + 2w = -20$
$2x - y + z + w = 11$
$3x - 2y + z - 2w = 27$

$$\begin{bmatrix} 1 & 1 & -1 & 2 & -20 \\ 2 & -1 & 1 & 1 & 11 \\ 3 & -2 & 1 & -2 & 27 \end{bmatrix}$$

$-2R_1 + R_2 \rightarrow R_2$
$-3R_1 + R_3 \rightarrow R_3$

$$\begin{bmatrix} 1 & 1 & -1 & 2 & -20 \\ 0 & -3 & 3 & -3 & 51 \\ 0 & -5 & 4 & -8 & 87 \end{bmatrix}$$

$-\frac{1}{3}R_2 \rightarrow R_2$

$$\begin{bmatrix} 1 & 1 & -1 & 2 & -20 \\ 0 & 1 & -1 & 1 & -17 \\ 0 & -5 & 4 & -8 & 87 \end{bmatrix}$$

$-1R_2 + R_1 \rightarrow R_1$

$5R_2 + R_3 \rightarrow R_3$

$$\begin{bmatrix} 1 & 0 & 0 & 1 & -3 \\ 0 & 1 & -1 & 1 & -17 \\ 0 & 0 & -1 & -3 & 2 \end{bmatrix}$$

$-1R_3 \rightarrow R_3$

$$\begin{bmatrix} 1 & 0 & 0 & 1 & -3 \\ 0 & 1 & -1 & 1 & -17 \\ 0 & 0 & 1 & 3 & -2 \end{bmatrix}$$

$R_3 + R_2 \rightarrow R_2$

$$\begin{bmatrix} 1 & 0 & 0 & 1 & -3 \\ 0 & 1 & 0 & 4 & -19 \\ 0 & 0 & 1 & 3 & -2 \end{bmatrix}$$

This is as far as we can go using row operations. To complete the solution, write the equations that correspond to the matrix.

$x + w = -3$
$y + 4w = -19$
$z + 3w = -2$

Let w be the parameter and express x, y, and z in terms of w. From the equations above, $x = -w - 3$, $y = -4w - 19$, and $z = -3w - 2$.

The solution is $(-w - 3, -4w - 19, -3w - 2, w)$.

41. $10.47x + 3.52y + 2.58z - 6.42w = 218.65$
$8.62x - 4.93y - 1.75z + 2.83w = 157.03$
$4.92x + 6.83y - 2.97z + 2.65w = 462.3$
$2.86x + 19.10y - 6.24z - 8.73w = 398.4$

Write the augmented matrix of the system.

$$\begin{bmatrix} 10.47 & 3.52 & 2.58 & -6.42 & 218.65 \\ 8.62 & -4.93 & -1.75 & 2.83 & 157.03 \\ 4.92 & 6.83 & -2.97 & 2.65 & 462.3 \\ 2.86 & 19.10 & -6.24 & -8.73 & 398.4 \end{bmatrix}$$

This exercise should be solved by graphing calculator or computer methods. The solution, which may vary slightly, is $x \approx 28.9436$, $y \approx 36.6326$, $z \approx 9.6390$, and $w \approx 37.1036$, or

$$(28.9436, 36.6326, 9.6390, 37.1036).$$

43. Let $x =$ the number of hours to hire the Garcia firm
and $y =$ the number of hours to hire the Wong firm.

The system to be solved is

$10x + 20y = 500$ 　 (1)
$30x + 10y = 750$ 　 (2)
$5x + 10y = 250.$ 　 (3)

Write the augmented matrix of the system.

$$\begin{bmatrix} 10 & 20 & 500 \\ 30 & 10 & 750 \\ 5 & 10 & 250 \end{bmatrix}$$

$\frac{1}{10}R_1 \rightarrow R_1$
$\frac{1}{10}R_2 \rightarrow R_2$
$\frac{1}{5}R_3 \rightarrow R_3$

$$\begin{bmatrix} 1 & 2 & 50 \\ 3 & 1 & 75 \\ 1 & 2 & 50 \end{bmatrix}$$

$-3R_1 + R_2 \rightarrow R_2$
$-1R_1 + R_3 \rightarrow R_3$

$$\begin{bmatrix} 1 & 2 & 50 \\ 0 & -5 & -75 \\ 0 & 0 & 0 \end{bmatrix}$$

$-\frac{1}{5}R_2 \rightarrow R_2$

$$\begin{bmatrix} 1 & 2 & 50 \\ 0 & 1 & 15 \\ 0 & 0 & 0 \end{bmatrix}$$

$-2R_2 + R_1 \rightarrow R_1$

$$\begin{bmatrix} 1 & 0 & 20 \\ 0 & 1 & 15 \\ 0 & 0 & 0 \end{bmatrix}$$

The solution is $(20, 15)$. Hire the Garcia firm for 20 hr and the Wong firm for 15 hr.

45. Let $x =$ the number of chairs produced each week,

$y =$ the number of cabinets produced each week, and

$z =$ the number of buffets produced each week.

Make a table to organize the information.

	Chair	Cabinet	Buffet	Totals
Cutting	.2	.5	.3	1950
Assembly	.3	.4	.1	1490
Finishing	.1	.6	.4	2160

The system to be solved is

$$.2x + .5y + .3z = 1950$$
$$.3x + .4y + .1z = 1490$$
$$.1x + .6y + .4z = 2160.$$

Write the augmented matrix of the system.

$$\begin{bmatrix} .2 & .5 & .3 & | & 1950 \\ .3 & .4 & .1 & | & 1490 \\ .1 & .6 & .4 & | & 2160 \end{bmatrix}$$

$$\begin{matrix} 10R_1 \rightarrow R_1 \\ 10R_2 \rightarrow R_2 \\ 10R_3 \rightarrow R_3 \end{matrix} \begin{bmatrix} 2 & 5 & 3 & | & 19,500 \\ 3 & 4 & 1 & | & 14,900 \\ 1 & 6 & 4 & | & 21,600 \end{bmatrix}$$

Interchange rows 1 and 3.

$$\begin{bmatrix} 1 & 6 & 4 & | & 21,600 \\ 3 & 4 & 1 & | & 14,900 \\ 2 & 5 & 3 & | & 19,500 \end{bmatrix}$$

$$\begin{matrix} -3R_1 + R_2 \rightarrow R_2 \\ -2R_1 + R_3 \rightarrow R_3 \end{matrix} \begin{bmatrix} 1 & 6 & 4 & | & 21,600 \\ 0 & -14 & -11 & | & -49,900 \\ 0 & -7 & -5 & | & -23,700 \end{bmatrix} \left(-\tfrac{1}{14}\right)$$

$$-\tfrac{1}{14}R_2 \rightarrow R_2 \begin{bmatrix} 1 & 6 & 4 & | & 21,600 \\ 0 & 1 & \tfrac{11}{14} & | & \tfrac{24,950}{7} \\ 0 & -7 & -5 & | & -23,700 \end{bmatrix}$$

$$\begin{matrix} -6R_2 + R_1 \rightarrow R_1 \\ \\ 7R_2 + R_3 \rightarrow R_3 \end{matrix} \begin{bmatrix} 1 & 0 & -\tfrac{5}{7} & | & \tfrac{1500}{7} \\ 0 & 1 & \tfrac{11}{14} & | & \tfrac{24,950}{7} \\ 0 & 0 & \tfrac{1}{2} & | & 1250 \end{bmatrix}$$

$$2R_3 \rightarrow R_3 \begin{bmatrix} 1 & 0 & -\tfrac{5}{7} & | & \tfrac{1500}{7} \\ 0 & 1 & \tfrac{11}{14} & | & \tfrac{24,950}{7} \\ 0 & 0 & 1 & | & 2500 \end{bmatrix}$$

$$\begin{matrix} \tfrac{5}{7}R_3 + R_1 \rightarrow R_1 \\ -\tfrac{11}{14}R_3 + R_2 \rightarrow R_2 \end{matrix} \begin{bmatrix} 1 & 0 & 0 & | & 2000 \\ 0 & 1 & 0 & | & 1600 \\ 0 & 0 & 1 & | & 2500 \end{bmatrix}$$

The solution is $(2000, 1600, 2500)$. Therefore, 2000 chairs, 1600 cabinets, and 2500 buffets should be produced.

47. Let $x =$ the number of trucks,

$y =$ the number of vans, and

$z =$ the number of station wagons.

(a) Make a table.

	Truck	Van	Station Wagon	Totals
A	2	1	1	15
B	1	3	3	20
C	3	2	1	22

The system to be solved is

$$2x + 1y + 1z = 15$$
$$1x + 3y + 3z = 20$$
$$3x + 2y + 1z = 22.$$

Write the augmented matrix of the system.

$$\begin{bmatrix} 2 & 1 & 1 & | & 15 \\ 1 & 3 & 3 & | & 20 \\ 3 & 2 & 1 & | & 22 \end{bmatrix}$$

Interchange rows 1 and 2.

$$\begin{bmatrix} 1 & 3 & 3 & | & 20 \\ 2 & 1 & 1 & | & 15 \\ 3 & 2 & 1 & | & 22 \end{bmatrix}$$

$$\begin{matrix} -2R_1 + R_2 \rightarrow R_2 \\ -3R_1 + R_3 \rightarrow R_3 \end{matrix} \begin{bmatrix} 1 & 3 & 3 & | & 20 \\ 0 & -5 & -5 & | & -25 \\ 0 & -7 & -8 & | & -38 \end{bmatrix}$$

$$-\tfrac{1}{5}R_2 \rightarrow R_2 \begin{bmatrix} 1 & 3 & 3 & | & 20 \\ 0 & 1 & 1 & | & 5 \\ 0 & -7 & -8 & | & -38 \end{bmatrix}$$

$$\begin{matrix} -3R_2 + R_1 \rightarrow R_1 \\ \\ 7R_2 + R_3 \rightarrow R_3 \end{matrix} \begin{bmatrix} 1 & 0 & 0 & | & 5 \\ 0 & 1 & 1 & | & 5 \\ 0 & 0 & -1 & | & -3 \end{bmatrix}$$

$$-1R_3 \rightarrow R_3 \begin{bmatrix} 1 & 0 & 0 & | & 5 \\ 0 & 1 & 1 & | & 5 \\ 0 & 0 & 1 & | & 3 \end{bmatrix}$$

$$-1R_3 + R_2 \rightarrow R_2 \begin{bmatrix} 1 & 0 & 0 & | & 5 \\ 0 & 1 & 0 & | & 2 \\ 0 & 0 & 1 & | & 3 \end{bmatrix}$$

The solution is $(5, 2, 3)$. Use 5 trucks, 2 vans, and 3 station wagons.

(b) Look at the table in part (a). Increase the numbers of model B, i.e., 1, 3, 3, by one to 2, 4, 4. There are now 16 boxes of model A instead of 15 and 22 boxes of model B instead of 20. With no equation for model C, the system of equations is

$$2x + 1y + 1z = 16$$
$$2x + 4y + 4z = 22.$$

Write the augmented matrix of the system.

$$\begin{bmatrix} 2 & 1 & 1 & | & 16 \\ 2 & 4 & 4 & | & 22 \end{bmatrix}$$

$$\tfrac{1}{2}R_2 \to R_2 \quad \begin{bmatrix} 2 & 1 & 1 & | & 16 \\ 1 & 2 & 2 & | & 11 \end{bmatrix}$$

Interchange rows 1 and 2.

$$\begin{bmatrix} 1 & 2 & 2 & | & 11 \\ 2 & 1 & 1 & | & 16 \end{bmatrix}$$

$$-2R_1 + R_2 \to R_2 \quad \begin{bmatrix} 1 & 2 & 2 & | & 11 \\ 0 & -3 & -3 & | & -6 \end{bmatrix}$$

$$-\tfrac{1}{3}R_2 \to R_2 \quad \begin{bmatrix} 1 & 2 & 2 & | & 11 \\ 0 & 1 & 1 & | & 2 \end{bmatrix}$$

$$(-2)R_2 + R_1 \to R_1 \quad \begin{bmatrix} 1 & 0 & 0 & | & 7 \\ 0 & 1 & 1 & | & 2 \end{bmatrix}$$

The system associated with this matrix is

$$x = 7$$
$$y + z = 2.$$

Solve the second equation for y in terms of z.

$$y = 2 - z$$

Therefore, the solution is $(7, 2 - z, z)$. The numbers in the solution must be nonnegative integers. Therefore, $0 \leq z \leq 2$.

Values of z:	7	$2 - z$	z	Solutions
0	7	2	0	$(7, 2, 0)$
1	7	1	1	$(7, 1, 1)$
2	7	0	2	$(7, 0, 2)$

There are three possible solutions:

1. 7 trucks, 2 vans, and no station wagons;

2. 7 trucks, 1 van, and 1 station wagon; or

3. 7 trucks, no vans, and 2 station wagons.

49. Let $x =$ the amount borrowed at 13%,

         $y =$ the amount borrowed at 14%,

and $z =$ the amount borrowed at 12%.

(a) The system to be solved is

$$x + y + z = 25{,}000 \qquad (1)$$
$$.13x + .14y + .12z = 3240 \qquad (2)$$
$$y = \frac{1}{2}x + 2000. \qquad (3)$$

Multiply equation (2) by 100 and equation (3) by 2. Then rewrite the system.

$$x + y + z = 25{,}000 \qquad (1)$$
$$13x + 14y + 12z = 324{,}000 \qquad (4)$$
$$-x + 2y = 4000 \qquad (5)$$

Write the augmented matrix of the system.

$$\begin{bmatrix} 1 & 1 & 1 & | & 25{,}000 \\ 13 & 14 & 12 & | & 324{,}000 \\ -1 & 2 & 0 & | & 4000 \end{bmatrix}$$

$$\begin{matrix} -13R_1 + R_2 \to R_2 \\ R_1 + R_3 \to R_3 \end{matrix} \quad \begin{bmatrix} 1 & 1 & 1 & | & 25{,}000 \\ 0 & 1 & -1 & | & -1000 \\ 0 & 3 & 1 & | & 29{,}000 \end{bmatrix}$$

$$\begin{matrix} -1R_2 + R_1 \to R_1 \\ \\ -3R_2 + R_3 \to R_3 \end{matrix} \quad \begin{bmatrix} 1 & 0 & 2 & | & 26{,}000 \\ 0 & 1 & -1 & | & -1000 \\ 0 & 0 & 4 & | & 32{,}000 \end{bmatrix}$$

$$\tfrac{1}{4}R_3 \to R_3 \quad \begin{bmatrix} 1 & 0 & 2 & | & 26{,}000 \\ 0 & 1 & -1 & | & -1000 \\ 0 & 0 & 1 & | & 8000 \end{bmatrix}$$

$$\begin{matrix} -2R_3 + R_1 \to R_1 \\ R_3 + R_2 \to R_2 \end{matrix} \quad \begin{bmatrix} 1 & 0 & 0 & | & 10{,}000 \\ 0 & 1 & 0 & | & 7000 \\ 0 & 0 & 1 & | & 8000 \end{bmatrix}$$

The solution is $(10{,}000, 7000, 8000)$. Borrow \$10,000 at 13%, \$7000 at 14%, and \$8000 at 12%.

(b) If the condition is dropped, refer to the first two rows of the fourth augmented matrix of part (a).

$$\begin{bmatrix} 1 & 0 & 2 & | & 26{,}000 \\ 0 & 1 & -1 & | & -1000 \end{bmatrix}$$

This gives

$$x = 26{,}000 - 2z$$
$$y = z - 1000.$$

Since all values must be nonnegative,

$$26{,}000 - 2z \geq 0 \qquad \text{and} \qquad z - 1000 \geq 0$$
$$z \leq 13{,}000 \qquad \text{and} \qquad z \geq 1000.$$

Therefore, the amount borrowed at 12% must be between $1000 and $13,000. If $z = 5000$, then

$$x = 26,000 - 2(5000) = 16,000 \text{ and}$$
$$y = 5000 - 1000 = 4000.$$

Therefore, $16,000 is borrowed at 13% and $4000 at 14%.

(c) Substitute $z = 6000$ into equations (1), (4), and (5) from part (a) to obtain the system

$$\begin{array}{rll} x + y + 6000 = & 25,000 & (6) \\ 13x + 14y + 12(6000) = & 324,000 & (7) \\ -x + 2y = & 4000. & (5) \end{array}$$

This gives the system

$$\begin{array}{rll} x + y = & 19,000 & (8) \\ 13x + 14y = & 252,000 & (9) \\ -x + 2y = & 4000. & (5) \end{array}$$

The augmented matrix for this system is

$$\begin{bmatrix} 1 & 1 & | & 19,000 \\ 13 & 14 & | & 252,000 \\ -1 & 2 & | & 4000 \end{bmatrix}.$$

$$\begin{array}{l} -13R_1 + R_2 \to R_2 \\ R_1 + R_3 \to R_3 \end{array} \begin{bmatrix} 1 & 1 & | & 19,000 \\ 0 & 1 & | & 5000 \\ 0 & 3 & | & 23,000 \end{bmatrix}.$$

Row 2 indicates that $y = 5000$, while row 3 indicates that $y = \frac{23,000}{3}$. Therefore, there is no solution if $6000 is borrowed at 12%.

51. Let $x_1 =$ the number of units from first supplier for Roseville,
$x_2 =$ the number of units from first supplier for Akron,
$x_3 =$ the number of units from second supplier for Roseville, and
$x_4 =$ the number of units from second supplier for Akron.

	Roseville	Akron
I	x_1	x_2
II	x_3	x_4

The first supplier will provide 75 units, so

$$x_1 + x_2 = 75.$$

The second supplier will supply 40 units, so

$$x_3 + x_4 = 40.$$

Roseville needs 40 units, so

$$x_1 + x_3 = 40.$$

Akron requires 75 units, so

$$x_2 + x_4 = 75.$$

The total shipping cost is $10,750, so

$$70x_1 + 90x_2 + 80x_3 + 120x_4 = 10,750.$$

The system to be solved is

$$\begin{array}{rl} x_1 + x_2 & = 75 \\ x_3 + x_4 & = 40 \\ x_1 + x_3 & = 40 \\ x_2 + x_4 & = 75 \\ 70x_1 + 90x_2 + 80x_3 + 120x_4 & = 10,750. \end{array}$$

Write the augmented matrix and use row operations.

$$\begin{bmatrix} 1 & 1 & 0 & 0 & | & 75 \\ 0 & 0 & 1 & 1 & | & 40 \\ 1 & 0 & 1 & 0 & | & 40 \\ 0 & 1 & 0 & 1 & | & 75 \\ 70 & 90 & 80 & 120 & | & 10,750 \end{bmatrix}$$

$$\begin{array}{l} -1R_1 + R_3 \to R_3 \\ \\ -70R_1 + R_5 \to R_5 \end{array} \begin{bmatrix} 1 & 1 & 0 & 0 & | & 75 \\ 0 & 0 & 1 & 1 & | & 40 \\ 0 & -1 & 1 & 0 & | & -35 \\ 0 & 1 & 0 & 1 & | & 75 \\ 0 & 20 & 80 & 120 & | & 5500 \end{bmatrix}$$

Interchange rows 2 and 4.

$$\begin{bmatrix} 1 & 1 & 0 & 0 & | & 75 \\ 0 & 1 & 0 & 1 & | & 75 \\ 0 & -1 & 1 & 0 & | & -35 \\ 0 & 0 & 1 & 1 & | & 40 \\ 0 & 20 & 80 & 120 & | & 5500 \end{bmatrix}$$

$$\begin{array}{l} -1R_2 + R_1 \to R_1 \\ \\ R_2 + R_3 \to R_3 \\ \\ -20R_2 + R_5 \to R_5 \end{array} \begin{bmatrix} 1 & 0 & 0 & -1 & | & 0 \\ 0 & 1 & 0 & 1 & | & 75 \\ 0 & 0 & 1 & 1 & | & 40 \\ 0 & 0 & 1 & 1 & | & 40 \\ 0 & 0 & 80 & 100 & | & 4000 \end{bmatrix}$$

$$\begin{array}{l} -1R_3 + R_4 \to R_4 \\ -80R_3 + R_5 \to R_5 \end{array} \begin{bmatrix} 1 & 0 & 0 & -1 & | & 0 \\ 0 & 1 & 0 & 1 & | & 75 \\ 0 & 0 & 1 & 1 & | & 40 \\ 0 & 0 & 0 & 0 & | & 0 \\ 0 & 0 & 0 & 20 & | & 800 \end{bmatrix}$$

Interchange rows 4 and 5.

$$\begin{bmatrix} 1 & 0 & 0 & -1 & | & 0 \\ 0 & 1 & 0 & 1 & | & 75 \\ 0 & 0 & 1 & 1 & | & 40 \\ 0 & 0 & 0 & 20 & | & 800 \\ 0 & 0 & 0 & 0 & | & 0 \end{bmatrix}$$

$\frac{1}{20}R_4 \to R_4$
$$\begin{bmatrix} 1 & 0 & 0 & -1 & | & 0 \\ 0 & 1 & 0 & 1 & | & 75 \\ 0 & 0 & 1 & 1 & | & 40 \\ 0 & 0 & 0 & 1 & | & 40 \\ 0 & 0 & 0 & 0 & | & 0 \end{bmatrix}$$

$R_1 + R_4 \to R_1$
$R_2 + (-1)R_4 \to R_2$
$R_3 + (-1)R_4 \to R_3$
$$\begin{bmatrix} 1 & 0 & 0 & 0 & | & 40 \\ 0 & 1 & 0 & 0 & | & 35 \\ 0 & 0 & 1 & 0 & | & 0 \\ 0 & 0 & 0 & 1 & | & 40 \\ 0 & 0 & 0 & 0 & | & 0 \end{bmatrix}$$

Each of the original variables has a value, so the last row of all zeros may be ignored. The solution of the system is $x_1 = 40$, $x_2 = 35$, $x_3 = 0$, $x_4 = 40$, or $(40, 35, 0, 40)$.

The manufacturer should purchase 40 units for Roseville from the first supplier, 35 units for Akron from the first supplier, 0 units for Roseville from the second supplier, and 40 units for Akron from the second supplier.

53. (a) The other two equations are

$$x_2 + x_3 = 700$$
$$x_3 + x_4 = 600.$$

(b) The augmented matrix is

$$\begin{bmatrix} 1 & 0 & 0 & 1 & | & 1000 \\ 1 & 1 & 0 & 0 & | & 1100 \\ 0 & 1 & 1 & 0 & | & 700 \\ 0 & 0 & 1 & 1 & | & 600 \end{bmatrix}.$$

$-1R_1 + R_2 \to R_2$
$$\begin{bmatrix} 1 & 0 & 0 & 1 & | & 1000 \\ 0 & 1 & 0 & -1 & | & 100 \\ 0 & 1 & 1 & 0 & | & 700 \\ 0 & 0 & 1 & 1 & | & 600 \end{bmatrix}$$

$-1R_2 + R_3 \to R_3$
$$\begin{bmatrix} 1 & 0 & 0 & 1 & | & 1000 \\ 0 & 1 & 0 & -1 & | & 100 \\ 0 & 0 & 1 & 1 & | & 600 \\ 0 & 0 & 1 & 1 & | & 600 \end{bmatrix}$$

$$\begin{bmatrix} 1 & 0 & 0 & 1 & | & 1000 \\ 0 & 1 & 0 & -1 & | & 100 \\ 0 & 0 & 1 & 1 & | & 600 \\ 0 & 0 & 0 & 0 & | & 0 \end{bmatrix}$$

$-1R_3 + R_4 \to R_4$

Let x_4 be arbitrary. Solve the first three equations for x_1, x_2, and x_3.

$$x_1 = 1000 - x_4$$
$$x_2 = 100 + x_4$$
$$x_3 = 600 - x_4$$

The solution is $(1000 - x_4, 100 + x_4, 600 - x_4, x_4)$.

(c) For x_4, we see that $x_4 \geq 0$ and $x_4 \leq 600$ since $600 - x_4$ must be nonnegative. Therefore, $0 \leq x_4 \leq 600$.

(d) x_1:   If $x_4 = 0$, then $x_1 = 1000$.
   If $x_4 = 600$, then $x_1 = 1000 - 600 = 400$.

Therefore, $400 \leq x_1 \leq 1000$.

   x_2:   If $x_4 = 0$, then $x_2 = 100$.
   If $x_4 = 600$, then $x_2 = 100 + 600 = 700$.

Therefore, $100 \leq x_2 \leq 700$.

   x_3:   If $x_4 = 0$, then $x_3 = 600$.
   If $x_4 = 600$, then $x_3 = 600 - 600 = 0$.

Therefore, $0 \leq x_3 \leq 600$.

(e) If you know the number of cars entering or leaving three of the intersections, then the number entering or leaving the fourth is automatically determined because the number leaving must equal the number entering.

55. Let $x_1 = $ the number of cases of Brand A,
$x_2 = $ the number of cases of Brand B,
$x_3 = $ the number of cases of Brand C, and
$x_4 = $ the number of cases of Brand D.

$$25x_1 + 50x_2 + 75x_3 + 100x_4 = 1200$$
$$30x_1 + 30x_2 + 30x_3 + 60x_4 = 600$$
$$30x_1 + 20x_2 + 20x_3 + 30x_4 = 400$$

The augmented matrix of the system is

$$\begin{bmatrix} 25 & 50 & 75 & 100 & | & 1200 \\ 30 & 30 & 30 & 60 & | & 600 \\ 30 & 20 & 20 & 30 & | & 400 \end{bmatrix}.$$

$\frac{1}{5}R_1 \to R_1$
$\frac{1}{30}R_2 \to R_2$
$\frac{1}{10}R_3 \to R_3$
$$\begin{bmatrix} 5 & 10 & 15 & 20 & | & 240 \\ 1 & 1 & 1 & 2 & | & 20 \\ 3 & 2 & 2 & 3 & | & 40 \end{bmatrix}$$

Interchange rows 1 and 2.

$$\begin{bmatrix} 1 & 1 & 1 & 2 & 20 \\ 5 & 10 & 15 & 20 & 240 \\ 3 & 2 & 2 & 3 & 40 \end{bmatrix}$$

$$\begin{array}{c} -5R_1 + R_2 \to R_2 \\ -3R_1 + R_3 \to R_3 \end{array} \begin{bmatrix} 1 & 1 & 1 & 2 & 20 \\ 0 & 5 & 10 & 10 & 140 \\ 0 & -1 & -1 & -3 & -20 \end{bmatrix}$$

$$\tfrac{1}{5}R_2 \to R_2 \begin{bmatrix} 1 & 1 & 1 & 2 & 20 \\ 0 & 1 & 2 & 2 & 28 \\ 0 & -1 & -1 & -3 & -20 \end{bmatrix}$$

$$\begin{array}{c} -1R_2 + R_1 \to R_1 \\ \\ R_2 + R_3 \to R_3 \end{array} \begin{bmatrix} 1 & 0 & -1 & 0 & -8 \\ 0 & 1 & 2 & 2 & 28 \\ 0 & 0 & 1 & -1 & 8 \end{bmatrix}$$

$$\begin{array}{c} R_3 + R_1 \to R_1 \\ -2R_3 + R_2 \to R_2 \end{array} \begin{bmatrix} 1 & 0 & 0 & -1 & 0 \\ 0 & 1 & 0 & 4 & 12 \\ 0 & 0 & 1 & -1 & 8 \end{bmatrix}$$

We cannot change the values in column 4 further without changing the form of the other three columns. Therefore, let x_4 be arbitrary. This matrix gives the equations

$$\begin{aligned} x_1 - x_4 &= 0 \quad \text{or} \quad x_1 = x_4, \\ x_2 + 4x_4 &= 12 \quad \text{or} \quad x_2 = 12 - 4x_4, \\ x_3 - x_4 &= 8 \quad \text{or} \quad x_3 = 8 + x_4. \end{aligned}$$

The solution is $(x_4, 12 - 4x_4, 8 + x_4, x_4)$. Since all solutions must be nonnegative,

$$12 - 4x_4 \geq 0$$
$$x_4 \leq 3.$$

If $x_4 = 0$, then $x_1 = 0$, $x_2 = 12$, and $x_3 = 8$.
If $x_4 = 1$, then $x_1 = 1$, $x_2 = 8$, and $x_3 = 9$.
If $x_4 = 2$, then $x_1 = 2$, $x_2 = 4$, and $x_3 = 10$.
If $x_4 = 3$, then $x_1 = 3$, $x_2 = 0$, and $x_3 = 11$.

Therefore, there are four possible solutions. The breeder should mix

1. 0 cases of A, 12 cases of B, 8 cases of C, and 0 cases of D;

2. 1 case of A, 8 cases of B, 9 cases of C, and 1 case of D;

3. 2 cases of A, 4 cases of B, 10 cases of C, and 2 cases of D; or

4. 3 cases of A, 0 cases of B, 11 cases of C, and 3 cases of D.

57. Let $x =$ the number of the first species,
$y =$ the number of the second species, and
$z =$ the number of the third species.

$$\begin{aligned} 1.3x + 1.1y + 8.1z &= 16{,}000 \\ 1.3x + 2.4y + 2.9z &= 28{,}000 \\ 2.3x + 3.7y + 5.1z &= 44{,}000 \end{aligned}$$

Write the augmented matrix of the system.

$$\begin{bmatrix} 1.3 & 1.1 & 8.1 & 16{,}000 \\ 1.3 & 2.4 & 2.9 & 28{,}000 \\ 2.3 & 3.7 & 5.1 & 44{,}000 \end{bmatrix}$$

This exercise should be solved by graphing calculator or computer methods. The solution, which may vary slightly, is 2340 of the first species, 10,128 of the second species, and 224 of the third species. (All of these are rounded to the nearest whole number.)

59. Let $x =$ the number of acres for honeydews,
$y =$ the number of acres for yellow onions,
and $z =$ the number of acres for lettuce.

(a)
$$\begin{aligned} x + y + z &= 220 \\ 120x + 150y + 180z &= 29{,}100 \\ 180x + 80y + 80z &= 32{,}600 \\ 4.97x + 4.45y + 4.65z &= 480 \end{aligned}$$

Write the augmented matrix for this system.

$$\begin{bmatrix} 1 & 1 & 1 & 220 \\ 120 & 150 & 180 & 29{,}100 \\ 180 & 80 & 80 & 32{,}600 \\ 4.97 & 4.45 & 4.65 & 480 \end{bmatrix}$$

Using graphing calculator or computer methods, we obtain

$$\begin{bmatrix} 1 & 0 & 0 & 150 \\ 0 & 1 & 0 & 50 \\ 0 & 0 & 1 & 20 \\ 0 & 0 & 0 & -58 \end{bmatrix}.$$

There is no solution to the system. Therefore, it is not possible to utilize all resources completely.

(b) If 1061 hr of labor are available, the augmented matrix becomes,

$$\begin{bmatrix} 1 & 1 & 1 & 220 \\ 120 & 150 & 180 & 29{,}100 \\ 180 & 80 & 80 & 32{,}600 \\ 4.97 & 4.45 & 4.65 & 1061 \end{bmatrix}.$$

Again, using graphing calculator or computer methods we obtain

$$\begin{bmatrix} 1 & 0 & 0 & 150 \\ 0 & 1 & 0 & 50 \\ 0 & 0 & 1 & 20 \\ 0 & 0 & 0 & 0 \end{bmatrix}.$$

The solution is $(150, 50, 20)$. Therefore, allot 150 acres for honeydews, 50 acres for onions, and 20 acres for lettuce.

2.3   Addition and Subtraction of Matrices

1. $\begin{bmatrix} 1 & 3 \\ 5 & 7 \end{bmatrix} = \begin{bmatrix} 1 & 5 \\ 3 & 7 \end{bmatrix}$

This statement is false, since not all corresponding elements are equal.

3. $\begin{bmatrix} x \\ y \end{bmatrix} = \begin{bmatrix} 3 \\ 5 \end{bmatrix}$ if $x = 3$ and $y = 5$.

This statement is true. The matrices are the same size and corresponding elements are equal.

5. $\begin{bmatrix} 1 & 9 & -4 \\ 3 & 7 & 2 \\ -1 & 1 & 0 \end{bmatrix}$ is a square matrix.

This statement is true. The matrix has 3 rows and 3 columns.

7. $\begin{bmatrix} -4 & 8 \\ 2 & 3 \end{bmatrix}$ is a 2×2 square matrix.

Its additive inverse is $\begin{bmatrix} 4 & -8 \\ -2 & -3 \end{bmatrix}$.

9. $\begin{bmatrix} -6 & 8 & 0 & 0 \\ 4 & 1 & 9 & 2 \\ 3 & -5 & 7 & 1 \end{bmatrix}$ is a 3×4 matrix.

Its additive inverse is
$$\begin{bmatrix} 6 & -8 & 0 & 0 \\ -4 & -1 & -9 & -2 \\ -3 & 5 & -7 & -1 \end{bmatrix}.$$

11. $\begin{bmatrix} 2 \\ 4 \end{bmatrix}$ is a 2×1 column matrix.

Its additive inverse is
$$\begin{bmatrix} -2 \\ -4 \end{bmatrix}.$$

13. The sum of an $n \times m$ matrix and its additive inverse is the $n \times m$ zero matrix.

15. $\begin{bmatrix} 2 & 1 \\ 4 & 8 \end{bmatrix} = \begin{bmatrix} x & 1 \\ y & z \end{bmatrix}$

Corresponding elements must be equal for the matrices to be equal. Therefore, $x = 2$, $y = 4$, and $z = 8$.

17. $\begin{bmatrix} x+6 & y+2 \\ 8 & 3 \end{bmatrix} = \begin{bmatrix} -9 & 7 \\ 8 & k \end{bmatrix}$

Corresponding elements must be equal.

$$x + 6 = -9 \qquad y + 2 = 7 \qquad k = 3$$
$$x = -15 \qquad\quad y = 5$$

Thus, $x = -15$, $y = 5$, and $k = 3$.

19. $\begin{bmatrix} -7+z & 4r & 8s \\ 6p & 2 & 5 \end{bmatrix} + \begin{bmatrix} -9 & 8r & 3 \\ 2 & 5 & 4 \end{bmatrix}$

$$= \begin{bmatrix} 2 & 36 & 27 \\ 20 & 7 & 12a \end{bmatrix}$$

Add the two matrices on the left side of this equation to obtain

$$\begin{bmatrix} -7+z & 4r & 8s \\ 6p & 2 & 5 \end{bmatrix} + \begin{bmatrix} -9 & 8r & 3 \\ 2 & 5 & 4 \end{bmatrix}$$

$$= \begin{bmatrix} (-7+z)+(-9) & 4r+8r & 8s+3 \\ 6p+2 & 7 & 9 \end{bmatrix}$$

$$= \begin{bmatrix} -16+z & 12r & 8s+3 \\ 6p+2 & 7 & 9 \end{bmatrix}.$$

Corresponding elements of this matrix and the matrix on the right side of the original equation must be equal.

$$-16 + z = 2 \qquad 12r = 36 \qquad 8s + 3 = 27$$
$$z = 18 \qquad\quad r = 3 \qquad\quad s = 3$$

$$6p + 2 = 20 \qquad 9 = 12a$$

$$p = 3 \qquad a = \frac{3}{4}$$

Thus, $z = 18$, $r = 3$, $s = 3$, $p = 3$, and $a = \frac{3}{4}$.

21. $\begin{bmatrix} 1 & 2 & 5 & -1 \\ 3 & 0 & 2 & -4 \end{bmatrix} + \begin{bmatrix} 8 & 10 & -5 & 3 \\ -2 & -1 & 0 & 0 \end{bmatrix}$

$$= \begin{bmatrix} 1+8 & 2+10 & 5+(-5) & -1+3 \\ 3+(-2) & 0+(-1) & 2+0 & -4+0 \end{bmatrix}$$

$$= \begin{bmatrix} 9 & 12 & 0 & 2 \\ 1 & -1 & 2 & -4 \end{bmatrix}$$

23. $\begin{bmatrix} 1 & 3 & -2 \\ 4 & 7 & 1 \end{bmatrix} + \begin{bmatrix} 3 & 0 \\ 6 & 4 \\ -5 & 2 \end{bmatrix}$

These matrices cannot be added since the first matrix has size 2×3, while the second has size 3×2. Only matrices that are the same size can be added.

25. The matrices have the same size, so the subtraction can be done. Let A and B represent the given matrices. Using the definition of subtraction, we have

$$A - B = A + (-B)$$

$$= \begin{bmatrix} 2 & 8 & 12 & 0 \\ 7 & 4 & -1 & 5 \\ 1 & 2 & 0 & 10 \end{bmatrix} + \begin{bmatrix} -1 & -3 & -6 & -9 \\ -2 & 3 & 3 & -4 \\ -8 & 0 & 2 & -17 \end{bmatrix}$$

$$= \begin{bmatrix} 1 & 5 & 6 & -9 \\ 5 & 7 & 2 & 1 \\ -7 & 2 & 2 & -7 \end{bmatrix}.$$

27. $\begin{bmatrix} 2 & 3 \\ -2 & 4 \end{bmatrix} + \begin{bmatrix} 4 & 3 \\ 7 & 8 \end{bmatrix} - \begin{bmatrix} 3 & 2 \\ 1 & 4 \end{bmatrix}$

$$= \begin{bmatrix} 2+4-3 & 3+3-2 \\ -2+7-1 & 4+8-4 \end{bmatrix} = \begin{bmatrix} 3 & 4 \\ 4 & 8 \end{bmatrix}$$

29. $\begin{bmatrix} 1 & 5 \\ -3 & 7 \end{bmatrix} - \begin{bmatrix} 6 & 3 \\ 2 & 4 \end{bmatrix} + \begin{bmatrix} 8 & 10 \\ -1 & 0 \end{bmatrix}$

$$= \begin{bmatrix} 1-6+8 & 5-3+10 \\ -3-2+(-1) & 7-4+0 \end{bmatrix}$$

$$= \begin{bmatrix} 3 & 12 \\ -6 & 3 \end{bmatrix}$$

31. $\begin{bmatrix} -4x+2y & -3x+y \\ 6x-3y & 2x-5y \end{bmatrix} + \begin{bmatrix} -8x+6y & 2x \\ 3y-5x & 6x+4y \end{bmatrix}$

$$= \begin{bmatrix} (-4x+2y)+(-8x+6y) & (-3x+y)+2x \\ (6x-3y)+(3y-5x) & (2x-5y)+(6x+4y) \end{bmatrix}$$

$$= \begin{bmatrix} -12x+8y & -x+y \\ x & 8x-y \end{bmatrix}$$

33. The additive inverse of

$$X = \begin{bmatrix} x & y \\ z & w \end{bmatrix}$$

is

$$-X = \begin{bmatrix} -x & -y \\ -z & -w \end{bmatrix}.$$

35. Show that $X + (T + P) = (X + T) + P$.

On the left side, the sum $T + P$ is obtained first, and then

$$X + (T + P).$$

This gives the matrix

$$\begin{bmatrix} x+(r+m) & y+(s+n) \\ z+(t+p) & w+(u+q) \end{bmatrix}.$$

For the right side, first the sum $X + T$ is obtained, and then

$$(X + T) + P.$$

This gives the matrix

$$\begin{bmatrix} (x+r)+m & (y+s)+n \\ (z+t)+p & (w+u)+q \end{bmatrix}.$$

Comparing corresponding elements, we see that they are equal by the associative property of addition of real numbers. Thus,

$$X + (T + P) = (X + T) + P.$$

37. Show that $P + O = P$.

$$P + O = \begin{bmatrix} m & n \\ p & q \end{bmatrix} + \begin{bmatrix} 0 & 0 \\ 0 & 0 \end{bmatrix}$$

$$= \begin{bmatrix} m+0 & n+0 \\ p+0 & q+0 \end{bmatrix}$$

$$= \begin{bmatrix} m & n \\ p & q \end{bmatrix}$$

$$= P$$

Thus, $P + O = P$.

39. (a) The production cost matrix for Chicago is

	Phones	Calculators
Material	4.05	7.01
Labor	3.27	3.51

The production cost matrix for Seattle is

	Phones	Calculators
Material	4.40	6.90
Labor	3.54	3.76

(b) The new production cost matrix for Chicago is

	Phones	Calculators
Material	4.05 + .37	7.01 + .42
Labor	3.27 + .11	3.51 + .11

or $\begin{bmatrix} 4.42 & 7.43 \\ 3.38 & 3.62 \end{bmatrix}.$

41. (a) There are four food groups and three meals. To represent the data by a 3×4 matrix, we must use the rows to correspond to the meals, breakfast, lunch, and dinner, and the columns to correspond to the four food groups. Thus, we obtain the matrix

$$\begin{bmatrix} 2 & 1 & 2 & 1 \\ 3 & 2 & 2 & 1 \\ 4 & 3 & 2 & 1 \end{bmatrix}.$$

(b) There are four food groups. These will correspond to the four rows. There are three components in each food group: fat, carbohydrates, and protein. These will correspond to the three columns. The matrix is

$$\begin{bmatrix} 5 & 0 & 7 \\ 0 & 10 & 1 \\ 0 & 15 & 2 \\ 10 & 12 & 8 \end{bmatrix}.$$

(c) The matrix is

$$\begin{bmatrix} 8 \\ 4 \\ 5 \end{bmatrix}.$$

43.

	Obtained Pain Relief	
	Yes	No
Painfree	22	3
Placebo	8	17

(a) Of the 25 patients who took the placebo, 8 got relief.

(b) Of the 25 patients who took Painfree, 3 got no relief.

(c) $\begin{bmatrix} 22 & 3 \\ 8 & 17 \end{bmatrix} + \begin{bmatrix} 21 & 4 \\ 6 & 19 \end{bmatrix} + \begin{bmatrix} 19 & 6 \\ 10 & 15 \end{bmatrix} + \begin{bmatrix} 23 & 2 \\ 3 & 22 \end{bmatrix}$

$= \begin{bmatrix} 85 & 15 \\ 27 & 73 \end{bmatrix}$

(d) Yes, it appears that Painfree is effective. Of the 100 patients who took the medication, 85% got relief.

45. (a)

	M	J	Ca	Cl
M	1	1	1	1
J	0	1	0	0
Ca	0	1	1	1
Cl	0	1	1	1

(b) Rows 1 and 2 will stay the same. Since the cats now like Musk, the zeros in rows 3 and 4 change to ones.

	M	J	Ca	Cl
M	1	1	1	1
J	0	1	0	0
Ca	1	1	1	1
Cl	1	1	1	1

2.4 Multiplication of Matrices

In Exercises 1-5, let

$$A = \begin{bmatrix} -2 & 4 \\ 0 & 3 \end{bmatrix} \text{ and } B = \begin{bmatrix} -6 & 2 \\ 4 & 0 \end{bmatrix}.$$

1. $2A = 2 \begin{bmatrix} -2 & 4 \\ 0 & 3 \end{bmatrix} = \begin{bmatrix} -4 & 8 \\ 0 & 6 \end{bmatrix}$

3. $-4B = -4 \begin{bmatrix} -6 & 2 \\ 4 & 0 \end{bmatrix} = \begin{bmatrix} 24 & -8 \\ -16 & 0 \end{bmatrix}$

5. $-4A + 5B = -4 \begin{bmatrix} -2 & 4 \\ 0 & 3 \end{bmatrix} + 5 \begin{bmatrix} -6 & 2 \\ 4 & 0 \end{bmatrix}$

$= \begin{bmatrix} 8 & -16 \\ 0 & -12 \end{bmatrix} + \begin{bmatrix} -30 & 10 \\ 20 & 0 \end{bmatrix}$

$= \begin{bmatrix} -22 & -6 \\ 20 & -12 \end{bmatrix}$

7.

Matrix A size	Matrix B size
2 × **2**	**2** × 2

The number of columns of A is the same as the number of rows of B, so the product AB exists. The size of the matrix AB is 2×2.

Matrix B size	Matrix A size
2 × **2**	**2** × 2

Since the number of columns of B is the same as the number of rows of A, the product BA also exists and has size 2×2.

9.

Matrix A size	Matrix B size
3 × **5**	**5** × 2

Since matrix A has 5 columns and matrix B has 5 rows, the product AB exists and has size 3×2.

Matrix B size	Matrix A size
5 × **2**	**3** × 5

Since B has 2 columns and A has 3 rows, the product BA does not exist.

11.

Matrix A size	Matrix B size
4 × **2**	**3** × 4

The number of columns of A is not the same as the number of rows of B, so the product AB does not exist.

Matrix B size   Matrix A size

$3 \times \underline{4}$ $\underline{4} \times 2$

The number of columns of B is the same as the number of rows of A, so the product BA exists and has size 3×2.

13. To find the product matrix AB, the number of *columns* of A must be the same as the number of *rows* of B.

15. Call the first matrix A and the second matrix B. The product matrix AB will have size 2×1.

Step 1: Multiply the elements of the first row of A by the corresponding elements of the column of B and add.

$$\begin{bmatrix} 1 & 2 \\ 3 & 4 \end{bmatrix} \begin{bmatrix} -1 \\ 7 \end{bmatrix} \qquad 1(-1) + 2(7) = 13$$

Therefore, 13 is the first row entry of the product matrix AB.

Step 2: Multiply the elements of the second row of A by the corresponding elements of the column of B and add.

$$\begin{bmatrix} 1 & 2 \\ 3 & 4 \end{bmatrix} \begin{bmatrix} -1 \\ 7 \end{bmatrix} \qquad 3(-1) + 4(7) = 25$$

The second row entry of the product is 25.

Step 3: Write the product using the two entries found above.

$$AB = \begin{bmatrix} 1 & 2 \\ 3 & 4 \end{bmatrix} \begin{bmatrix} -1 \\ 7 \end{bmatrix} = \begin{bmatrix} 13 \\ 25 \end{bmatrix}$$

17. $\begin{bmatrix} 1 & 5 & 3 \\ -1 & 2 & 7 \end{bmatrix} \begin{bmatrix} 4 \\ 2 \\ -3 \end{bmatrix}$

$= \begin{bmatrix} 1 \cdot 4 + 5 \cdot 2 + 3(-3) \\ -1(4) + 2 \cdot 2 + 7(-3) \end{bmatrix}$

$= \begin{bmatrix} 5 \\ -21 \end{bmatrix}$

19. $\begin{bmatrix} 5 & 1 \\ 2 & 3 \end{bmatrix} \begin{bmatrix} 3 & -1 & 0 \\ 1 & 0 & 2 \end{bmatrix}$

$= \begin{bmatrix} 5 \cdot 3 + 1 \cdot 1 & 5(-1) + 1 \cdot 0 & 5 \cdot 0 + 1 \cdot 2 \\ 2 \cdot 3 + 3 \cdot 1 & 2(-1) + 3 \cdot 0 & 2 \cdot 0 + 3 \cdot 2 \end{bmatrix}$

$= \begin{bmatrix} 16 & -5 & 2 \\ 9 & -2 & 6 \end{bmatrix}$

21. $\begin{bmatrix} 2 & 2 & -1 \\ 3 & 0 & 1 \end{bmatrix} \begin{bmatrix} 0 & 2 \\ -1 & 4 \\ 0 & 2 \end{bmatrix}$

$= \begin{bmatrix} 2 \cdot 0 + 2(-1) + (-1)0 & 2 \cdot 2 + 2 \cdot 4 + (-1)2 \\ 3 \cdot 0 + 0(-1) + 1(0) & 3 \cdot 2 + 0 \cdot 4 + 1 \cdot 2 \end{bmatrix}$

$= \begin{bmatrix} -2 & 10 \\ 0 & 8 \end{bmatrix}$

23. $\begin{bmatrix} 1 & 2 \\ 3 & 4 \end{bmatrix} \begin{bmatrix} -1 & 5 \\ 7 & 0 \end{bmatrix}$

$= \begin{bmatrix} 1(-1) + 2 \cdot 7 & 1 \cdot 5 + 2 \cdot 0 \\ 3(-1) + 4 \cdot 7 & 3 \cdot 5 + 4 \cdot 0 \end{bmatrix}$

$= \begin{bmatrix} 13 & 5 \\ 25 & 15 \end{bmatrix}$

25. $\begin{bmatrix} -2 & -3 & 7 \\ 1 & 5 & 6 \end{bmatrix} \begin{bmatrix} 1 \\ 2 \\ 3 \end{bmatrix}$

$= \begin{bmatrix} -2(1) + (-3)2 + 7 \cdot 3 \\ 1 \cdot 1 + 5 \cdot 2 + 6 \cdot 3 \end{bmatrix}$

$= \begin{bmatrix} 13 \\ 29 \end{bmatrix}$

27. $\left(\begin{bmatrix} 4 & 3 \\ 1 & 2 \\ 0 & -5 \end{bmatrix} \begin{bmatrix} 2 & -2 \\ 1 & -1 \end{bmatrix} \right) \begin{bmatrix} 10 \\ 0 \end{bmatrix}$

$= \begin{bmatrix} 11 & -11 \\ 4 & -4 \\ -5 & 5 \end{bmatrix} \begin{bmatrix} 10 \\ 0 \end{bmatrix} = \begin{bmatrix} 110 \\ 40 \\ -50 \end{bmatrix}$

29. $\begin{bmatrix} 2 & -2 \\ 1 & -1 \end{bmatrix} \left(\begin{bmatrix} 4 & 3 \\ 1 & 2 \end{bmatrix} + \begin{bmatrix} 7 & 0 \\ -1 & 5 \end{bmatrix} \right)$

$= \begin{bmatrix} 2 & -2 \\ 1 & -1 \end{bmatrix} \begin{bmatrix} 11 & 3 \\ 0 & 7 \end{bmatrix}$

$= \begin{bmatrix} 22 & -8 \\ 11 & -4 \end{bmatrix}$

31. (a) $AB = \begin{bmatrix} -2 & 4 \\ 1 & 3 \end{bmatrix} \begin{bmatrix} -2 & 1 \\ 3 & 6 \end{bmatrix} = \begin{bmatrix} 16 & 22 \\ 7 & 19 \end{bmatrix}$

(b) $BA = \begin{bmatrix} -2 & 1 \\ 3 & 6 \end{bmatrix} \begin{bmatrix} -2 & 4 \\ 1 & 3 \end{bmatrix} = \begin{bmatrix} 5 & -5 \\ 0 & 30 \end{bmatrix}$

(c) No, AB and BA are not equal here.

(d) No, AB does not always equal BA.

33. Verify that $P(X + T) = PX + PT$.
Find $P(X+T)$ and $PX+PT$ separately and compare their values to see if they are the same.

$P(X + T)$

$$= \begin{bmatrix} m & n \\ p & q \end{bmatrix} \left(\begin{bmatrix} x & y \\ z & w \end{bmatrix} + \begin{bmatrix} r & s \\ t & u \end{bmatrix} \right)$$

$$= \begin{bmatrix} m & n \\ p & q \end{bmatrix} \left(\begin{bmatrix} x+r & y+s \\ z+t & w+u \end{bmatrix} \right)$$

$$= \begin{bmatrix} m(x+r) + n(z+t) & m(y+s) + n(w+u) \\ p(x+r) + q(z+t) & p(y+s) + q(w+u) \end{bmatrix}$$

$$= \begin{bmatrix} mx + mr + nz + nt & my + ms + nw + nu \\ px + pr + qz + qt & py + ps + qw + qu \end{bmatrix}$$

$PX + PT$

$$= \begin{bmatrix} m & n \\ p & q \end{bmatrix} \begin{bmatrix} x & y \\ z & w \end{bmatrix} + \begin{bmatrix} m & n \\ p & q \end{bmatrix} \begin{bmatrix} r & s \\ t & u \end{bmatrix}$$

$$= \begin{bmatrix} mx + nz & my + nw \\ px + qz & py + qw \end{bmatrix} + \begin{bmatrix} mr + nt & ms + nu \\ pr + qt & ps + qu \end{bmatrix}$$

$$= \begin{bmatrix} (mx + nz) + (mr + nt) & (my + nw) + (ms + nu) \\ (px + qz) + (pr + qt) & (py + qw) + (ps + qu) \end{bmatrix}$$

$$= \begin{bmatrix} mx + nz + mr + nt & my + nw + ms + nu \\ px + qz + pr + qt & py + qw + ps + qu \end{bmatrix}$$

$$= \begin{bmatrix} mx + mr + nz + nt & my + ms + nw + nu \\ px + pr + qz + qt & py + ps + qw + qu \end{bmatrix}$$

Observe that the two results are identical. Thus, $P(X + T) = PX + PT$.

35. Verify that $(k+h)P = kP + hP$ for any real numbers k and h.

$$(k + h)P = (k + h) \begin{bmatrix} m & n \\ p & q \end{bmatrix}$$

$$= \begin{bmatrix} (k+h)m & (k+h)n \\ (k+h)p & (k+h)q \end{bmatrix}$$

$$= \begin{bmatrix} km + hm & kn + hn \\ kp + hp & kq + hq \end{bmatrix}$$

$$= \begin{bmatrix} km & kn \\ kp & kq \end{bmatrix} + \begin{bmatrix} hm & hn \\ hp & hq \end{bmatrix}$$

$$= k \begin{bmatrix} m & n \\ p & q \end{bmatrix} + h \begin{bmatrix} m & n \\ p & q \end{bmatrix}$$

$$= kP + hP$$

Thus, $(k + h)P = kP + hP$ for any real numbers k and h.

37.
$$\begin{bmatrix} 2 & 3 & 1 \\ 1 & -4 & 5 \end{bmatrix} \begin{bmatrix} x_1 \\ x_2 \\ x_3 \end{bmatrix} = \begin{bmatrix} 2x_1 + 3x_2 + x_3 \\ x_1 - 4x_2 + 5x_3 \end{bmatrix},$$

and $\begin{bmatrix} 2x_1 + 3x_2 + x_3 \\ x_1 - 4x_2 + 5x_3 \end{bmatrix} = \begin{bmatrix} 5 \\ 8 \end{bmatrix}.$

This is equivalent to

$$2x_1 + 3x_2 + x_3 = 5$$
$$x_1 - 4x_2 + 5x_3 = 8$$

since corresponding elements of equal matrices must be equal. Reversing this, observe that the given system of linear equations can be written as the matrix equation

$$\begin{bmatrix} 2 & 3 & 1 \\ 1 & -4 & 5 \end{bmatrix} \begin{bmatrix} x_1 \\ x_2 \\ x_3 \end{bmatrix} = \begin{bmatrix} 5 \\ 8 \end{bmatrix}.$$

39. **(a)** Use a graphing calculator or a computer to find the product matrix. The answer is

$$AC = \begin{bmatrix} 6 & 106 & 158 & 222 & 28 \\ 120 & 139 & 64 & 75 & 115 \\ -146 & -2 & 184 & 144 & -129 \\ 106 & 94 & 24 & 116 & 110 \end{bmatrix}.$$

(b) CA does not exist.

(c) AC and CA are clearly not equal, since CA does not even exist.

41. Use a graphing calculator or computer to find the matrix products and sums. The answers are as follows.

(a) $C + D = \begin{bmatrix} -1 & 5 & 9 & 13 & -1 \\ 7 & 17 & 2 & -10 & 6 \\ 18 & 9 & -12 & 12 & 22 \\ 9 & 4 & 18 & 10 & -3 \\ 1 & 6 & 10 & 28 & 5 \end{bmatrix}$

(b) $(C + D)B = \begin{bmatrix} -2 & -9 & 90 & 77 \\ -42 & -63 & 127 & 62 \\ 413 & 76 & 180 & -56 \\ -29 & -44 & 198 & 85 \\ 137 & 20 & 162 & 103 \end{bmatrix}$

(c) $CB = \begin{bmatrix} -56 & -1 & 1 & 45 \\ -156 & -119 & 76 & 122 \\ 315 & 86 & 118 & -91 \\ -17 & -17 & 116 & 51 \\ 118 & 19 & 125 & 77 \end{bmatrix}$

(d) $DB = \begin{bmatrix} 54 & -8 & 89 & 32 \\ 114 & 56 & 51 & -60 \\ 98 & -10 & 62 & 35 \\ -12 & -27 & 82 & 34 \\ 19 & 1 & 37 & 26 \end{bmatrix}$

(e) $CB + DB = \begin{bmatrix} -2 & -9 & 90 & 77 \\ -42 & -63 & 127 & 62 \\ 413 & 76 & 180 & -56 \\ -29 & -44 & 198 & 85 \\ 137 & 20 & 162 & 103 \end{bmatrix}$

(f) Yes, $(C+D)B$ and $CB+DB$ are equal, as can be seen by observing that the answers to parts (b) and (e) are identical.

43. (a) $\begin{bmatrix} 10 & 4 & 3 & 5 & 6 \\ 7 & 2 & 2 & 3 & 8 \\ 4 & 5 & 1 & 0 & 10 \\ 0 & 3 & 4 & 5 & 5 \end{bmatrix} \begin{bmatrix} 2 & 3 \\ 1 & 1 \\ 4 & 3 \\ 3 & 3 \\ 1 & 2 \end{bmatrix}$

$$= \begin{array}{c} \\ \text{Dept. 1} \\ \text{Dept. 2} \\ \text{Dept. 3} \\ \text{Dept. 4} \end{array} \begin{array}{c} \text{A} \quad\; \text{B} \\ \begin{bmatrix} 57 & 70 \\ 41 & 54 \\ 27 & 40 \\ 39 & 40 \end{bmatrix} \end{array}$$

(b) The total cost to buy from supplier A is $57 + 41 + 27 + 39 = \$164$, and the total cost to buy from supplier B is $70 + 54 + 40 + 40 = \$204$. The company should make the purchase from supplier A, since $\$164$ is a lower total cost than $\$204$.

45. (a) To find the average, add the matrices. Then multiply the resulting matrix by $\frac{1}{3}$. (Multiplying by $\frac{1}{3}$ is the same as dividing by 3.)

$$\frac{1}{3}\left(\begin{bmatrix} 4.27 & 6.94 \\ 3.45 & 3.65 \end{bmatrix} + \begin{bmatrix} 4.05 & 7.01 \\ 3.27 & 3.51 \end{bmatrix} + \begin{bmatrix} 4.40 & 6.90 \\ 3.54 & 3.76 \end{bmatrix} \right)$$

$$= \frac{1}{3} \begin{bmatrix} 12.72 & 20.85 \\ 10.26 & 10.92 \end{bmatrix} = \begin{bmatrix} 4.24 & 6.95 \\ 3.42 & 3.64 \end{bmatrix}$$

(b) To find the new average, add the new matrix for the Chicago plant and the matrix for the Seattle plant. Since there are only two matrices now, multiply the resulting matrix by $\frac{1}{2}$ to get the average. (Multiplying by $\frac{1}{2}$ is the same as dividing by 2.)

$$\frac{1}{2}\left(\begin{bmatrix} 4.42 & 7.43 \\ 3.38 & 3.62 \end{bmatrix} + \begin{bmatrix} 4.40 & 6.90 \\ 3.54 & 3.76 \end{bmatrix} \right)$$

$$= \frac{1}{2} \begin{bmatrix} 8.82 & 14.33 \\ 6.92 & 7.38 \end{bmatrix} = \begin{bmatrix} 4.41 & 7.17 \\ 3.46 & 3.69 \end{bmatrix}$$

47. (a)

$$P = \begin{array}{c} \\ \text{Sal's} \\ \text{Fred's} \end{array} \begin{array}{c} \text{Sh} \quad \text{Sa} \quad \text{B} \\ \begin{bmatrix} 80 & 40 & 120 \\ 60 & 30 & 150 \end{bmatrix} \end{array}$$

(b)

$$F = \begin{array}{c} \\ \text{Sh} \\ \text{Sa} \\ \text{B} \end{array} \begin{array}{c} \text{CA} \quad\; \text{AR} \\ \begin{bmatrix} \frac{1}{2} & \frac{1}{5} \\ \frac{1}{4} & \frac{1}{5} \\ \frac{1}{4} & \frac{3}{5} \end{bmatrix} \end{array}$$

(c) PF

$$= \begin{bmatrix} 80 & 40 & 120 \\ 60 & 30 & 150 \end{bmatrix} \begin{bmatrix} \frac{1}{2} & \frac{1}{5} \\ \frac{1}{4} & \frac{1}{5} \\ \frac{1}{4} & \frac{3}{5} \end{bmatrix}$$

$$= \begin{bmatrix} 80\left(\frac{1}{2}\right)+40\left(\frac{1}{4}\right)+120\left(\frac{1}{4}\right) & 80\left(\frac{1}{5}\right)+40\left(\frac{1}{5}\right)+120\left(\frac{3}{5}\right) \\ 60\left(\frac{1}{2}\right)+30\left(\frac{1}{4}\right)+150\left(\frac{1}{4}\right) & 60\left(\frac{1}{5}\right)+30\left(\frac{1}{5}\right)+150\left(\frac{3}{5}\right) \end{bmatrix}$$

$$= \begin{bmatrix} 80 & 96 \\ 75 & 108 \end{bmatrix}$$

The rows give the average price per pair of footwear sold by each store, and the columns give the state.

49. (a) The matrices are

$$S = \begin{bmatrix} .027 & .009 \\ .030 & .007 \\ .015 & .009 \\ .013 & .011 \\ .019 & .011 \end{bmatrix} \quad \text{and}$$

$$P = \begin{bmatrix} 1596 & 218 & 199 & 425 & 214 \\ 1996 & 286 & 226 & 460 & 243 \\ 2440 & 365 & 252 & 484 & 266 \\ 2906 & 455 & 277 & 499 & 291 \end{bmatrix}.$$

(b)

$$PS = \begin{array}{c} \\ 1960 \\ 1970 \\ 1980 \\ 1990 \end{array} \begin{array}{c} \text{Births} \quad\; \text{Deaths} \\ \begin{bmatrix} 62.208 & 24.710 \\ 76.459 & 29.733 \\ 91.956 & 35.033 \\ 108.28 & 40.522 \end{bmatrix} \end{array}$$

This product matrix gives the total number of births and deaths (in millions) in each year.

2.5 Matrix Inverses

1. $\begin{bmatrix} 2 & 3 \\ 1 & 1 \end{bmatrix} \begin{bmatrix} -1 & 3 \\ 1 & -2 \end{bmatrix} = \begin{bmatrix} 1 & 0 \\ 0 & 1 \end{bmatrix} = I$

$\begin{bmatrix} -1 & 3 \\ 1 & -2 \end{bmatrix} \begin{bmatrix} 2 & 3 \\ 1 & 1 \end{bmatrix} = \begin{bmatrix} 1 & 0 \\ 0 & 1 \end{bmatrix} = I$

Yes, these matrices are inverses of each other since their product matrix (both ways) is I.

3. $\begin{bmatrix} 2 & 1 \\ 3 & 2 \end{bmatrix} \begin{bmatrix} 2 & 1 \\ -3 & 2 \end{bmatrix} = \begin{bmatrix} 1 & 4 \\ 0 & 7 \end{bmatrix} \neq I$

No, these matrices are not inverses of each other since their product matrix is not I.

5. $\begin{bmatrix} 1 & 2 & 0 \\ 0 & 1 & 0 \\ 0 & 1 & 0 \end{bmatrix} \begin{bmatrix} 1 & -2 & 0 \\ 0 & 1 & 0 \\ 0 & -1 & 1 \end{bmatrix} = \begin{bmatrix} 1 & 0 & 0 \\ 0 & 1 & 0 \\ 0 & 1 & 0 \end{bmatrix} \neq I$

No, these matrices are not inverses of each other.

7. $\begin{bmatrix} 1 & 3 & 3 \\ 1 & 4 & 3 \\ 1 & 3 & 4 \end{bmatrix} \begin{bmatrix} 7 & -3 & -3 \\ -1 & 1 & 0 \\ -1 & 0 & 1 \end{bmatrix} = \begin{bmatrix} 1 & 0 & 0 \\ 0 & 1 & 0 \\ 0 & 0 & 1 \end{bmatrix} = I$

$\begin{bmatrix} 7 & -3 & -3 \\ -1 & 1 & 0 \\ -1 & 0 & 1 \end{bmatrix} \begin{bmatrix} 1 & 3 & 3 \\ 1 & 4 & 3 \\ 1 & 3 & 4 \end{bmatrix} = \begin{bmatrix} 1 & 0 & 0 \\ 0 & 1 & 0 \\ 0 & 0 & 1 \end{bmatrix} = I$

Yes, these matrices are inverses of each other.

9. No, a matrix with a row of all zeros does not have an inverse; the row of all zeros makes it impossible to get all the 1's in the main diagonal of the identity matrix.

11. Let $A = \begin{bmatrix} 1 & -1 \\ 2 & 0 \end{bmatrix}$.

Form the augmented matrix $[A|I]$.

$[A|I] = \begin{bmatrix} 1 & -1 & 1 & 0 \\ 2 & 0 & 0 & 1 \end{bmatrix}$

Perform row operations on $[A|I]$ to get a matrix of the form $[I|B]$.

$\begin{bmatrix} 1 & -1 & 1 & 0 \\ 2 & 0 & 0 & 1 \end{bmatrix}$

$-2R_1 + R_2 \to R_2 \quad \begin{bmatrix} 1 & -1 & 1 & 0 \\ 0 & 2 & -2 & 1 \end{bmatrix}$

$2R_1 + R_2 \to R_1 \quad \begin{bmatrix} 2 & 0 & 0 & 1 \\ 0 & 2 & -2 & 1 \end{bmatrix}$

$\frac{1}{2}R_1 \to R_1 \quad \begin{bmatrix} 1 & 0 & 0 & \frac{1}{2} \\ 0 & 1 & -1 & \frac{1}{2} \end{bmatrix} = [I|B]$
$\frac{1}{2}R_2 \to R_2$

The matrix B in the last transformation is the desired multiplicative inverse.

$A^{-1} = \begin{bmatrix} 0 & \frac{1}{2} \\ -1 & \frac{1}{2} \end{bmatrix}$

This answer may be checked by showing that $AA^{-1} = I$ and $A^{-1}A = I$.

13. Let $A = \begin{bmatrix} 3 & -1 \\ -5 & 2 \end{bmatrix}$.

$[A|I] = \begin{bmatrix} 3 & -1 & 1 & 0 \\ -5 & 2 & 0 & 1 \end{bmatrix}$

$5R_1 + 3R_2 \to R_2 \quad \begin{bmatrix} 3 & -1 & 1 & 0 \\ 0 & 1 & 5 & 3 \end{bmatrix}$

$R_1 + R_2 \to R_1 \quad \begin{bmatrix} 3 & 0 & 6 & 3 \\ 0 & 1 & 5 & 3 \end{bmatrix}$

$\frac{1}{3}R_1 \to R_1 \quad \begin{bmatrix} 1 & 0 & 2 & 1 \\ 0 & 1 & 5 & 3 \end{bmatrix} = [I|B]$

The desired inverse is

$A^{-1} = \begin{bmatrix} 2 & 1 \\ 5 & 3 \end{bmatrix}$.

15. Let $A = \begin{bmatrix} -6 & 4 \\ -3 & 2 \end{bmatrix}$.

$[A|I] = \begin{bmatrix} -6 & 4 & 1 & 0 \\ -3 & 2 & 0 & 1 \end{bmatrix}$

$R_1 + (-2)R_2 \to R_2 \quad \begin{bmatrix} -6 & 4 & 1 & 0 \\ 0 & 0 & 1 & -2 \end{bmatrix}$

Because the last row has all zeros to the left of the vertical bar, there is no way to complete the desired transformation. A has no inverse.

17. Let $A = \begin{bmatrix} 1 & 0 & 0 \\ 0 & -1 & 0 \\ 1 & 0 & 1 \end{bmatrix}$.

$[A|I] = \begin{bmatrix} 1 & 0 & 0 & 1 & 0 & 0 \\ 0 & -1 & 0 & 0 & 1 & 0 \\ 1 & 0 & 1 & 0 & 0 & 1 \end{bmatrix}$

$-1R_1 + R_3 \to R_3 \quad \begin{bmatrix} 1 & 0 & 0 & 1 & 0 & 0 \\ 0 & -1 & 0 & 0 & 1 & 0 \\ 0 & 0 & 1 & -1 & 0 & 1 \end{bmatrix}$

$-1R_2 \to R_2 \quad \begin{bmatrix} 1 & 0 & 0 & 1 & 0 & 0 \\ 0 & 1 & 0 & 0 & -1 & 0 \\ 0 & 0 & 1 & -1 & 0 & 1 \end{bmatrix}$

$A^{-1} = \begin{bmatrix} 1 & 0 & 0 \\ 0 & -1 & 0 \\ -1 & 0 & 1 \end{bmatrix}$

19. Let $A = \begin{bmatrix} -1 & -1 & -1 \\ 4 & 5 & 0 \\ 0 & 1 & -3 \end{bmatrix}$.

$[A|I] = \begin{bmatrix} -1 & -1 & -1 & | & 1 & 0 & 0 \\ 4 & 5 & 0 & | & 0 & 1 & 0 \\ 0 & 1 & -3 & | & 0 & 0 & 1 \end{bmatrix}$

$4R_1 + R_2 \to R_2 \quad \begin{bmatrix} -1 & -1 & -1 & | & 1 & 0 & 0 \\ 0 & 1 & -4 & | & 4 & 1 & 0 \\ 0 & 1 & -3 & | & 0 & 0 & 1 \end{bmatrix}$

$\begin{matrix} R_2 + R_1 \to R_1 \\ -1R_2 + R_3 \to R_3 \end{matrix} \quad \begin{bmatrix} -1 & 0 & -5 & | & 5 & 1 & 0 \\ 0 & 1 & -4 & | & 4 & 1 & 0 \\ 0 & 0 & 1 & | & -4 & -1 & 1 \end{bmatrix}$

$\begin{matrix} 5R_3 + R_1 \to R_1 \\ 4R_3 + R_2 \to R_2 \end{matrix} \quad \begin{bmatrix} -1 & 0 & 0 & | & -15 & -4 & 5 \\ 0 & 1 & 0 & | & -12 & -3 & 4 \\ 0 & 0 & 1 & | & -4 & -1 & 1 \end{bmatrix}$

$-1R_1 \to R_1 \quad \begin{bmatrix} 1 & 0 & 0 & | & 15 & 4 & -5 \\ 0 & 1 & 0 & | & -12 & -3 & 4 \\ 0 & 0 & 1 & | & -4 & -1 & 1 \end{bmatrix}$

$A^{-1} = \begin{bmatrix} 15 & 4 & -5 \\ -12 & -3 & 4 \\ -4 & -1 & 1 \end{bmatrix}$

21. Let $A = \begin{bmatrix} 1 & 2 & 3 \\ -3 & -2 & -1 \\ -1 & 0 & 1 \end{bmatrix}$.

$[A|I] = \begin{bmatrix} 1 & 2 & 3 & | & 1 & 0 & 0 \\ -3 & -2 & -1 & | & 0 & 1 & 0 \\ -1 & 0 & 1 & | & 0 & 0 & 1 \end{bmatrix}$

$\begin{matrix} 3R_1 + R_2 \to R_2 \\ R_1 + R_3 \to R_3 \end{matrix} \quad \begin{bmatrix} 1 & 2 & 3 & | & 1 & 0 & 0 \\ 0 & 4 & 8 & | & 3 & 1 & 0 \\ 0 & 2 & 4 & | & 1 & 0 & 1 \end{bmatrix}$

$\begin{matrix} R_2 + (-2R_1) \to R_1 \\ R_2 + (-2R_3) \to R_3 \end{matrix} \quad \begin{bmatrix} -2 & 0 & 2 & | & 1 & 1 & 0 \\ 0 & 4 & 8 & | & 3 & 1 & 0 \\ 0 & 0 & 0 & | & 1 & 1 & -2 \end{bmatrix}$

Because the last row has all zeros to the left of the vertical bar, there is no way to complete the desired transformation. A has no inverse.

23. Let $A = \begin{bmatrix} 2 & 4 & 6 \\ -1 & -4 & -3 \\ 0 & 1 & -1 \end{bmatrix}$.

$[A|I] = \begin{bmatrix} 2 & 4 & 6 & | & 1 & 0 & 0 \\ -1 & -4 & -3 & | & 0 & 1 & 0 \\ 0 & 1 & -1 & | & 0 & 0 & 1 \end{bmatrix}$

$R_1 + 2R_2 \to R_2 \quad \begin{bmatrix} 2 & 4 & 6 & | & 1 & 0 & 0 \\ 0 & -4 & 0 & | & 1 & 2 & 0 \\ 0 & 1 & -1 & | & 0 & 0 & 1 \end{bmatrix}$

$\begin{matrix} R_2 + R_1 \to R_1 \\ R_2 + 4R_3 \to R_3 \end{matrix} \quad \begin{bmatrix} 2 & 0 & 6 & | & 2 & 2 & 0 \\ 0 & -4 & 0 & | & 1 & 2 & 0 \\ 0 & 0 & -4 & | & 1 & 2 & 4 \end{bmatrix}$

$6R_3 + 4R_1 \to R_1 \quad \begin{bmatrix} 8 & 0 & 0 & | & 14 & 20 & 24 \\ 0 & -4 & 0 & | & 1 & 2 & 0 \\ 0 & 0 & -4 & | & 1 & 2 & 4 \end{bmatrix}$

$\begin{matrix} \frac{1}{8}R_1 \to R_1 \\ -\frac{1}{4}R_2 \to R_2 \\ -\frac{1}{4}R_3 \to R_3 \end{matrix} \quad \begin{bmatrix} 1 & 0 & 0 & | & \frac{7}{4} & \frac{5}{2} & 3 \\ 0 & 1 & 0 & | & -\frac{1}{4} & -\frac{1}{2} & 0 \\ 0 & 0 & 1 & | & -\frac{1}{4} & -\frac{1}{2} & -1 \end{bmatrix}$

$A^{-1} = \begin{bmatrix} \frac{7}{4} & \frac{5}{2} & 3 \\ -\frac{1}{4} & -\frac{1}{2} & 0 \\ -\frac{1}{4} & -\frac{1}{2} & -1 \end{bmatrix}$

25. Let

$A = \begin{bmatrix} 1 & -2 & 3 & 0 \\ 0 & 1 & -1 & 1 \\ -2 & 2 & -2 & 4 \\ 0 & 2 & -3 & 1 \end{bmatrix}$.

$[A|I] = \begin{bmatrix} 1 & -2 & 3 & 0 & | & 1 & 0 & 0 & 0 \\ 0 & 1 & -1 & 1 & | & 0 & 1 & 0 & 0 \\ -2 & 2 & -2 & 4 & | & 0 & 0 & 1 & 0 \\ 0 & 2 & -3 & 1 & | & 0 & 0 & 0 & 1 \end{bmatrix}$

$2R_1 + R_3 \to R_3 \quad \begin{bmatrix} 1 & -2 & 3 & 0 & | & 1 & 0 & 0 & 0 \\ 0 & 1 & -1 & 1 & | & 0 & 1 & 0 & 0 \\ 0 & -2 & 4 & 4 & | & 2 & 0 & 1 & 0 \\ 0 & 2 & -3 & 1 & | & 0 & 0 & 0 & 1 \end{bmatrix}$

$\begin{matrix} 2R_2 + R_1 \to R_1 \\ \\ 2R_2 + R_3 \to R_3 \\ -2R_2 + R_4 \to R_4 \end{matrix} \begin{bmatrix} 1 & 0 & 1 & 2 & | & 1 & 2 & 0 & 0 \\ 0 & 1 & -1 & 1 & | & 0 & 1 & 0 & 0 \\ 0 & 0 & 2 & 6 & | & 2 & 2 & 1 & 0 \\ 0 & 0 & -1 & -1 & | & 0 & -2 & 0 & 1 \end{bmatrix}$

$\begin{matrix} R_3 + (-2)R_1 \to R_1 \\ R_3 + 2R_2 \to R_2 \\ \\ R_3 + 2R_4 \to R_4 \end{matrix} \begin{bmatrix} -2 & 0 & 0 & 0 & | & 0 & -2 & 1 & 0 \\ 0 & 2 & 0 & 8 & | & 2 & 4 & 1 & 0 \\ 0 & 0 & 2 & 6 & | & 2 & 2 & 1 & 0 \\ 0 & 0 & 0 & 4 & | & 2 & -2 & 1 & 2 \end{bmatrix}$

$\begin{matrix} -2R_1 + R_4 \to R_1 \\ R_2 + (-2)R_4 \to R_2 \\ 2R_3 + (-3)R_4 \to R_3 \end{matrix} \begin{bmatrix} 4 & 0 & 0 & 0 & | & 2 & 2 & -1 & 2 \\ 0 & 2 & 0 & 0 & | & -2 & 8 & -1 & -4 \\ 0 & 0 & 4 & 0 & | & -2 & 10 & -1 & -6 \\ 0 & 0 & 0 & 4 & | & 2 & -2 & 1 & 2 \end{bmatrix}$

$$\begin{array}{l}\frac{1}{4}R_1 \to R_1 \\ \frac{1}{2}R_2 \to R_2 \\ \frac{1}{4}R_3 \to R_3 \\ \frac{1}{4}R_4 \to R_4\end{array}\left[\begin{array}{cccc|cccc} 1 & 0 & 0 & 0 & \frac{1}{2} & \frac{1}{2} & -\frac{1}{4} & \frac{1}{2} \\ 0 & 1 & 0 & 0 & -1 & 4 & -\frac{1}{2} & -2 \\ 0 & 0 & 1 & 0 & -\frac{1}{2} & \frac{5}{2} & -\frac{1}{4} & -\frac{3}{2} \\ 0 & 0 & 0 & 1 & \frac{1}{2} & -\frac{1}{2} & \frac{1}{4} & \frac{1}{2} \end{array}\right]$$

$$A^{-1} = \begin{bmatrix} \frac{1}{2} & \frac{1}{2} & -\frac{1}{4} & \frac{1}{2} \\ -1 & 4 & -\frac{1}{2} & -2 \\ -\frac{1}{2} & \frac{5}{2} & -\frac{1}{4} & -\frac{3}{2} \\ \frac{1}{2} & -\frac{1}{2} & \frac{1}{4} & \frac{1}{2} \end{bmatrix}.$$

27. $2x + 3y = 10$
$\quad\; x - \;\; y = -5$

First write the system in matrix form.

$$\begin{bmatrix} 2 & 3 \\ 1 & -1 \end{bmatrix}\begin{bmatrix} x \\ y \end{bmatrix} = \begin{bmatrix} 10 \\ -5 \end{bmatrix}$$

Let $A = \begin{bmatrix} 2 & 3 \\ 1 & -1 \end{bmatrix}$, $X = \begin{bmatrix} x \\ y \end{bmatrix}$, $B = \begin{bmatrix} 10 \\ -5 \end{bmatrix}$.

The system is (in matrix form) $AX = B$.
Now use row operations to find A^{-1}.

$$[A|I] = \begin{bmatrix} 2 & 3 & 1 & 0 \\ 1 & -1 & 0 & 1 \end{bmatrix}$$

$$R_1 + (-2)R_2 \to R_2 \qquad \begin{bmatrix} 2 & 3 & 1 & 0 \\ 0 & 5 & 1 & -2 \end{bmatrix}$$

$$-3R_2 + 5R_1 \to R_1 \qquad \begin{bmatrix} 10 & 0 & 2 & 6 \\ 0 & 5 & 1 & -2 \end{bmatrix}$$

$$\begin{array}{l}\frac{1}{10}R_1 \to R_1 \\ \frac{1}{5}R_2 \to R_2\end{array}\qquad \begin{bmatrix} 1 & 0 & \frac{1}{5} & \frac{3}{5} \\ 0 & 1 & \frac{1}{5} & -\frac{2}{5} \end{bmatrix}$$

$$A^{-1} = \begin{bmatrix} \frac{1}{5} & \frac{3}{5} \\ \frac{1}{5} & -\frac{2}{5} \end{bmatrix}.$$

Next, find the product $A^{-1}B$.

$$A^{-1}B = \begin{bmatrix} \frac{1}{5} & \frac{3}{5} \\ \frac{1}{5} & -\frac{2}{5} \end{bmatrix}\begin{bmatrix} 10 \\ -5 \end{bmatrix} = \begin{bmatrix} -1 \\ 4 \end{bmatrix}$$

Since $X = A^{-1}B$,

$$X = \begin{bmatrix} x \\ y \end{bmatrix} = \begin{bmatrix} -1 \\ 4 \end{bmatrix}.$$

Thus, the solution is $(-1, 4)$.

29. $2x + \;\; y = \;\; 5$
$\quad\; 5x + 3y = 13$

Let $A = \begin{bmatrix} 2 & 1 \\ 5 & 3 \end{bmatrix}$, $X = \begin{bmatrix} x \\ y \end{bmatrix}$, $B = \begin{bmatrix} 5 \\ 13 \end{bmatrix}$.

Use row operations to obtain

$$A^{-1} = \begin{bmatrix} 3 & -1 \\ -5 & 2 \end{bmatrix}.$$

$$X = A^{-1}B = \begin{bmatrix} 3 & -1 \\ -5 & 2 \end{bmatrix}\begin{bmatrix} 5 \\ 13 \end{bmatrix} = \begin{bmatrix} 2 \\ 1 \end{bmatrix}$$

The solution is $(2, 1)$.

31. $-x + y = 1$
$\quad\;\; 2x - y = 1$

Let $A = \begin{bmatrix} -1 & 1 \\ 2 & -1 \end{bmatrix}$, $X = \begin{bmatrix} x \\ y \end{bmatrix}$, $B = \begin{bmatrix} 1 \\ 1 \end{bmatrix}$.

Use row operations to obtain

$$A^{-1} = \begin{bmatrix} 1 & 1 \\ 2 & 1 \end{bmatrix}.$$

$$X = A^{-1}B = \begin{bmatrix} 1 & 1 \\ 2 & 1 \end{bmatrix}\begin{bmatrix} 1 \\ 1 \end{bmatrix} = \begin{bmatrix} 2 \\ 3 \end{bmatrix}$$

The solution is $(2, 3)$.

33. $-x - \;\; 8y = \;\; 12$
$\quad\;\; 3x + 24y = -36$

Let $A = \begin{bmatrix} -1 & -8 \\ 3 & 24 \end{bmatrix}$, $X = \begin{bmatrix} x \\ y \end{bmatrix}$, $B = \begin{bmatrix} 12 \\ -36 \end{bmatrix}$.

Using row operations on $[A|I]$ leads to the matrix

$$\begin{bmatrix} 1 & -8 & -1 & 0 \\ 0 & 0 & 3 & 1 \end{bmatrix},$$

but the zeros in the second row indicate that matrix A does not have an inverse. We cannot complete the solution by this method.

Since the second equation is a multiple of the first, the equations are dependent. Solve the first equation of the system for x.

$$\begin{aligned} -x - 8y &= 12 \\ -x &= 8y + 12 \\ x &= -8y - 12 \end{aligned}$$

The solution is $(-8y - 12, y)$.

35. $-x - \;\; y - \;\; z = \;\; 1$
$\quad\;\; 4x + 5y \qquad\;\; = -2$
$\qquad\qquad y - 3z = \;\; 3$

has coefficient matrix

$$A = \begin{bmatrix} -1 & -1 & -1 \\ 4 & 5 & 0 \\ 0 & 1 & -3 \end{bmatrix}.$$

In Exercise 19, it was found that

$$A^{-1} = \begin{bmatrix} -1 & -1 & -1 \\ 4 & 5 & 0 \\ 0 & 1 & 3 \end{bmatrix}^{-1}$$

$$= \begin{bmatrix} 15 & 4 & -5 \\ -12 & -3 & 4 \\ -4 & -1 & 1 \end{bmatrix}.$$

Since $X = A^{-1}B$,

$$\begin{bmatrix} x \\ y \\ z \end{bmatrix} = \begin{bmatrix} 15 & 4 & -5 \\ -12 & -3 & 4 \\ -4 & -1 & 1 \end{bmatrix} \begin{bmatrix} 1 \\ -2 \\ 3 \end{bmatrix} = \begin{bmatrix} -8 \\ 6 \\ 1 \end{bmatrix}.$$

The solution is $(-8, 6, 1)$.

37. $\begin{aligned} 2x + 4y + 6z &= 4 \\ -x - 4y - 3z &= 8 \\ y - z &= -4 \end{aligned}$

has coefficient matrix

$$A = \begin{bmatrix} 2 & 4 & 6 \\ -1 & -4 & -3 \\ 0 & 1 & -1 \end{bmatrix}.$$

In Exercise 23, it was found that

$$A^{-1} = \begin{bmatrix} 2 & 4 & 6 \\ -1 & -4 & -3 \\ 0 & 1 & -1 \end{bmatrix}^{-1}$$

$$= \begin{bmatrix} \frac{7}{4} & \frac{5}{2} & 3 \\ -\frac{1}{4} & -\frac{1}{2} & 0 \\ -\frac{1}{4} & -\frac{1}{2} & -1 \end{bmatrix}.$$

Since $X = A^{-1}B$,

$$\begin{bmatrix} x \\ y \\ z \end{bmatrix} = \begin{bmatrix} \frac{7}{4} & \frac{5}{2} & 3 \\ -\frac{1}{4} & -\frac{1}{2} & 0 \\ -\frac{1}{4} & -\frac{1}{2} & -1 \end{bmatrix} \begin{bmatrix} 4 \\ 8 \\ -4 \end{bmatrix} = \begin{bmatrix} 15 \\ -5 \\ -1 \end{bmatrix}.$$

The solution is $(15, -5, -1)$.

39. $\begin{aligned} 2x - 2y &= 5 \\ 4y + 8z &= 7 \\ x + 2z &= 1 \end{aligned}$

has coefficient matrix

$$A = \begin{bmatrix} 2 & -2 & 0 \\ 0 & 4 & 8 \\ 1 & 0 & 2 \end{bmatrix}.$$

However, using row operations on $[A|I]$ shows that A does not have an inverse, so another method must be used.

Try the Gauss-Jordan method. The augmented

matrix is

$$\begin{bmatrix} 2 & -2 & 0 & 5 \\ 0 & 4 & 8 & 7 \\ 1 & 0 & 2 & 1 \end{bmatrix}.$$

After several row operations, we obtain the matrix

$$\begin{bmatrix} 1 & 0 & 2 & \frac{17}{4} \\ 0 & 1 & 2 & \frac{7}{4} \\ 0 & 0 & 0 & 13 \end{bmatrix}.$$

The bottom row of this matrix shows that the system has no solution, since $0 = 13$ is a false statement.

41. $\begin{aligned} x - 2y + 3z &= 4 \\ y - z + w &= -8 \\ -2x + 2y - 2z + 4w &= 12 \\ 2y - 3z + w &= -4 \end{aligned}$

has coefficient matrix

$$A = \begin{bmatrix} 1 & -2 & 3 & 0 \\ 0 & 1 & -1 & 1 \\ -2 & 2 & -2 & 4 \\ 0 & 2 & -3 & 1 \end{bmatrix}.$$

In Exercise 25, it was found that

$$A^{-1} = \begin{bmatrix} \frac{1}{2} & \frac{1}{2} & -\frac{1}{4} & \frac{1}{2} \\ -1 & 4 & -\frac{1}{2} & -2 \\ -\frac{1}{2} & \frac{5}{2} & -\frac{1}{4} & -\frac{3}{2} \\ \frac{1}{2} & -\frac{1}{2} & \frac{1}{4} & \frac{1}{2} \end{bmatrix}.$$

Since $X = A^{-1}B$,

$$\begin{bmatrix} x \\ y \\ z \\ w \end{bmatrix} = \begin{bmatrix} \frac{1}{2} & \frac{1}{2} & -\frac{1}{4} & \frac{1}{2} \\ -1 & 4 & -\frac{1}{2} & -2 \\ -\frac{1}{2} & \frac{5}{2} & -\frac{1}{4} & -\frac{3}{2} \\ \frac{1}{2} & -\frac{1}{2} & \frac{1}{4} & \frac{1}{2} \end{bmatrix} \begin{bmatrix} 4 \\ -8 \\ 12 \\ -4 \end{bmatrix} = \begin{bmatrix} -7 \\ -34 \\ -19 \\ 7 \end{bmatrix}.$$

The solution is $(-7, -34, -19, 7)$.

In Exercises 43-47, let $A = \begin{bmatrix} a & b \\ c & d \end{bmatrix}$.

43. $IA = \begin{bmatrix} 1 & 0 \\ 0 & 1 \end{bmatrix} \begin{bmatrix} a & b \\ c & d \end{bmatrix} = \begin{bmatrix} a & b \\ c & d \end{bmatrix} = A$

Thus, $IA = A$.

45. $A \cdot 0 = \begin{bmatrix} a & b \\ c & d \end{bmatrix} \begin{bmatrix} 0 & 0 \\ 0 & 0 \end{bmatrix} = \begin{bmatrix} 0 & 0 \\ 0 & 0 \end{bmatrix} = 0$

Thus, $A \cdot 0 = 0$.

47. In Exercise 46, it was found that

$$A^{-1} = \frac{1}{ad - bc} \begin{bmatrix} d & -b \\ -c & a \end{bmatrix}.$$

$$A^{-1}A = \left(\frac{1}{ad - bc} \begin{bmatrix} d & -b \\ -c & a \end{bmatrix} \right) \begin{bmatrix} a & b \\ c & d \end{bmatrix}$$

$$= \frac{1}{ad - bc} \left(\begin{bmatrix} d & -b \\ -c & a \end{bmatrix} \begin{bmatrix} a & b \\ c & d \end{bmatrix} \right)$$

$$= \frac{1}{ad - bc} \begin{bmatrix} ad - bc & 0 \\ 0 & ad - bc \end{bmatrix}$$

$$= \begin{bmatrix} 1 & 0 \\ 0 & 1 \end{bmatrix} = I$$

Thus, $A^{-1}A = I$.

49.
$$AB = O$$
$$A^{-1}(AB) = A^{-1} \cdot O$$
$$(A^{-1}A)B = O$$
$$I \cdot B = O$$
$$B = O$$

Thus, if $AB = O$ and A^{-1} exists, then $B = O$.

51. This exercise should be solved by graphing calculator or computer methods. The solution, which may vary slightly, is

$$C^{-1} = \begin{bmatrix} -.0477 & -.0230 & .0292 & .0895 & -.0402 \\ .0921 & .0150 & .0321 & .0209 & -.0276 \\ -.0678 & .0315 & -.0404 & .0326 & .0373 \\ .0171 & -.0248 & .0069 & -.0003 & .0246 \\ -.0208 & .0740 & .0096 & -.1018 & .0646 \end{bmatrix}.$$

(Entries are rounded to 4 places.)

53. This exercise should be solved by graphing calculator or computer methods. The solution, which may vary slightly, is

$$D^{-1} = \begin{bmatrix} .0394 & -.0880 & .0033 & .0530 & -.1499 \\ -.1492 & .0289 & .0187 & .1033 & .1668 \\ -.1330 & -.0543 & .0356 & .1768 & .1055 \\ .1407 & .0175 & -.0453 & -.1344 & .0655 \\ .0102 & -.0653 & .0993 & .0085 & -.0388 \end{bmatrix}.$$

(Entries are rounded to 4 places.)

55. This exercise should be solved by graphing calculator or computer methods. The solution may vary slightly.

The answer is, yes, $D^{-1}C^{-1} = (CD)^{-1}$.

57. This exercise should be solved by graphing calculator or computer methods. The solution, which

may vary slightly, is

$$\begin{bmatrix} 1.51482 \\ .053479 \\ -.637242 \\ .462629 \end{bmatrix}.$$

59. (a) The matrix is $B = \begin{bmatrix} 72 \\ 48 \\ 60 \end{bmatrix}$.

(b) The matrix equation is

$$\begin{bmatrix} 2 & 4 & 2 \\ 2 & 1 & 2 \\ 2 & 1 & 3 \end{bmatrix} \begin{bmatrix} x_1 \\ x_2 \\ x_3 \end{bmatrix} = \begin{bmatrix} 72 \\ 48 \\ 60 \end{bmatrix}.$$

(c) To solve the system, begin by using row operations to find A^{-1}.

$$[A|I] = \begin{bmatrix} 2 & 4 & 2 & | & 1 & 0 & 0 \\ 2 & 1 & 2 & | & 0 & 1 & 0 \\ 2 & 1 & 3 & | & 0 & 0 & 1 \end{bmatrix}$$

$$\begin{matrix} \\ R_1 - 1R_2 \to R_2 \\ R_1 - 1R_3 \to R_3 \end{matrix} \begin{bmatrix} 2 & 4 & 2 & | & 1 & 0 & 0 \\ 0 & 3 & 0 & | & 1 & -1 & 0 \\ 0 & 3 & -1 & | & 1 & 0 & -1 \end{bmatrix}$$

$$\begin{matrix} -4R_2 + 3R_1 \to R_1 \\ \\ R_2 - 1R_3 \to R_3 \end{matrix} \begin{bmatrix} 6 & 0 & 6 & | & -1 & 4 & 0 \\ 0 & 3 & 0 & | & 1 & -1 & 0 \\ 0 & 0 & 1 & | & 0 & -1 & 1 \end{bmatrix}$$

$$-6R_3 + R_1 \to R_1 \begin{bmatrix} 6 & 0 & 0 & | & -1 & 10 & -6 \\ 0 & 3 & 0 & | & 1 & -1 & 0 \\ 0 & 0 & 1 & | & 0 & -1 & 1 \end{bmatrix}$$

$$\begin{matrix} \frac{1}{6}R_1 \to R_1 \\ \frac{1}{3}R_2 \to R_2 \end{matrix} \begin{bmatrix} 1 & 0 & 0 & | & -\frac{1}{6} & \frac{5}{3} & -1 \\ 0 & 1 & 0 & | & \frac{1}{3} & -\frac{1}{3} & 0 \\ 0 & 0 & 1 & | & 0 & -1 & 1 \end{bmatrix}$$

The inverse matrix is

$$A^{-1} = \begin{bmatrix} -\frac{1}{6} & \frac{5}{3} & -1 \\ \frac{1}{3} & -\frac{1}{3} & 0 \\ 0 & -1 & 1 \end{bmatrix}.$$

Since $X = A^{-1}B$,

$$\begin{bmatrix} x_1 \\ x_2 \\ x_3 \end{bmatrix} = \begin{bmatrix} -\frac{1}{6} & \frac{5}{3} & -1 \\ \frac{1}{3} & -\frac{1}{3} & 0 \\ 0 & -1 & 1 \end{bmatrix} \begin{bmatrix} 72 \\ 48 \\ 60 \end{bmatrix} = \begin{bmatrix} 8 \\ 8 \\ 12 \end{bmatrix}.$$

There are 8 daily orders for type I, 8 for type II, and 12 for type III.

61. Let $x_1 =$ amount invested in AAA bonds,
$x_2 =$ amount invested in A bonds, and
$x_3 =$ amount invested in B bonds.

The total investment is

$$x_1 + x_2 + x_3.$$

The annual return is

$$.06x_1 + .07x_2 + .10x_3.$$

Since twice as much is invested in AAA bonds as in B bonds,

$$x_1 = 2x_3.$$

(a) The system to be solved is

$$\begin{array}{rcrcrcr} x_1 &+& x_2 &+& x_3 &=& 25{,}000 \\ .06x_1 &+& .07x_2 &+& .10x_3 &=& 1810 \\ x_1 & & & - & 2x_3 &=& 0. \end{array}$$

Let $A = \begin{bmatrix} 1 & 1 & 1 \\ .06 & .07 & .10 \\ 1 & 0 & -2 \end{bmatrix}$, $X = \begin{bmatrix} x_1 \\ x_2 \\ x_3 \end{bmatrix}$,

$$B = \begin{bmatrix} 25{,}000 \\ 1810 \\ 0 \end{bmatrix}.$$

Use row operations to obtain

$$A^{-1} = \begin{bmatrix} -14 & 200 & 3 \\ 22 & -300 & -4 \\ -7 & 100 & 1 \end{bmatrix}.$$

Since $X = A^{-1}B$,

$$\begin{bmatrix} x_1 \\ x_2 \\ x_3 \end{bmatrix} = \begin{bmatrix} -14 & 200 & 3 \\ 22 & -300 & -4 \\ -7 & 100 & 1 \end{bmatrix}\begin{bmatrix} 25{,}000 \\ 1810 \\ 0 \end{bmatrix} = \begin{bmatrix} 12{,}000 \\ 7000 \\ 6000 \end{bmatrix}.$$

$12{,}000 should be invested in AAA bonds at 6%, $7000 in A bonds at 7%, and $6000 in B bonds at 10%.

(b) The matrix of constants is changed to

$$B = \begin{bmatrix} 30{,}000 \\ 2150 \\ 0 \end{bmatrix}.$$

$$X = A^{-1}B = \begin{bmatrix} -14 & 200 & 3 \\ 22 & -300 & -4 \\ -7 & 100 & 1 \end{bmatrix}\begin{bmatrix} 30{,}000 \\ 2150 \\ 0 \end{bmatrix}$$

$$= \begin{bmatrix} 10{,}000 \\ 15{,}000 \\ 5000 \end{bmatrix}$$

$10{,}000 should be invested in AAA bonds at 6%, $15{,}000 in A bonds at 7%, and $5000 in B bonds at 10%.

(c) The matrix of constants is changed to

$$B = \begin{bmatrix} 40{,}000 \\ 2900 \\ 0 \end{bmatrix}.$$

$$X = A^{-1}B = \begin{bmatrix} -14 & 200 & 3 \\ 22 & -300 & -4 \\ -7 & 100 & 1 \end{bmatrix}\begin{bmatrix} 40{,}000 \\ 2900 \\ 0 \end{bmatrix}$$

$$= \begin{bmatrix} 20{,}000 \\ 10{,}000 \\ 10{,}000 \end{bmatrix}$$

$20{,}000 should be invested in AAA bonds at 6%, $10{,}000 in A bonds at 7%, and $10{,}000 in B bonds at 10%.

63. Let $x =$ the number of Super Vim tablets,
$y =$ the number of Multitab tablets, and
$z =$ the number of Mighty Mix tablets.

The total number of vitamins is

$$x + y + z.$$

The total amount of niacin is

$$15x + 20y + 25z.$$

The total amount of Vitamin E is

$$12x + 15y + 35z.$$

(a) The system to be solved is

$$\begin{array}{rcrcrcr} x &+& y &+& z &=& 225 \\ 15x &+& 20y &+& 25z &=& 4750 \\ 12x &+& 15y &+& 35z &=& 5225. \end{array}$$

Let $A = \begin{bmatrix} 1 & 1 & 1 \\ 15 & 20 & 25 \\ 12 & 15 & 35 \end{bmatrix}$, $X = \begin{bmatrix} x \\ y \\ z \end{bmatrix}$, $B = \begin{bmatrix} 225 \\ 4750 \\ 5225 \end{bmatrix}$.

Thus, $AX = B$ and

$$\begin{bmatrix} 1 & 1 & 1 \\ 15 & 20 & 25 \\ 12 & 15 & 35 \end{bmatrix}\begin{bmatrix} x \\ y \\ z \end{bmatrix} = \begin{bmatrix} 225 \\ 4750 \\ 5225 \end{bmatrix}.$$

Use row operations to obtain the inverse of the coefficient matrix.

$$A^{-1} = \begin{bmatrix} \frac{65}{17} & -\frac{4}{17} & \frac{1}{17} \\ -\frac{45}{17} & \frac{23}{85} & -\frac{2}{17} \\ -\frac{3}{17} & -\frac{3}{85} & \frac{1}{17} \end{bmatrix}$$

Since $X = A^{-1}B$,

$$\begin{bmatrix} x \\ y \\ z \end{bmatrix} = \begin{bmatrix} \frac{65}{17} & -\frac{4}{17} & \frac{1}{17} \\ -\frac{45}{17} & \frac{23}{85} & -\frac{2}{17} \\ -\frac{3}{17} & -\frac{3}{85} & \frac{1}{17} \end{bmatrix} \begin{bmatrix} 225 \\ 4750 \\ 5225 \end{bmatrix} = \begin{bmatrix} 50 \\ 75 \\ 100 \end{bmatrix}.$$

There are 50 Super Vim tablets, 50 Multitab tablets, and 100 Mighty Mix tablets.

(b) The matrix of constants is changed to

$$B = \begin{bmatrix} 185 \\ 3625 \\ 3750 \end{bmatrix}.$$

$$\begin{bmatrix} x \\ y \\ z \end{bmatrix} = \begin{bmatrix} \frac{65}{17} & -\frac{4}{17} & \frac{1}{17} \\ -\frac{45}{17} & \frac{23}{85} & -\frac{2}{17} \\ -\frac{3}{17} & -\frac{3}{85} & \frac{1}{17} \end{bmatrix} \begin{bmatrix} 185 \\ 3625 \\ 3750 \end{bmatrix} = \begin{bmatrix} 75 \\ 50 \\ 60 \end{bmatrix}.$$

There are 75 Super Vim tablets, 50 Multitab tablets, and 60 Mighty Mix tablets.

(c) The matrix of constants is changed to

$$B = \begin{bmatrix} 230 \\ 4450 \\ 4210 \end{bmatrix}.$$

$$\begin{bmatrix} x \\ y \\ z \end{bmatrix} = \begin{bmatrix} \frac{65}{17} & -\frac{4}{17} & \frac{1}{17} \\ -\frac{45}{17} & \frac{23}{85} & -\frac{2}{17} \\ -\frac{3}{17} & -\frac{3}{85} & \frac{1}{17} \end{bmatrix} \begin{bmatrix} 230 \\ 4450 \\ 4210 \end{bmatrix} = \begin{bmatrix} 80 \\ 100 \\ 50 \end{bmatrix}.$$

There are 80 Super Vim tablets, 100 Multitab tablets, and 50 Mighty Mix tablets.

2.6 Input-Output Models

1. $A = \begin{bmatrix} .5 & .4 \\ .25 & .2 \end{bmatrix}$, $D = \begin{bmatrix} 2 \\ 4 \end{bmatrix}$

To find the production matrix, first calculate $I - A$.

$$I - A = \begin{bmatrix} 1 & 0 \\ 0 & 1 \end{bmatrix} - \begin{bmatrix} .5 & .4 \\ .25 & .2 \end{bmatrix} = \begin{bmatrix} .5 & -.4 \\ -.25 & .8 \end{bmatrix}$$

Using row operations, find the inverse of $I - A$.

$$[I - A|I] = \begin{bmatrix} .5 & -.4 & 1 & 0 \\ -.25 & .8 & 0 & 1 \end{bmatrix}$$

$$\begin{matrix} 10R_1 \to R_1 \\ 100R_2 \to R_2 \end{matrix} \quad \begin{bmatrix} 5 & -4 & 10 & 0 \\ -25 & 80 & 0 & 100 \end{bmatrix}$$

$$5R_1 + R_2 \to R_2 \quad \begin{bmatrix} 5 & -4 & 10 & 0 \\ 0 & 60 & 50 & 100 \end{bmatrix}$$

$$R_2 + 15R_1 \to R_1 \quad \begin{bmatrix} 75 & 0 & 200 & 100 \\ 0 & 60 & 50 & 100 \end{bmatrix}$$

$$\begin{matrix} \frac{1}{75}R_1 \to R_1 \\ \frac{1}{60}R_2 \to R_2 \end{matrix} \quad \begin{bmatrix} 1 & 0 & \frac{8}{3} & \frac{4}{3} \\ 0 & 1 & \frac{5}{6} & \frac{5}{3} \end{bmatrix}$$

$$(I - A)^{-1} = \begin{bmatrix} \frac{8}{3} & \frac{4}{3} \\ \frac{5}{6} & \frac{5}{3} \end{bmatrix} \approx \begin{bmatrix} 2.67 & 1.33 \\ .83 & 1.67 \end{bmatrix}$$

Since $X = (I - A)^{-1}D$, the production matrix is

$$X = \begin{bmatrix} 2.67 & 1.33 \\ .83 & 1.67 \end{bmatrix} \begin{bmatrix} 2 \\ 4 \end{bmatrix} = \begin{bmatrix} 10.67 \\ 8.33 \end{bmatrix}.$$

3. $A = \begin{bmatrix} .1 & .03 \\ .07 & .6 \end{bmatrix}$, $D = \begin{bmatrix} 5 \\ 10 \end{bmatrix}$

First, calculate $I - A$.

$$I - A = \begin{bmatrix} .9 & -.03 \\ -.07 & .4 \end{bmatrix}$$

Use row operations to find the inverse of $I - A$, which is

$$(I - A)^{-1} \approx \begin{bmatrix} 1.118 & .084 \\ .195 & 2.515 \end{bmatrix}.$$

Since $X = (I - A)^{-1}D$, the production matrix is

$$X = \begin{bmatrix} 1.118 & .084 \\ .195 & 2.515 \end{bmatrix} \begin{bmatrix} 5 \\ 10 \end{bmatrix} = \begin{bmatrix} 6.43 \\ 26.12 \end{bmatrix}.$$

5. $A = \begin{bmatrix} .4 & 0 & .3 \\ 0 & .8 & .1 \\ 0 & .2 & .4 \end{bmatrix}$, $D = \begin{bmatrix} 1 \\ 3 \\ 2 \end{bmatrix}$

First, calculate $I - A$.

$$I - A = \begin{bmatrix} 1 & 0 & 0 \\ 0 & 1 & 0 \\ 0 & 0 & 1 \end{bmatrix} - \begin{bmatrix} .4 & 0 & .3 \\ 0 & .8 & .1 \\ 0 & .2 & .4 \end{bmatrix}$$

$$= \begin{bmatrix} .6 & 0 & -.3 \\ 0 & .2 & -.1 \\ 0 & -.2 & .6 \end{bmatrix}$$

Now use row operations to find $(I - A)^{-1}$.

$$[I - A|I] = \begin{bmatrix} .6 & 0 & -.3 & 1 & 0 & 0 \\ 0 & .2 & -.1 & 0 & 1 & 0 \\ 0 & -.2 & .6 & 0 & 0 & 1 \end{bmatrix}$$

$R_2 + R_3 \to R_3$
$$\begin{bmatrix} .6 & 0 & -.3 & 1 & 0 & 0 \\ 0 & .2 & -.1 & 0 & 1 & 0 \\ 0 & 0 & .5 & 0 & 1 & 1 \end{bmatrix}$$

$3R_3 + 5R_1 \to R_1$
$R_3 + 5R_2 \to R_2$
$$\begin{bmatrix} 3 & 0 & 0 & 5 & 3 & 3 \\ 0 & 1 & 0 & 0 & 6 & 1 \\ 0 & 0 & .5 & 0 & 1 & 1 \end{bmatrix}$$

$\frac{1}{3}R_1 \to R_1$
$2R_3 \to R_3$
$$\begin{bmatrix} 1 & 0 & 0 & \frac{5}{3} & 1 & 1 \\ 0 & 1 & 0 & 0 & 6 & 1 \\ 0 & 0 & 1 & 0 & 2 & 2 \end{bmatrix}$$

$$(I - A)^{-1} \approx \begin{bmatrix} 1.67 & 1 & 1 \\ 0 & 6 & 1 \\ 0 & 2 & 2 \end{bmatrix}.$$

Since $X = (I - A)^{-1}D$, the production matrix is

$$X = \begin{bmatrix} 1.67 & 1 & 1 \\ 0 & 6 & 1 \\ 0 & 2 & 2 \end{bmatrix} \begin{bmatrix} 1 \\ 3 \\ 2 \end{bmatrix} = \begin{bmatrix} 6.67 \\ 20 \\ 10 \end{bmatrix}.$$

7.

$$\begin{array}{cccc} & A & B & C \\ A & \\ B & \\ C & \end{array} \begin{bmatrix} .3 & .1 & .8 \\ .5 & .6 & .1 \\ .2 & .3 & .1 \end{bmatrix} = A$$

$$I - A = \begin{bmatrix} .7 & -.1 & -.8 \\ -.5 & .4 & -.1 \\ -.2 & -.3 & .9 \end{bmatrix}$$

Set $(I - A)X = O$ to obtain the following.

$$\begin{bmatrix} .7 & -.1 & -.8 \\ -.5 & .4 & -.1 \\ -.2 & -.3 & .9 \end{bmatrix} \begin{bmatrix} x_1 \\ x_2 \\ x_3 \end{bmatrix} = \begin{bmatrix} 0 \\ 0 \\ 0 \end{bmatrix}$$

$$\begin{bmatrix} .7x_1 - .1x_2 - .8x_3 \\ -.5x_1 + .4x_2 - .1x_3 \\ -.2x_1 - .3x_2 + .9x_3 \end{bmatrix} = \begin{bmatrix} 0 \\ 0 \\ 0 \end{bmatrix}$$

Rewrite this matrix equation as a system of equations.

$$.7x_1 - .1x_2 - .8x_3 = 0$$
$$-.5x_1 + .4x_2 - .1x_3 = 0$$
$$-.2x_1 - .3x_2 + .9x_3 = 0$$

Rewrite the equations without decimals.

$$7x_1 - x_2 - 8x_3 = 0 \quad (1)$$
$$-5x_1 + 4x_2 - x_3 = 0 \quad (2)$$
$$-2x_1 - 3x_2 + 9x_3 = 0 \quad (3)$$

Use row operations to solve this system of equations. Begin by eliminating x_1 in equations (2) and (3).

$$7x_1 - x_2 - 8x_3 = 0 \quad (1)$$
$5R_1 + 7R_2 \to R_2$ $\qquad 23x_2 - 47x_3 = 0 \quad (4)$
$2R_1 + 7R_3 \to R_3$ $\qquad -23x_2 + 47x_3 = 0 \quad (5)$

Eliminate x_2 in equations (1) and (5).

$23R_1 + R_2 \to R_1$ $161x_1 \quad - 231x_3 = 0 \quad (6)$
$\qquad\qquad\qquad 23x_2 - 47x_3 = 0 \quad (4)$
$\qquad\qquad\qquad\qquad\qquad 0 = 0 \quad (7)$

The true statement in equation (7) indicates that the equations are dependent. Solve equation (6) for x_1 and equation (4) for x_2, each in terms of x_3.

$$x_1 = \frac{231}{161}x_3 = \frac{33}{23}x_3$$

$$x_2 = \frac{47}{23}x_3$$

The solution of the system is

$$\left(\frac{33}{23}x_3, \frac{47}{23}x_3, x_3 \right).$$

If $x_3 = 23$, then $x_1 = 33$ and $x_2 = 47$, so the production of the three commodities should be in the ratio 33:47:23.

9. Use a graphing calculator or a computer to find the production matrix $X = (I - A)^{-1}D$. The answer is

$$X = \begin{bmatrix} 7697 \\ 4205 \\ 6345 \\ 4106 \end{bmatrix}.$$

Values have been rounded.

11. In Example 4, it was found that

$$(I - A)^{-1} \approx \begin{bmatrix} 1.39 & .13 \\ .51 & 1.17 \end{bmatrix}.$$

Since $X = (I - A)^{-1}D$, the production matrix is

$$X = \begin{bmatrix} 1.39 & .13 \\ .51 & 1.17 \end{bmatrix} \begin{bmatrix} 690 \\ 920 \end{bmatrix} = \begin{bmatrix} 1078.7 \\ 1428.3 \end{bmatrix}.$$

Thus, about 1079 metric tons of wheat and 1428 metric tons of oil should be produced.

13. In Example 3, it was found that

$$(I - A)^{-1} \approx \begin{bmatrix} 1.40 & .50 & .59 \\ .84 & 1.36 & .62 \\ .56 & .47 & 1.30 \end{bmatrix}.$$

Since $X = (I - A)^{-1}D$, the production matrix is

$$X = \begin{bmatrix} 1.40 & .50 & .59 \\ .84 & 1.36 & .62 \\ .56 & .47 & 1.30 \end{bmatrix} \begin{bmatrix} 516 \\ 516 \\ 516 \end{bmatrix} = \begin{bmatrix} 1284.8 \\ 1455.1 \\ 1202.3 \end{bmatrix}.$$

Thus, about 1285 units of agriculture, 1455 units of manufacturing, and 1202 units of transportation should be produced.

15. From the given data, we get the input-output matrix

$$A = \begin{bmatrix} 0 & \frac{1}{2} & \frac{1}{4} \\ \frac{1}{4} & 0 & \frac{1}{4} \\ \frac{1}{2} & \frac{1}{4} & 0 \end{bmatrix}.$$

$$I - A = \begin{bmatrix} 1 & -\frac{1}{2} & -\frac{1}{4} \\ -\frac{1}{4} & 1 & -\frac{1}{4} \\ -\frac{1}{2} & -\frac{1}{4} & 1 \end{bmatrix}$$

Use row operations to find the inverse of $I - A$, which is

$$(I - A)^{-1} \approx \begin{bmatrix} 1.538 & .923 & .615 \\ .615 & 1.436 & .513 \\ .923 & .821 & 1.436 \end{bmatrix}.$$

Since $X = (I - A)^{-1}D$, the production matrix is

$$X = \begin{bmatrix} 1.538 & .923 & .615 \\ .615 & 1.436 & .513 \\ .923 & .821 & 1.436 \end{bmatrix} \begin{bmatrix} 1000 \\ 1000 \\ 1000 \end{bmatrix}$$

$$\approx \begin{bmatrix} 3077 \\ 2564 \\ 3179 \end{bmatrix}.$$

Thus, the production should be about 3077 units of agriculture, 2564 units of manufacturing, and 3179 units of transportation.

17. From the given data, we get the input-output matrix

$$A = \begin{bmatrix} \frac{1}{4} & \frac{1}{6} \\ \frac{1}{2} & 0 \end{bmatrix}.$$

$$I - A = \begin{bmatrix} \frac{3}{4} & -\frac{1}{6} \\ -\frac{1}{2} & 1 \end{bmatrix}$$

Use row operations to find the inverse of $I - A$, which is

$$(I - A)^{-1} = \begin{bmatrix} \frac{3}{2} & \frac{1}{4} \\ \frac{3}{4} & \frac{9}{8} \end{bmatrix}.$$

(a) The production matrix is

$$X = (I - A)^{-1}D = \begin{bmatrix} \frac{3}{2} & \frac{1}{4} \\ \frac{3}{4} & \frac{9}{8} \end{bmatrix} \begin{bmatrix} 1 \\ 1 \end{bmatrix} = \begin{bmatrix} \frac{7}{4} \\ \frac{15}{8} \end{bmatrix}.$$

Thus, $\frac{7}{4}$ bushels of yams and $\frac{15}{8} \approx 2$ pigs should be produced.

(b) The production matrix is

$$X = (I - A)^{-1}D = \begin{bmatrix} \frac{3}{2} & \frac{1}{4} \\ \frac{3}{4} & \frac{9}{8} \end{bmatrix} \begin{bmatrix} 100 \\ 70 \end{bmatrix} = \begin{bmatrix} 167.5 \\ 153.75 \end{bmatrix}.$$

Thus, 167.5 bushels of yams and $153.75 \approx 154$ pigs should be produced.

19. Use a graphing calculator or a computer to find the production matrix $X = (I - A)^{-1}D$. The answer is

$$\begin{bmatrix} 848 \\ 516 \\ 2970 \end{bmatrix}.$$

Values have been rounded.

Produce 848 units of agriculture, 516 units of manufacturing, and 2970 units of households.

21. Use a graphing calculator or a computer to find the production matrix $X = (I - A)^{-1}D$. The answer is

$$\begin{bmatrix} 195,492 \\ 25,933 \\ 13,580 \end{bmatrix}.$$

Values have been rounded. Change from thousands of pounds to millions of pounds.

Produce about 195 million lb of agriculture, 26 million lb of manufacturing, and 13.6 million lb of energy.

23. Find the value of $I - A$, then set $(I - A)X = O$.

$$(I - A)X = \left(\begin{bmatrix} 1 & 0 \\ 0 & 1 \end{bmatrix} - \begin{bmatrix} \frac{1}{4} & \frac{1}{2} \\ \frac{3}{4} & \frac{1}{2} \end{bmatrix} \right) \begin{bmatrix} x_1 \\ x_2 \end{bmatrix}$$

$$= \begin{bmatrix} \frac{3}{4} & -\frac{1}{2} \\ -\frac{3}{4} & \frac{1}{2} \end{bmatrix} \begin{bmatrix} x_1 \\ x_2 \end{bmatrix}$$

$$= \begin{bmatrix} \frac{3}{4}x_1 - \frac{1}{2}x_2 \\ -\frac{3}{4}x_1 + \frac{1}{2}x_2 \end{bmatrix} = \begin{bmatrix} 0 \\ 0 \end{bmatrix}$$

Thus,

$$\frac{3}{4}x_1 - \frac{1}{2}x_2 = 0$$

$$\frac{3}{4}x_1 = \frac{1}{2}x_2$$

$$x_1 = \frac{2}{3}x_2.$$

If $x_2 = 3$, then $x_1 = 2$. Therefore, produce 2 units of yams for every 3 units of pigs.

25. Find the value of $I - A$, then set $(I - A)X = O$.

$$(I - A)X = \left(\begin{bmatrix} 1 & 0 & 0 \\ 0 & 1 & 0 \\ 0 & 0 & 1 \end{bmatrix} - \begin{bmatrix} \frac{1}{3} & \frac{1}{2} & 0 \\ \frac{1}{3} & \frac{1}{4} & \frac{1}{4} \\ \frac{1}{3} & \frac{1}{4} & \frac{3}{4} \end{bmatrix} \right) \begin{bmatrix} x_1 \\ x_2 \\ x_3 \end{bmatrix}$$

$$= \begin{bmatrix} \frac{2}{3} & -\frac{1}{2} & 0 \\ -\frac{1}{3} & \frac{3}{4} & -\frac{1}{4} \\ -\frac{1}{3} & -\frac{1}{4} & \frac{1}{4} \end{bmatrix} \begin{bmatrix} x_1 \\ x_2 \\ x_3 \end{bmatrix}$$

$$= \begin{bmatrix} \frac{2}{3}x_1 - \frac{1}{2}x_2 \\ -\frac{1}{3}x_1 + \frac{3}{4}x_2 - \frac{1}{4}x_3 \\ -\frac{1}{3}x_1 - \frac{1}{4}x_2 + \frac{1}{4}x_3 \end{bmatrix} = \begin{bmatrix} 0 \\ 0 \\ 0 \end{bmatrix}$$

The system to be solved is

$$\frac{2}{3}x_1 - \frac{1}{2}x_2 \qquad = 0$$
$$-\frac{1}{3}x_1 + \frac{3}{4}x_2 - \frac{1}{4}x_3 = 0$$
$$-\frac{1}{3}x_1 - \frac{1}{4}x_2 + \frac{1}{4}x_3 = 0.$$

Write the augmented matrix of the system.

$$\begin{bmatrix} \frac{2}{3} & -\frac{1}{2} & 0 & \Big| & 0 \\ -\frac{1}{3} & \frac{3}{4} & -\frac{1}{4} & \Big| & 0 \\ -\frac{1}{3} & -\frac{1}{4} & \frac{1}{4} & \Big| & 0 \end{bmatrix}$$

$\begin{matrix} \frac{3}{2}R_1 \to R_1 \\ 12R_2 \to R_2 \\ 12R_3 \to R_3 \end{matrix}$ $\begin{bmatrix} 1 & -\frac{3}{4} & 0 & \Big| & 0 \\ -4 & 9 & -3 & \Big| & 0 \\ -4 & -3 & 3 & \Big| & 0 \end{bmatrix}$

$\begin{matrix} 4R_1 + R_2 \to R_2 \\ 4R_1 + R_3 \to R_3 \end{matrix}$ $\begin{bmatrix} 1 & -\frac{3}{4} & 0 & \Big| & 0 \\ 0 & 6 & -3 & \Big| & 0 \\ 0 & -6 & 3 & \Big| & 0 \end{bmatrix}$

$\frac{1}{6}R_2 \to R_2$ $\begin{bmatrix} 1 & -\frac{3}{4} & 0 & \Big| & 0 \\ 0 & 1 & -\frac{1}{2} & \Big| & 0 \\ 0 & -6 & 3 & \Big| & 0 \end{bmatrix}$

$\begin{matrix} \frac{3}{4}R_2 + R_1 \to R_1 \\ \\ 6R_2 + R_3 \to R_3 \end{matrix}$ $\begin{bmatrix} 1 & 0 & -\frac{3}{8} & \Big| & 0 \\ 0 & 1 & -\frac{1}{2} & \Big| & 0 \\ 0 & 0 & 0 & \Big| & 0 \end{bmatrix}$

Use x_3 as the parameter. Therefore, $x_1 = \frac{3}{8}x_3$ and $x_2 = \frac{1}{2}x_3$, and the solution is $\left(\frac{3}{8}x_3, \frac{1}{2}x_3, x_3 \right)$. If $x_3 = 8$, then $x_1 = 3$ and $x_2 = 4$.

Produce 3 units of agriculture to every 4 units of manufacturing and 8 units of transportation.

Chapter 2 Review Exercises

3. $2x + 3y = 10 \quad (1)$
$-3x + y = 18 \quad (2)$

Eliminate x in equation (2).

$\qquad\qquad 2x + 3y = 10 \quad (1)$
$3R_2 + 2R_1 \to R_2 \qquad 11y = 66 \quad (3)$

Make each leading coefficient equal 1.

$\frac{1}{2}R_1 \to R_1 \quad x + \frac{3}{2}y = 5 \quad (4)$
$\frac{1}{11}R_2 \to R_2 \qquad\quad y = 6 \quad (5)$

Substitute 6 for y in equation (4) to get $x = -4$.

The solution is $(-4, 6)$.

5. $2x - 3y + z = -5 \quad (1)$
$x + 4y + 2z = 13 \quad (2)$
$5x + 5y + 3z = 14 \quad (3)$

Eliminate x in equations (2) and (3).

$\qquad\qquad\qquad 2x - 3y + z = -5 \quad (1)$
$-2R_2 + R_1 \to R_2 \qquad -11y - 3z = -31 \quad (4)$
$5R_1 + (-2)R_3 \to R_3 \qquad -25y - z = -53 \quad (5)$

Eliminate y in equation (5).

$\qquad\qquad\qquad 2x - 3y + z = -5 \quad (1)$
$\qquad\qquad\qquad -11y - 3z = -31 \quad (6)$
$-25R_2 + R_3 \to R_3 \qquad 64z = 192 \quad (7)$

Make each leading coefficient equal 1.

$\frac{1}{2}R_1 \to R_1 \quad x - \frac{3}{2}y + \frac{1}{2}z = -\frac{5}{2} \quad (8)$
$-\frac{1}{11}R_2 \to R_2 \qquad y + \frac{3}{11}z = \frac{31}{11} \quad (9)$
$\frac{1}{64}R_3 \to R_3 \qquad\qquad z = 3 \quad (10)$

Substitute 3 for z in equation (9) to get $y = 2$. Substitute 3 for z and 2 for y in equation (8) to get $x = -1$.

The solution is $(-1, 2, 3)$.

7. $2x + 4y = -6$
$-3x - 5y = 12$

Write the augmented matrix and use row operations.

$$\begin{bmatrix} 2 & 4 & | & -6 \\ -3 & -5 & | & 12 \end{bmatrix}$$

$3R_1 + 2R_2 \to R_2$ $\begin{bmatrix} 2 & 4 & | & -6 \\ 0 & 2 & | & 6 \end{bmatrix}$

$-2R_2 + R_1 \to R_1$ $\begin{bmatrix} 2 & 0 & | & -18 \\ 0 & 2 & | & 6 \end{bmatrix}$

$\frac{1}{2}R_1 \to R_1$
$\frac{1}{2}R_2 \to R_2$ $\begin{bmatrix} 1 & 0 & | & -9 \\ 0 & 1 & | & 3 \end{bmatrix}$

The solution is $(-9, 3)$.

9. $x - y + 3z = 13$
$4x + y + 2z = 17$
$3x + 2y + 2z = 1$

Write the augmented matrix and use row operations.

$$\begin{bmatrix} 1 & -1 & 3 & | & 13 \\ 4 & 1 & 2 & | & 17 \\ 3 & 2 & 2 & | & 1 \end{bmatrix}$$

$-4R_1 + R_2 \to R_2$
$-3R_1 + R_3 \to R_3$ $\begin{bmatrix} 1 & -1 & 3 & | & 13 \\ 0 & 5 & -10 & | & -35 \\ 0 & 5 & -7 & | & -38 \end{bmatrix}$

$R_2 + 5R_1 \to R_1$
$-1R_2 + R_3 \to R_3$ $\begin{bmatrix} 5 & 0 & 5 & | & 30 \\ 0 & 5 & -10 & | & -35 \\ 0 & 0 & 3 & | & -3 \end{bmatrix}$

$5R_3 + (-3R_1) \to R_1$
$10R_3 + 3R_2 \to R_2$ $\begin{bmatrix} -15 & 0 & 0 & | & -105 \\ 0 & 15 & 0 & | & -135 \\ 0 & 0 & 3 & | & -3 \end{bmatrix}$

$-\frac{1}{15}R_1 \to R_1$
$\frac{1}{15}R_2 \to R_2$
$\frac{1}{3}R_3 \to R_3$ $\begin{bmatrix} 1 & 0 & 0 & | & 7 \\ 0 & 1 & 0 & | & -9 \\ 0 & 0 & 1 & | & -1 \end{bmatrix}$

The solution is $(7, -9, -1)$.

11. $3x - 6y + 9z = 12$
$-x + 2y - 3z = -4$
$x + y + 2z = 7$

Write the augmented matrix and use row operations.

$$\begin{bmatrix} 3 & -6 & 9 & | & 12 \\ -1 & 2 & -3 & | & -4 \\ 1 & 1 & 2 & | & 7 \end{bmatrix}$$

$R_1 + 3R_2 \to R_2$
$-1R_1 + 3R_3 \to R_3$ $\begin{bmatrix} 3 & -6 & 9 & | & 12 \\ 0 & 0 & 0 & | & 0 \\ 0 & 9 & -3 & | & 9 \end{bmatrix}$

The zero in row 2, column 2 is an obstacle. To proceed, interchange the second and third rows.

$$\begin{bmatrix} 3 & -6 & 9 & | & 12 \\ 0 & 9 & -3 & | & 9 \\ 0 & 0 & 0 & | & 0 \end{bmatrix}$$

$3R_1 + 2R_2 \to R_1$ $\begin{bmatrix} 9 & 0 & 21 & | & 54 \\ 0 & 9 & -3 & | & 9 \\ 0 & 0 & 0 & | & 0 \end{bmatrix}$

$\frac{1}{9}R_1 \to R_1$
$\frac{1}{9}R_2 \to R_2$ $\begin{bmatrix} 1 & 0 & \frac{7}{3} & | & 6 \\ 0 & 1 & -\frac{1}{3} & | & 1 \\ 0 & 0 & 0 & | & 0 \end{bmatrix}$

The row of zeros indicates dependent equations. Solve the first two equations respectively for x and y in terms of z to obtain

$$x = 6 - \frac{7}{3}z \quad \text{and} \quad y = 1 + \frac{1}{3}z.$$

The solution of the system is

$$\left(6 - \frac{7}{3}z, 1 + \frac{1}{3}z, z\right).$$

13. $\begin{bmatrix} 2 & x \\ y & 6 \\ 5 & z \end{bmatrix} = \begin{bmatrix} a & -1 \\ 4 & 6 \\ p & 7 \end{bmatrix}$

The size of these matrices is 3×2. For matrices to be equal, corresponding elements must be equal, so $a = 2$, $x = -1$, $y = 4$, $p = 5$, and $z = 7$.

15. $\begin{bmatrix} a+5 & 3b & 6 \\ 4c & 2+d & -3 \\ -1 & 4p & q-1 \end{bmatrix} = \begin{bmatrix} -7 & b+2 & 2k-3 \\ 3 & 2d-1 & 4\ell \\ m & 12 & 8 \end{bmatrix}$

These are 3×3 square matrices. Since corresponding elements must be equal,

$a + 5 = -7$, so $a = -12$;
$3b = b + 2$, so $b = 1$;
$6 = 2k - 3$, so $k = \frac{9}{2}$;
$4c = 3$, so $c = \frac{3}{4}$;
$2 + d = 2d - 1$, so $d = 3$;
$-3 = 4\ell$, so $\ell = -\frac{3}{4}$;
$m = -1$;
$4p = 12$, so $p = 3$; and
$q - 1 = 8$, so $q = 9$.

17. $2G - 4F = 2\begin{bmatrix} 2 & 5 \\ 1 & 6 \end{bmatrix} - 4\begin{bmatrix} -1 & 4 \\ 3 & 7 \end{bmatrix}$

$= \begin{bmatrix} 4 & 10 \\ 2 & 12 \end{bmatrix} + \begin{bmatrix} 4 & -16 \\ -12 & -28 \end{bmatrix}$

$= \begin{bmatrix} 8 & -6 \\ -10 & -16 \end{bmatrix}$

19. Since B is a 3×3 matrix, and A is a 3×2 matrix, the calculation of $B - A$ is not possible.

21. A has size 3×2 and F has 2×2, so AF will have size 3×2.

$AF = \begin{bmatrix} 4 & 10 \\ -2 & -3 \\ 6 & 9 \end{bmatrix}\begin{bmatrix} -1 & 4 \\ 3 & 7 \end{bmatrix} = \begin{bmatrix} 26 & 86 \\ -7 & -29 \\ 21 & 87 \end{bmatrix}$

23. D has size 3×1 and E has size 1×3, so DE will have size 3×3.

$DE = \begin{bmatrix} 6 \\ 1 \\ 0 \end{bmatrix}\begin{bmatrix} 1 & 3 & -4 \end{bmatrix} = \begin{bmatrix} 6 & 18 & -24 \\ 1 & 3 & -4 \\ 0 & 0 & 0 \end{bmatrix}$

25. B has size 3×3 and D has size 3×1, so BD will have size 3×1.

$BD = \begin{bmatrix} 2 & 3 & -2 \\ 2 & 4 & 0 \\ 0 & 1 & 2 \end{bmatrix}\begin{bmatrix} 6 \\ 1 \\ 0 \end{bmatrix} = \begin{bmatrix} 15 \\ 16 \\ 1 \end{bmatrix}$

27. $F = \begin{bmatrix} -1 & 4 \\ 3 & 7 \end{bmatrix}$

$[F|I] = \begin{bmatrix} -1 & 4 & | & 1 & 0 \\ 3 & 7 & | & 0 & 1 \end{bmatrix}$

$3R_1 + R_2 \to R_2 \quad \begin{bmatrix} -1 & 4 & | & 1 & 0 \\ 0 & 19 & | & 3 & 1 \end{bmatrix}$

$4R_2 + (-19R_1) \to R_1 \quad \begin{bmatrix} 19 & 0 & | & -7 & 4 \\ 0 & 19 & | & 3 & 1 \end{bmatrix}$

$\frac{1}{19}R_1 \to R_1 \atop \frac{1}{19}R_2 \to R_2 \quad \begin{bmatrix} 1 & 0 & | & -\frac{7}{19} & \frac{4}{19} \\ 0 & 1 & | & \frac{3}{19} & \frac{1}{19} \end{bmatrix}$

$F^{-1} = \begin{bmatrix} -\frac{7}{19} & \frac{4}{19} \\ \frac{3}{19} & \frac{1}{19} \end{bmatrix}$

29. A and C are 3×2 matrices, so their sum $A + C$ is a 3×2 matrix. Only square matrices have inverses. Therefore, $(A + C)^{-1}$ does not exist.

31. Let $A = \begin{bmatrix} -4 & 2 \\ 0 & 3 \end{bmatrix}$.

$[A|I] = \begin{bmatrix} -4 & 2 & | & 1 & 0 \\ 0 & 3 & | & 0 & 1 \end{bmatrix}$

$2R_2 + (-3R_1) \to R_1 \quad \begin{bmatrix} 12 & 0 & | & -3 & 2 \\ 0 & 3 & | & 0 & 1 \end{bmatrix}$

$\frac{1}{12}R_1 \to R_1 \atop \frac{1}{3}R_2 \to R_2 \quad \begin{bmatrix} 1 & 0 & | & -\frac{1}{4} & \frac{1}{6} \\ 0 & 1 & | & 0 & \frac{1}{3} \end{bmatrix}$

$A^{-1} = \begin{bmatrix} -\frac{1}{4} & \frac{1}{6} \\ 0 & \frac{1}{3} \end{bmatrix}$

33. Let $A = \begin{bmatrix} 6 & 4 \\ 3 & 2 \end{bmatrix}$.

$[A|I] = \begin{bmatrix} 6 & 4 & | & 1 & 0 \\ 3 & 2 & | & 0 & 1 \end{bmatrix}$

$R_1 + (-2)R_2 \to R_2 \quad \begin{bmatrix} 6 & 4 & | & 1 & 0 \\ 0 & 0 & | & 1 & -2 \end{bmatrix}$

The zeros in the second row indicate that the original matrix has no inverse.

35. Let $A = \begin{bmatrix} 2 & 0 & 4 \\ 1 & -1 & 0 \\ 0 & 1 & -2 \end{bmatrix}$.

$[A|I] = \begin{bmatrix} 2 & 0 & 4 & | & 1 & 0 & 0 \\ 1 & -1 & 0 & | & 0 & 1 & 0 \\ 0 & 1 & -2 & | & 0 & 0 & 1 \end{bmatrix}$

$-2R_2 + R_1 \to R_2 \quad \begin{bmatrix} 2 & 0 & 4 & | & 1 & 0 & 0 \\ 0 & 2 & 4 & | & 1 & -2 & 0 \\ 0 & 1 & -2 & | & 0 & 0 & 1 \end{bmatrix}$

$-2R_3 + R_2 \to R_3 \quad \begin{bmatrix} 2 & 0 & 4 & | & 1 & 0 & 0 \\ 0 & 2 & 4 & | & 1 & -2 & 0 \\ 0 & 0 & 8 & | & 1 & -2 & -2 \end{bmatrix}$

$-1R_3 + 2R_1 \to R_1 \atop -1R_3 + 2R_2 \to R_2 \quad \begin{bmatrix} 4 & 0 & 0 & | & 1 & 2 & 2 \\ 0 & 4 & 0 & | & 1 & -2 & 2 \\ 0 & 0 & 8 & | & 1 & -2 & -2 \end{bmatrix}$

$\frac{1}{4}R_1 \to R_1 \atop \frac{1}{4}R_2 \to R_2 \atop \frac{1}{8}R_3 \to R_3 \quad \begin{bmatrix} 1 & 0 & 0 & | & \frac{1}{4} & \frac{1}{2} & \frac{1}{2} \\ 0 & 1 & 0 & | & \frac{1}{4} & -\frac{1}{2} & \frac{1}{2} \\ 0 & 0 & 1 & | & \frac{1}{8} & -\frac{1}{4} & -\frac{1}{4} \end{bmatrix}$

$A^{-1} = \begin{bmatrix} \frac{1}{4} & \frac{1}{2} & \frac{1}{2} \\ \frac{1}{4} & -\frac{1}{2} & \frac{1}{2} \\ \frac{1}{8} & -\frac{1}{4} & -\frac{1}{4} \end{bmatrix}.$

37. Let

$$A = \begin{bmatrix} 2 & 3 & 5 \\ -2 & -3 & -5 \\ 1 & 4 & 2 \end{bmatrix}.$$

$$[A|I] = \begin{bmatrix} 2 & 3 & 5 & | & 1 & 0 & 0 \\ -2 & -3 & -5 & | & 0 & 1 & 0 \\ 1 & 4 & 2 & | & 0 & 0 & 1 \end{bmatrix}$$

$$\begin{matrix} R_1 + R_2 \to R_2 \\ R_1 + (-2R_3) \to R_3 \end{matrix} \begin{bmatrix} 2 & 3 & 5 & | & 1 & 0 & 0 \\ 0 & 0 & 0 & | & 1 & 1 & 0 \\ 0 & -5 & 1 & | & 1 & 0 & -2 \end{bmatrix}$$

The zeros in the second row to the left of the vertical bar indicate that the original matrix has no inverse.

39. $A = \begin{bmatrix} 1 & 2 \\ 2 & 4 \end{bmatrix}$, $B = \begin{bmatrix} 5 \\ 10 \end{bmatrix}$

Row operations may be used to see that matrix A has no inverse. The matrix equation $AX = B$ may be written as the system of equations

$$\begin{aligned} x + 2y &= 5 \quad (1) \\ 2x + 4y &= 10. \quad (2) \end{aligned}$$

Use the elimination method to solve this system. Begin by eliminating x in equation (2).

$$\begin{matrix} & x + 2y = 5 & (1) \\ -2R_1 + R_2 \to R_2 & 0 = 0 & (3) \end{matrix}$$

The true statement in equation (3) indicates that the equations are dependent. Solve equation (1) for x in terms of y.

$$x = -2y + 5$$

The solution is $(-2y + 5, y)$.

41. $A = \begin{bmatrix} 2 & 4 & 0 \\ 1 & -2 & 0 \\ 0 & 0 & 3 \end{bmatrix}$, $B = \begin{bmatrix} 72 \\ -24 \\ 48 \end{bmatrix}$

Use row operations to find the inverse of A, which is

$$A^{-1} = \begin{bmatrix} \frac{1}{4} & \frac{1}{2} & 0 \\ \frac{1}{8} & -\frac{1}{4} & 0 \\ 0 & 0 & \frac{1}{3} \end{bmatrix}.$$

Since $X = A^{-1}B$,

$$X = \begin{bmatrix} \frac{1}{4} & \frac{1}{2} & 0 \\ \frac{1}{8} & -\frac{1}{4} & 0 \\ 0 & 0 & \frac{1}{3} \end{bmatrix} \begin{bmatrix} 72 \\ -24 \\ 48 \end{bmatrix} = \begin{bmatrix} 6 \\ 15 \\ 16 \end{bmatrix}.$$

43. $\begin{aligned} 5x + 10y &= 80 \\ 3x - 2y &= 120 \end{aligned}$

Let $A = \begin{bmatrix} 5 & 10 \\ 3 & -2 \end{bmatrix}$, $X = \begin{bmatrix} x \\ y \end{bmatrix}$, $B \begin{bmatrix} 80 \\ 120 \end{bmatrix}$.

Use row operations to find the inverse of A, which is

$$A^{-1} = \begin{bmatrix} \frac{1}{20} & \frac{1}{4} \\ \frac{3}{40} & -\frac{1}{8} \end{bmatrix}.$$

Since $X = A^{-1}B$,

$$\begin{bmatrix} x \\ y \end{bmatrix} = \begin{bmatrix} \frac{1}{20} & \frac{1}{4} \\ \frac{3}{40} & -\frac{1}{8} \end{bmatrix} \begin{bmatrix} 80 \\ 120 \end{bmatrix} = \begin{bmatrix} 34 \\ -9 \end{bmatrix}.$$

The solution is $(34, -9)$.

45. $\begin{aligned} x + 4y - z &= 6 \\ 2x - y + z &= 3 \\ 3x + 2y + 3z &= 16 \end{aligned}$

Let $A = \begin{bmatrix} 1 & 4 & -1 \\ 2 & -1 & 1 \\ 3 & 2 & 3 \end{bmatrix}$, $X = \begin{bmatrix} x \\ y \\ z \end{bmatrix}$, $B = \begin{bmatrix} 6 \\ 3 \\ 16 \end{bmatrix}$.

Use row operations to find the inverse of A, which is

$$A^{-1} = \begin{bmatrix} \frac{5}{24} & \frac{7}{12} & -\frac{1}{8} \\ \frac{1}{8} & -\frac{1}{4} & \frac{1}{8} \\ -\frac{7}{24} & -\frac{5}{12} & \frac{3}{8} \end{bmatrix}.$$

Since $X = A^{-1}B$,

$$\begin{bmatrix} x \\ y \\ z \end{bmatrix} = \begin{bmatrix} \frac{5}{24} & \frac{7}{12} & -\frac{1}{8} \\ \frac{1}{8} & -\frac{1}{4} & \frac{1}{8} \\ -\frac{7}{24} & -\frac{5}{12} & \frac{3}{8} \end{bmatrix} \begin{bmatrix} 6 \\ 3 \\ 16 \end{bmatrix} = \begin{bmatrix} 1 \\ 2 \\ 3 \end{bmatrix}.$$

The solution is $(1, 2, 3)$.

47. $A = \begin{bmatrix} .2 & .1 & .3 \\ .1 & 0 & .2 \\ 0 & 0 & .4 \end{bmatrix}$, $D = \begin{bmatrix} 500 \\ 200 \\ 100 \end{bmatrix}$

$$X = (I - A)^{-1}D$$

$$I - A = \begin{bmatrix} .8 & -.1 & -.3 \\ -.1 & 1 & -.2 \\ 0 & 0 & .6 \end{bmatrix}$$

$$(I - A)^{-1} = \begin{bmatrix} 1.266 & .1266 & .6510 \\ .1266 & 1.0126 & .40084 \\ 0 & 0 & 1.6667 \end{bmatrix}$$

Since $X = (I - A)^{-1}D$,

$$X = \begin{bmatrix} 1.266 & .1266 & .6510 \\ .1266 & 1.0126 & .40084 \\ 0 & 0 & 1.6667 \end{bmatrix} \begin{bmatrix} 500 \\ 200 \\ 100 \end{bmatrix}$$

$$= \begin{bmatrix} 725.7 \\ 305.9 \\ 166.7 \end{bmatrix}.$$

49. Use a table to organize the information.

	Standard	Extra Large	Time Available
Hours Cutting	$\frac{1}{4}$	$\frac{1}{3}$	4
Hours Shaping	$\frac{1}{2}$	$\frac{1}{3}$	6

Let $x =$ the number of standard paper clips (in thousands),

and $y =$ the number of extra large paper clips (in thousands).

The given information leads to the system

$$\tfrac{1}{4}x + \tfrac{1}{3}y = 4$$
$$\tfrac{1}{2}x + \tfrac{1}{3}y = 6.$$

Solve this system by any method to get $x = 8$, $y = 6$. The manufacturer can make 8 thousand (8000) standard and 6 thousand (6000) extra large paper clips.

51. Let $x_1 =$ the number of blankets,
  $x_2 =$ the number of rugs, and
  $x_3 =$ the number of skirts.

The given information leads to the system

$$24x_1 + 30x_2 + 12x_3 = 306 \quad (1)$$
$$4x_1 + 5x_2 + 3x_3 = 59 \quad (2)$$
$$15x_1 + 18x_2 + 9x_3 = 201. \quad (3)$$

Simplify equations (1) and (3).

$$\begin{array}{ll} \tfrac{1}{6}R_1 \to R_1 & 4x_1 + 5x_2 + 2x_3 = 51 \quad (4) \\ & 4x_1 + 5x_2 + 3x_3 = 59 \quad (2) \\ \tfrac{1}{3}R_3 \to R_3 & 5x_1 + 6x_2 + 3x_3 = 67 \quad (6) \end{array}$$

Solve this system by the Gauss-Jordan method. Write the augmented matrix and use row operations.

$$\begin{bmatrix} 4 & 5 & 2 & | & 51 \\ 4 & 5 & 3 & | & 59 \\ 5 & 6 & 3 & | & 67 \end{bmatrix}$$

$$\begin{array}{l} -1R_1 + R_2 \to R_2 \\ -4R_3 + 5R_1 \to R_3 \end{array} \begin{bmatrix} 4 & 5 & 2 & | & 51 \\ 0 & 0 & 1 & | & 8 \\ 0 & 1 & -2 & | & -13 \end{bmatrix}$$

Interchange the second and third rows.

$$\begin{bmatrix} 4 & 5 & 2 & | & 51 \\ 0 & 1 & -2 & | & -13 \\ 0 & 0 & 1 & | & 8 \end{bmatrix}$$

$$-5R_2 + R_1 \to R_1 \begin{bmatrix} 4 & 0 & 12 & | & 116 \\ 0 & 1 & -2 & | & -13 \\ 0 & 0 & 1 & | & 8 \end{bmatrix}$$

$$\begin{array}{l} -12R_3 + R_1 \to R_1 \\ 2R_3 + R_2 \to R_2 \end{array} \begin{bmatrix} 4 & 0 & 0 & | & 20 \\ 0 & 1 & 0 & | & 3 \\ 0 & 0 & 1 & | & 8 \end{bmatrix}$$

$$\tfrac{1}{4}R_1 \to R_1 \begin{bmatrix} 1 & 0 & 0 & | & 5 \\ 0 & 1 & 0 & | & 3 \\ 0 & 0 & 1 & | & 8 \end{bmatrix}$$

The solution of the system is $x = 5$, $y = 3$, $z = 8$. 5 blankets, 3 rugs, and 8 skirts can be made.

53. The 4×5 matrix of stock reports is

$$\begin{bmatrix} 5 & 7 & 2532 & 52\tfrac{3}{8} & -\tfrac{1}{4} \\ 3 & 9 & 1464 & 56 & \tfrac{1}{8} \\ 2.50 & 5 & 4974 & 41 & -1\tfrac{1}{2} \\ 1.36 & 10 & 1754 & 18\tfrac{7}{8} & \tfrac{1}{2} \end{bmatrix}.$$

55. (a) The input-output matrix is

$$A = \begin{bmatrix} 0 & \tfrac{1}{2} \\ \tfrac{2}{3} & 0 \end{bmatrix}.$$

(b) $I - A = \begin{bmatrix} 1 & -\tfrac{1}{2} \\ -\tfrac{2}{3} & 1 \end{bmatrix}$, $D = \begin{bmatrix} 400 \\ 800 \end{bmatrix}$

Use row operations to find the inverse of $I - A$, which is

$$(I - A)^{-1} = \begin{bmatrix} \tfrac{3}{2} & \tfrac{3}{4} \\ 1 & \tfrac{3}{2} \end{bmatrix}.$$

Since $X = (I - A)^{-1}D$,

$$X = \begin{bmatrix} \tfrac{3}{2} & \tfrac{3}{4} \\ 1 & \tfrac{3}{2} \end{bmatrix} \begin{bmatrix} 400 \\ 800 \end{bmatrix} = \begin{bmatrix} 1200 \\ 1600 \end{bmatrix}.$$

The production required is 1200 units of cheese and 1600 units of goats.

57. (a) The X-ray passes through cells B and C, so the attenuation value for beam 3 is $b + c$.

(b)   Beam 1: $a + b = .8$
Beam 2: $a + c = .55$
Beam 3: $b + c = .65$

$$\begin{bmatrix} 1 & 1 & 0 \\ 1 & 0 & 1 \\ 0 & 1 & 1 \end{bmatrix} \begin{bmatrix} a \\ b \\ c \end{bmatrix} = \begin{bmatrix} .8 \\ .55 \\ .65 \end{bmatrix}$$

$$\begin{bmatrix} a \\ b \\ c \end{bmatrix} = \begin{bmatrix} 1 & 1 & 0 \\ 1 & 0 & 1 \\ 0 & 1 & 1 \end{bmatrix}^{-1} \begin{bmatrix} .8 \\ .55 \\ .65 \end{bmatrix}$$

$$= \begin{bmatrix} \frac{1}{2} & \frac{1}{2} & -\frac{1}{2} \\ \frac{1}{2} & -\frac{1}{2} & \frac{1}{2} \\ -\frac{1}{2} & \frac{1}{2} & \frac{1}{2} \end{bmatrix} \begin{bmatrix} .8 \\ .55 \\ .65 \end{bmatrix}$$

$$= \begin{bmatrix} .35 \\ .45 \\ .2 \end{bmatrix}$$

The solution is $(.35, .45, .2)$, so A is tumorous, B is bone, and C is healthy.

(c) For patient X,

$$\begin{bmatrix} a \\ b \\ c \end{bmatrix} = \begin{bmatrix} \frac{1}{2} & \frac{1}{2} & -\frac{1}{2} \\ \frac{1}{2} & -\frac{1}{2} & \frac{1}{2} \\ -\frac{1}{2} & \frac{1}{2} & \frac{1}{2} \end{bmatrix} \begin{bmatrix} .54 \\ .40 \\ .52 \end{bmatrix} = \begin{bmatrix} .21 \\ .33 \\ .19 \end{bmatrix}.$$

A and C are healthy; B is tumorous.

For patient Y,

$$\begin{bmatrix} a \\ b \\ c \end{bmatrix} = \begin{bmatrix} \frac{1}{2} & \frac{1}{2} & -\frac{1}{2} \\ \frac{1}{2} & -\frac{1}{2} & \frac{1}{2} \\ -\frac{1}{2} & \frac{1}{2} & \frac{1}{2} \end{bmatrix} \begin{bmatrix} .65 \\ .80 \\ .75 \end{bmatrix} = \begin{bmatrix} .35 \\ .3 \\ .45 \end{bmatrix}.$$

A and B are tumorous; C is bone.

For patient Z,

$$\begin{bmatrix} a \\ b \\ c \end{bmatrix} = \begin{bmatrix} \frac{1}{2} & \frac{1}{2} & -\frac{1}{2} \\ \frac{1}{2} & -\frac{1}{2} & \frac{1}{2} \\ -\frac{1}{2} & \frac{1}{2} & \frac{1}{2} \end{bmatrix} \begin{bmatrix} .51 \\ .49 \\ .44 \end{bmatrix} = \begin{bmatrix} .28 \\ .23 \\ .21 \end{bmatrix}.$$

A could be healthy or tumorous; B and C are healthy.

59. $\dfrac{\sqrt{3}}{2}(W_1 + W_2) = 100$ (1)

$W_1 - W_2 = 0$ (2)

Equation (2) gives $W_1 = W_2$. Substitute W_1 for

W_2 in equation (1).

$$\frac{\sqrt{3}}{2}(W_1 + W_1) = 100$$

$$\frac{\sqrt{3}}{2}(2W_1) = 100$$

$$\sqrt{3}W_1 = 100$$

$$W_1 = \frac{100}{\sqrt{3}} = \frac{100\sqrt{3}}{3} \approx 58$$

Therefore, $W_1 = W_2 \approx 58$ lb.

61. $C = at^2 + bt + c$

Use the values for C from the table.

(a) For 1958, $t = 0$, so

$$a(0)^2 + b(0) + c = 315.$$

Therefore, $c = 315$ and

$$C = at^2 + bt + 315.$$

For 1973, $t = 15$ $(1973 - 1958 = 15)$, so

$$a(15)^2 + b(15) + 315 = 325$$
$$225a + 15b = 10. \quad (1)$$

For 1988, $t = 30$ $(1988 - 1958 = 30)$, so

$$a(30)^2 + b(30) + 315 = 352$$
$$900a + 30b = 37. \quad (2)$$

Equations (1) and (2) give the system

$$225a + 15b = 10$$
$$900a + 30b = 37.$$

$$\begin{bmatrix} 225 & 15 \\ 900 & 30 \end{bmatrix} \begin{bmatrix} a \\ b \end{bmatrix} = \begin{bmatrix} 10 \\ 37 \end{bmatrix}$$

$$\begin{bmatrix} a \\ b \end{bmatrix} = \begin{bmatrix} 225 & 15 \\ 900 & 30 \end{bmatrix}^{-1} \begin{bmatrix} 10 \\ 37 \end{bmatrix}$$

$$\begin{bmatrix} a \\ b \end{bmatrix} = \begin{bmatrix} -\frac{1}{225} & \frac{1}{450} \\ \frac{2}{15} & -\frac{1}{30} \end{bmatrix} \begin{bmatrix} 10 \\ 37 \end{bmatrix}$$

$$\begin{bmatrix} a \\ b \end{bmatrix} = \begin{bmatrix} .038 \\ .1 \end{bmatrix}$$

Therefore, $C = .038t^2 + .1t + 315$

(b) We want to know when the 1958 CO_2 level, 315, will double, that is, will be $2(315) = 630$.

$$.038t^2 + .1t + 315 = 630$$
$$.038t^2 + .1t - 315 = 0$$
$$38t^2 + 100t - 315{,}000 = 0$$

Use the quadratic formula with $a = 38$, $b = 100$, and $c = -315{,}000$.

$$t = \frac{-b \pm \sqrt{b^2 - 4ac}}{2a}$$

$$t = \frac{-100 \pm \sqrt{100^2 - 4(38)(-315{,}000)}}{2(38)}$$

$$t = \frac{-100 \pm \sqrt{47{,}890{,}000}}{76}$$

$$t = \frac{-100 + \sqrt{47{,}890{,}000}}{76} \approx 90 \text{ or}$$

$$t = \frac{-100 - \sqrt{47{,}890{,}000}}{76} \approx -92$$

Ignore the negative value. If $t = 90$, then $1958 + 90 = 2048$. The 1958 CO_2 level will double in the year 2048.

Extended Application: Contagion

1. $PQ = \begin{bmatrix} 1 & 0 & 0 & 1 & 0 \\ 0 & 0 & 1 & 1 & 0 \\ 1 & 1 & 0 & 0 & 0 \end{bmatrix} \begin{bmatrix} 1 & 1 & 0 & 1 & 1 & 1 \\ 0 & 0 & 0 & 0 & 1 & 0 \\ 0 & 0 & 0 & 0 & 0 & 0 \\ 0 & 1 & 0 & 1 & 0 & 0 \\ 1 & 0 & 0 & 0 & 1 & 0 \end{bmatrix}$

$$= \begin{bmatrix} 1 & 2 & 0 & 2 & 1 & 1 \\ 0 & 1 & 0 & 1 & 0 & 0 \\ 1 & 1 & 0 & 1 & 2 & 1 \end{bmatrix}$$

2. In the product PQ, $a_{23} = 0$, so there were no contacts.

3. In the product PQ, column 3 has all zeros. The third person had no contacts with the first group.

4. In the product PQ, the entries in columns 2 and 4 both have a sum of 4. The second and fourth persons in the third group each had four contacts in all.

Chapter 2 Test

[2.1]

1. Solve the system using the echelon method.

$$3x + y = 11$$
$$x - 2y = -8$$

[2.2]

2. Solve the system using the Gauss-Jordan method.

$$x + 2y + 3z = 5$$
$$2x - y + z = 5$$
$$x + y + z = 2$$

3. Use the Gauss-Jordan method to find all solutions of the following system, given that all variables must be nonnegative integers.

$$2x + 2y - z = 30$$
$$3x + 2y - 2z = 41$$
$$x + 4y + z = 27$$

[2.1–2.2]

4. Use a system of equations to solve the following problem. Solve the system by the method of your choice.

   An investor has $80,000 and she would like to earn 7.675% per year by investing it in mutual funds, bonds, and certificates of deposit. She wants to invest $4000 more in bonds than the total investment in mutual funds and certificates of deposit. Mutual funds pay 10% per year, bonds pay 7% per year, and certificates of deposit pay 5% per year. How much should she invest in each of the three?

[2.3]

5. Find the values of the variables.

$$\begin{bmatrix} 6-x & 2y+1 \\ 3m & 5p+2 \end{bmatrix} = \begin{bmatrix} 8 & 10 \\ -5 & 3p-1 \end{bmatrix}$$

[2.3–2.4]

6. Given the following matrices, perform the indicated operations, if possible.

$$A = \begin{bmatrix} 1 & 2 & -1 \\ 0 & 1 & 1 \\ 1 & 0 & 1 \end{bmatrix} \qquad B = \begin{bmatrix} 1 & -2 \\ 1 & 1 \\ 0 & 1 \end{bmatrix} \qquad C = \begin{bmatrix} 2 & 1 & 3 \\ 0 & 4 & 1 \\ 1 & 1 & 1 \end{bmatrix}$$

   (a) AB **(b)** $2A - C$ **(c)** $A + 2B$

[2.5]

Find the inverse of each matrix which has an inverse.

7. $A = \begin{bmatrix} 1 & 0 & -1 \\ 2 & 1 & 1 \\ 1 & 1 & 5 \end{bmatrix}$

8. $B = \begin{bmatrix} 2 & -1 & 1 \\ 0 & 2 & 4 \\ 2 & 1 & 5 \end{bmatrix}$

9. For $A = \begin{bmatrix} 1 & 0 & 1 \\ 1 & 1 & 1 \\ 2 & 1 & 3 \end{bmatrix}$, $A^{-1} = \begin{bmatrix} 2 & 1 & -1 \\ -1 & 1 & 0 \\ -1 & -1 & 1 \end{bmatrix}$.

Use this inverse to solve the equation $AX = B$, where $B = \begin{bmatrix} 1 \\ 2 \\ -1 \end{bmatrix}$.

10. Solve the system using the inverse of the coefficient matrix.

$$x - 2y = 3$$
$$x + 3y = 5$$

11. Solve the system by using the inverse of the coefficient matrix. Use a graphing calculator.

$$.103x - .247y + .489z = .936$$
$$-.218x + .379y + .702z = .863$$
$$.315x - .742y - .913z = -.768$$

[2.6]

12. Find the production matrix, given the following input-output and demand matrices.

$$A = \begin{bmatrix} .2 & .1 & .3 \\ .1 & 0 & .2 \\ 0 & 0 & .4 \end{bmatrix}, \qquad D = \begin{bmatrix} 1000 \\ 2000 \\ 5000 \end{bmatrix}$$

Chapter 2 Test Answers

1. $(2, 5)$

2. $(1, -1, 2)$

3. $(11, 4, 0)$, $(13, 3, 2)$, $(15, 2, 4)$, $(17, 1, 6)$, $(19, 0, 8)$

4. $26,000 in mutual funds; $42,000 in bonds, $12,000 in certificates of deposit

5. $x = -2$, $y = \frac{9}{2}$, $m = -\frac{5}{3}$, $p = -\frac{3}{2}$

6. (a) $\begin{bmatrix} 3 & -1 \\ 1 & 2 \\ 1 & -1 \end{bmatrix}$
   (b) $\begin{bmatrix} 0 & 3 & -5 \\ 0 & -2 & 1 \\ 1 & -1 & 1 \end{bmatrix}$
   (c) Not possible

7. $A^{-1} = \begin{bmatrix} \frac{4}{3} & -\frac{1}{3} & \frac{1}{3} \\ -3 & 2 & -1 \\ \frac{1}{3} & -\frac{1}{3} & \frac{1}{3} \end{bmatrix}$

8. B^{-1} does not exist.

9. $X = \begin{bmatrix} 5 \\ 1 \\ -4 \end{bmatrix}$

10. $\left(\frac{19}{5}, \frac{2}{5} \right)$

11. $(-2.061, -1.683, 1.498)$

12. $\begin{bmatrix} 4895 \\ 4156 \\ 8333 \end{bmatrix}$

LINEAR PROGRAMMING: THE GRAPHICAL METHOD

3.1 Graphing Linear Inequalities

For Exercises 1-39, see the answer graphs in the back of the textbook.

1. $x + y \leq 2$

First graph the boundary line $x + y = 2$ using the points $(2,0)$ and $(0,2)$. Since the points on this line satisfy $x + y \leq 2$, draw a solid line. To find the correct region to shade, choose any point not on the line. If $(0,0)$ is used as the test point, we have

$$x + y \leq 2$$
$$0 + 0 \leq 2$$
$$0 \leq 2,$$

which is a true statement. Shade the half-plane containing $(0,0)$, or all points below the line.

3. $x \geq 3 + y$

First graph the boundary line $x = 3 + y$ using the points $(0,-3)$ and $(3,0)$. This will be a solid line. Choose $(0,0)$ as a test point.

$$x \geq 3 + y$$
$$0 \geq 3 + 0$$
$$0 \geq 3,$$

which is a false statement. Shade the half-plane that does not contain $(0,0)$, or all points below the line.

5. $4x - y < 6$

Graph $4x - y = 6$ as a dashed line, since the points on the line are not part of the solution; the line passes through the points $(0,-6)$ and $\left(\frac{3}{2},0\right)$. Using the test point $(0,0)$, $0 - 0 < 6$ or $0 < 6$, a true statement. Shade the half-plane containing $(0,0)$, or all points above the line.

7. $3x + y < 6$

Graph $3x + y = 6$ as a dashed line through $(2,0)$ and $(0,6)$. Using the test point $(0,0)$, we get $3(0) + 0 < 6$ or $0 < 6$, a true statement. Shade the half-plane containing $(0,0)$, or all points below the line.

9. $x + 3y \geq -2$

The graph includes the line $x + 3y = -2$, whose intercepts are the points $\left(0, -\frac{2}{3}\right)$ and $(-2,0)$. Graph $x + 3y = -2$ as a solid line and use the origin as a test point. Since $0 + 3(0) \geq -2$ is true, shade the half-plane containing $(0,0)$, or all points above the line.

11. $4x + 3y > -3$

Graph $4x + 3y = -3$ as a dashed line through the points $(0,-1)$ and $\left(-\frac{3}{4},0\right)$. Use the origin as a test point. Since $4(0) + 3(0) > -3$ is true, shade the half-plane containing $(0,0)$, or all points above the line.

13. $x \leq 5y$

Graph $x = 5y$ as a solid line through the points $(0,0)$ and $(5,1)$. Since this line contains the origin, some point other than $(0,0)$ must be used as a test point. If we use the point $(1,2)$, we obtain $1 \leq 5(2)$ or $1 \leq 10$, a true statement. Shade the half-plane containing $(1,2)$, or all points above the line.

15. $-3x < y$

Graph $y = -3x$ as a dashed line through the points $(0,0)$ and $(1,-3)$. Since this line contains the origin, use some point other than $(0,0)$ as a test point. If $(1,1)$ is used as a test point, we obtain $-3 < 3$, a true statement. Shade the half-plane containing $(1,1)$, or all points above the line.

17. $x + y \leq 0$

Graph $x + y = 0$ as a solid line through the points $(0,0)$ and $(1,-1)$. This line contains $(0,0)$. If we use $(-1,0)$ as a test point, we obtain $-1 + 0 \leq 0$ or $-1 \leq 0$, a true statement. Shade the half-plane containing $(-1,0)$, or all points below the line.

19. $y < x$

Graph $y = x$ as a dashed line through the points $(0,0)$ and $(1,1)$. Since this line contains the origin, choose a point other than $(0,0)$ as a test point. If we use $(2,3)$, we obtain $3 < 2$, which is false.

Shade the half-plane that does not contain $(2, 3)$, or all points below the line.

21. $x < 4$

Graph $x = 4$ as a dashed line. This is the vertical line crossing the x-axis at the point $(4, 0)$. Using $(0, 0)$ as a test point, we obtain $0 < 4$, which is true. Shade the half-plane containing $(0, 0)$, or all points to the left of the line.

23. $y \leq -2$

Graph $y = -2$ as a solid horizontal line through the point $(0, -2)$. Using the origin as a test point, we obtain $0 \leq -2$, which is false. Shade the half-plane that does not contain $(0, 0)$, or all points below the line.

25. $x + y \leq 1$
    $x - y \geq 2$

Graph the solid lines

$$x + y = 1 \text{ and}$$
$$x - y = 2.$$

$0 + 0 \leq 1$ is true, and $0 - 0 \geq 2$ is false. In each case, the graph is the region below the line. Shade the overlapping part of these two half-planes, which is the region below both lines. The shaded region is the feasible region for this system.

27. $x + 3y \leq 6$
    $2x + 4y \geq 7$

Graph the solid lines $x + 3y = 6$ and $2x + 4y = 7$. Use $(0, 0)$ as a test point. $0 + 0 \leq 6$ is true, and $0 + 0 \geq 7$ is false. Shade all points below $x + 3y = 6$ and above $2x + 4y = 7$. The feasible region is the overlap of the two half-planes.

29. $x + y \leq 4$
    $x - y \leq 5$
    $4x + y \leq -4$

The graph of $x + y \leq 4$ consists of the solid line $x + y = 4$ and all the points below it. The graph of $x - y \leq 5$ consists of the solid line $x - y = 5$ and all the points above it. The graph of $4x + y \leq -4$ consists of the solid line $4x + y = -4$ and all the points below it. The feasible region is the overlapping part of these three half-planes.

31. $-2 < x < 3$
    $-1 \leq y \leq 5$
    $2x + y < 6$

The graph of $-2 < x < 3$ is the region between the vertical lines $x = -2$ and $x = 3$, but not including the lines themselves (so the two vertical boundaries are drawn as dashed lines). The graph of $-1 \leq y \leq 5$ is the region between the horizontal lines $y = -1$ and $y = 5$, including the lines (so the two horizontal boundaries are drawn as solid lines). The graph of $2x + y < 6$ is the region below the line $2x + y = 6$ (so the boundary is drawn as a dashed line). Shade the region common to all three graphs to show the feasible region.

33. $2y + x \geq -5$
    $y \leq 3 + x$
    $x \geq 0$
    $y \geq 0$

The graph of $2y + x \geq -5$ consists of the boundary line $2y + x = 5$ and the region above it. The graph of $y \leq 3 + x$ consists of the boundary line $y = 3 + x$ and the region below it. The inequalities $x \geq 0$ and $y \geq 0$ restrict the feasible region to the first quadrant. Shade the region in the first quadrant where the first two graphs overlap to show the feasible region.

35. $3x + 4y > 12$
    $2x - 3y < 6$
    $0 \leq y \leq 2$
    $x \geq 0$

$3x + 4y > 12$ is the set of points above the dashed line $3x + 4y = 12$; $2x - 3y < 6$ is the set of points above the dashed line $2x - 3y = 6$; $0 \leq y \leq 2$ is the set of points lying on or between the horizontal lines $y = 0$ and $y = 2$; and $x \geq 0$ consists of all the points on or to the right of the y-axis. Shade the feasible region, which is the triangular region satisfying all of the inequalities.

37. $2x - 4y > 3$

Use a graphing calculator. The boundary line is the graph of $2x - 4y = 3$. Solve this equation for y.

$$-4y = -2x + 3$$

$$y = \frac{-2}{-4}x + \frac{3}{-4}$$

$$y = \frac{1}{2}x - \frac{3}{4}$$

Enter $y_1 = \frac{1}{2}x - \frac{3}{4}$ and graph it. Using the origin as a test point, we obtain $0 > 3$, which is false. Shade the region that does not contain the origin.

39. $3x - 4y < 6$
    $2x + 5y > 15$

Use a graphing calculator. One boundary line is the graph of $3x - 4y = 6$. Solve this equation for y.

$$-4y = -3x + 6$$

$$y = \frac{-3}{-4}x + \frac{6}{-4}$$

$$y = \frac{3}{4}x - \frac{3}{2}$$

Enter $y_1 = \frac{3}{4}x - \frac{3}{2}$ and graph it. Using the origin as a test point, we obtain $0 < 6$, which is true. Shade the region that contains the origin.

The other boundary line is the graph of $2x + 5y = 15$. Solve this equation for y.

$$5y = -2x + 15$$

$$y = -\frac{2}{5}x + 3$$

Enter $y_2 = -\frac{2}{5}x + 3$ and graph it. Using the origin as a test point, we obtain $0 > 15$, which is false. Shade the region that does not contain the origin. The overlap of the two graphs is the feasible region.

41. (a)

	Glazed	Unglazed	Maximum
Number Made	x	y	
Time on Wheel	$\frac{1}{2}$	1	8
Time in Kiln	1	6	20

(b) On the wheel, x glazed planters require $\frac{1}{2} \cdot x = \frac{1}{2}x$ hr and y unglazed planters require $1 \cdot y = y$ hr. Since the wheel is available for at most 8 hr per day,

$$\frac{1}{2}x + y \leq 8.$$

In the kiln, x glazed planters require $1 \cdot x = x$ hr and y unglazed planters require $6 \cdot y = 6y$ hr. Since the kiln is available for at most 20 hr per day,

$$x + 6y \leq 20.$$

Since it is not possible to produce a negative number of pots,

$$x \geq 0 \text{ and } y \geq 0.$$

Thus, we have the system

$$\frac{1}{2}x + y \leq 8$$
$$x + 6y \leq 20$$
$$x \geq 0$$
$$y \geq 0.$$

See the graph of the feasible region in the back of the textbook.

(c) Yes, 5 glazed and 2 unglazed planters can be made, since the point $(5, 2)$ lies within the feasible region.

From the graph, it looks like the point $(10, 2)$ might lie right on a boundary of the feasible region. However, $(10, 2)$ does not satisfy the inequality $x + 6y \leq 20$, so the point is definitely outside the feasible region. Therefore, 10 glazed and 2 unglazed planters cannot be made.

43. (a) $x \geq 3000$
             $y \geq 5000$
        $x + y \leq 10,000$

(b) The first inequality gives the half-plane to the right of the vertical line $x = 3000$, including the points on the line. The second inequality gives the half-plane above the horizontal line $y = 5000$, including the points on the line. The third inequality gives the half-plane below the line $x + y \leq 10,000$, including the points on the line. Shade the region where the three half-planes overlap to show the feasible region. See the graph of the feasible region in the back of the textbook.

45. (a) $x \geq 1000$
             $y \geq 800$
        $x + y \leq 2400$

(b) The first inequality gives the set of points on and to the right of the vertical line $x = 1000$. The second inequality gives the set of points on and above the horizontal line $y = 800$. The third inequality gives the set of points on and below the line $x + y = 2400$. Shade the region where the three graphs overlap to show the feasible region. See the graph of the feasible region in the back of the textbook.

3.2   Solving Linear Programming Problems Graphically

1. Make a table indicating the value of the objective function $z = 3x + 5y$ at each corner point.

Corner Point	Value of $z = 3x + 5y$
$(1,1)$	$3(1) + 5(1) = 8$ Minimum
$(2,7)$	$3(2) + 5(7) = 41$
$(5,10)$	$3(5) + 5(10) = 65$ Maximum
$(6,3)$	$3(6) + 5(3) = 33$

The maximum value of 65 occurs at $(5,10)$. The minimum value of 8 occurs at $(1,1)$.

3.

Corner Point	Value of $z = .40x + .75y$
$(0,0)$	$.40(0) + .75(0) = 0$ Minimum
$(0,12)$	$.40(0) + .75(12) = 9$ Maximum
$(4,8)$	$.40(4) + .75(8) = 7.6$
$(7,3)$	$.40(7) + .75(3) = 5.05$
$(8,0)$	$.40(8) + .75(0) = 3.2$

The maximum value is 9 at $(0,12)$; the minimum value is 0 at $(0,0)$.

5.

(a)

Corner Point	Value of $z = 4x + 2y$
$(0,8)$	$4(0) + 2(8) = 16$ Minimum
$(3,4)$	$4(3) + 2(4) = 20$
$\left(\frac{13}{2},2\right)$	$4\left(\frac{13}{2}\right) + 2(2) = 30$
$(12,0)$	$4(12) + 2(0) = 48$

The minimum value is 16 at $(0,8)$. Since the feasible region is unbounded, there is no maximum value.

(b)

Corner Point	Value of $z = 2x + 3y$
$(0,8)$	$2(0) + 3(8) = 24$
$(3,4)$	$2(3) + 3(4) = 18$ Minimum
$\left(\frac{13}{2},2\right)$	$2\left(\frac{13}{2}\right) + 3(2) = 19$
$(12,0)$	$2(12) + 3(0) = 24$

The minimum value is 18 at $(3,4)$; there is no maximum value since the feasible region is unbounded.

(c)

Corner Point	Value of $z = 2x + 4y$
$(0,8)$	$2(0) + 4(8) = 32$
$(3,4)$	$2(3) + 4(4) = 22$
$\left(\frac{13}{2},2\right)$	$2\left(\frac{13}{2}\right) + 4(2) = 21$ Minimum
$(12,0)$	$2(12) + 4(0) = 24$

The minimum value is 21 at $\left(\frac{13}{2},2\right)$; there is no maximum value since the feasible region is unbounded.

(d)

Corner Point	Value of $z = x + 4y$
$(0,8)$	$0 + 4(8) = 32$
$(3,4)$	$3 + 4(4) = 19$
$\left(\frac{13}{2},2\right)$	$\frac{13}{2} + 4(2) = \frac{29}{2}$
$(12,0)$	$12 + 4(0) = 12$ Minimum

The minimum value is 12 at $(12,0)$; there is no maximum value since the feasible region is unbounded.

7. Maximize $z = 5x + 2y$

subject to: $2x + 3y \le 6$
$4x + y \ge 6$
$x \ge 0$
$y \ge 0.$

Sketch the feasible region.

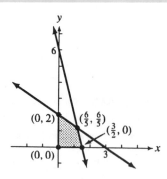

The graph shows that the feasible region is bounded. The corner points are $(0,0)$, $(0,2)$, $\left(\frac{3}{2},0\right)$, and $\left(\frac{6}{5},\frac{6}{5}\right)$. The corner point $\left(\frac{6}{5},\frac{6}{5}\right)$ can be found by solving the system

$$2x + 3y = 6$$
$$4x + y = 6.$$

Use the corner points to find the maximum value of the objective function.

Corner Point	Value of $z = 5x + 2y$
$(0,0)$	$5(0) + 2(0) = 0$
$(0,2)$	$5(0) + 2(2) = 4$
$\left(\frac{6}{5}, \frac{6}{5}\right)$	$5\left(\frac{6}{5}\right) + 2\left(\frac{6}{5}\right) = \frac{42}{5}$ Maximum
$\left(\frac{3}{2}, 0\right)$	$5\left(\frac{3}{2}\right) + 2(0) = \frac{15}{2}$

The maximum value is $\frac{42}{5}$ when $x = \frac{6}{5}$ and $y = \frac{6}{5}$.

9. Maximize $z = 2x + 2y$

   subject to: $3x - y \geq 12$
   $\qquad\qquad x + y \leq 15$
   $\qquad\qquad\quad x \geq 2$
   $\qquad\qquad\quad y \geq 5.$

   Sketch the feasible region.

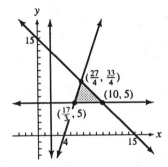

The graph shows that the feasible region is bounded. The corner points are: $\left(\frac{17}{3}, 5\right)$, which is the intersection of $y = 5$ and $3x - y = 12$; $\left(\frac{27}{4}, \frac{33}{4}\right)$, which is the intersection of $3x - y = 12$ and $x + y = 15$; and $(10, 5)$, which is the intersection of $y = 5$ and $x + y = 15$. Use the corner points to find the maximum value of the objective function.

Corner Point	Value of $z = 2x + 2y$
$\left(\frac{17}{3}, 5\right)$	$2\left(\frac{17}{3}\right) + 2(5) = \frac{64}{3}$
$\left(\frac{27}{4}, \frac{33}{4}\right)$	$2\left(\frac{27}{4}\right) + 2\left(\frac{33}{4}\right) = 30$ Maximum
$(10, 5)$	$2(10) + 2(5) = 30$ Maximum

The maximum value is 30 when $x = \frac{27}{4}$ and $y = \frac{33}{4}$, as well as when $x = 10$ and $y = 5$ and at all points in between.

11. Maximize $z = 4x + 2y$

    subject to: $x - y \leq 10$
    $\qquad\qquad 5x + 3y \leq 75$
    $\qquad\qquad\quad x + y \leq 20$
    $\qquad\qquad\qquad x \geq 0$
    $\qquad\qquad\qquad y \geq 0.$

Sketch the feasible region.

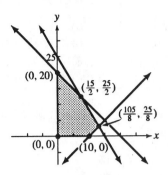

The region is bounded, with corner points $(0,0)$, $(0, 20)$, $\left(\frac{15}{2}, \frac{25}{2}\right)$, which is the intersection of $5x + 3y = 75$ and $x + y = 20$, $\left(\frac{105}{8}, \frac{25}{8}\right)$, which is the intersection of $x - y = 10$ and $5x + 3y = 75$, and $(10, 0)$.

Corner Point	Value of $z = 4x + 2y$
$(0,0)$	$4(0) + 2(0) = 0$
$(0, 20)$	$4(0) + 2(20) = 40$
$\left(\frac{15}{2}, \frac{25}{2}\right)$	$4\left(\frac{15}{2}\right) + 2\left(\frac{25}{2}\right) = 55$
$\left(\frac{105}{8}, \frac{25}{8}\right)$	$4\left(\frac{105}{8}\right) + 2\left(\frac{25}{8}\right) = \frac{235}{4} = 58\frac{3}{4}$ Maximum
$(10, 0)$	$4(10) + 2(0) = 40$

The maximum value is $\frac{235}{4}$ when $x = \frac{105}{8}$ and $y = \frac{25}{8}$.

13. Maximize $z = 10x + 12y$ subject to the following sets of constraints, with $x \geq 0$ and $y \geq 0$.

    (a) $x + y \leq 20$
    $\qquad x + 3y \leq 24$

Sketch the feasible region in the first quadrant, and identify the corner points at $(0,0)$, $(0,8)$, $(18,2)$, which is the intersection of $x + y = 20$ and $x + 3y = 24$, and $(20, 0)$.

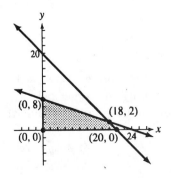

Corner Point	Value of $z = 10x + 12y$
$(0,0)$	$10(0) + 12(0) = 0$
$(0,8)$	$10(0) + 12(8) = 96$
$(18,2)$	$10(18) + 12(2) = 204$ Maximum
$(20,0)$	$10(20) + 12(0) = 200$

The maximum value of 204 occurs when $x = 18$ and $y = 2$.

(b) $3x + y \leq 15$
    $x + 2y \leq 18$

Sketch the feasible region in the first quadrant, and identify the corner points. The corner point $\left(\frac{12}{5}, \frac{39}{5}\right)$ can be found by solving the system

$$3x + y = 15$$
$$x + 2y = 18.$$

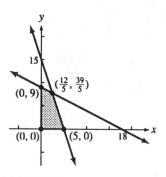

Corner Point	Value of $z = 10x + 12y$
$(0,0)$	$10(0) + 12(0) = 0$
$(0,9)$	$10(0) + 12(9) = 108$
$\left(\frac{12}{5}, \frac{39}{5}\right)$	$10\left(\frac{12}{5}\right) + 12\left(\frac{39}{5}\right) = \frac{588}{5} = 117\frac{3}{5}$ Maximum
$(5,0)$	$10(5) + 12(0) = 50$

The maximum value of $\frac{588}{5}$ occurs when $x = \frac{12}{5}$ and $y = \frac{39}{5}$.

(c) $2x + 5y \geq 22$
    $4x + 3y \leq 28$
    $2x + 2y \leq 17$

Sketch the feasible region in the first quadrant, and identify the corner points. The corner point $\left(\frac{5}{2}, 6\right)$ can be found by solving the system

$$4x + 3y = 28$$
$$2x + 2y = 17,$$

and the corner point $\left(\frac{37}{7}, \frac{16}{7}\right)$ can be found by solving the system

$$2x + 5y = 22$$
$$4x + 3y = 28.$$

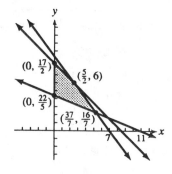

Corner Point	Value of $z = 10x + 12y$
$\left(0, \frac{22}{5}\right)$	$10(0) + 12\left(\frac{22}{5}\right) = \frac{264}{5} = 52.8$
$\left(0, \frac{17}{2}\right)$	$10(0) + 12\left(\frac{17}{2}\right) = 102$ Maximum
$\left(\frac{5}{2}, 6\right)$	$10\left(\frac{5}{2}\right) + 12(6) = 97$
$\left(\frac{37}{7}, \frac{16}{7}\right)$	$10\left(\frac{37}{7}\right) + 12\left(\frac{16}{7}\right) = \frac{562}{7} \approx 80.3$

The maximum value of 102 occurs when $x = 0$ and $y = \frac{17}{2}$.

15. Maximize $z = c_1 x_1 + c_2 x_2$

subject to: $2x_1 + x_2 \leq 11$
        $-x_1 + 2x_2 \leq 2$
        $x_1 \geq 0, \ x_2 \geq 0.$

Sketch the feasible region.

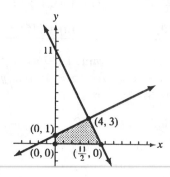

The region is bounded, with corner points $(0,0)$, $(0,1)$, $(4,3)$, and $\left(\frac{11}{2}, 0\right)$.

Corner Point	Value of $z = c_1 x_1 + c_2 x_2$
$(0,0)$	$c_1(0) + c_2(0) = 0$
$(0,1)$	$c_1(0) + c_2(1) = c_2$
$(4,3)$	$c_1(4) + c_2(3) = 4c_1 + 3c_2$
$\left(\frac{11}{2}, 0\right)$	$c_1\left(\frac{11}{2}\right) + c_2(0) = \frac{11}{2}c_1$

If we are to have $(x_1, x_2) = (4, 3)$ as an optimal solution, then it must be true that both $4c_1 + 3c_2 \geq c_2$ and $4c_1 + 3c_2 \geq \frac{11}{2}c_1$, because the value of z at $(4, 3)$ cannot be smaller than the other values of z in the table. Manipulate the symbols in these two inequalities in order to isolate $\frac{c_1}{c_2}$ in each; keep in mind the given information that $c_2 > 0$ when performing division by c_2. First,

$$4c_1 + 3c_2 \geq c_2$$
$$4c_1 \geq -2c_2$$
$$\frac{4c_1}{4c_2} \geq \frac{-2c_2}{4c_2}$$
$$\frac{c_1}{c_2} \geq -\frac{1}{2}.$$

Then,

$$4c_1 + 3c_2 \geq \frac{11}{2}c_1$$
$$-\frac{3}{2}c_1 + 3c_2 \geq 0$$
$$3c_1 - 6c_2 \leq 0$$
$$3c_1 \leq 6c_2$$
$$\frac{3c_1}{3c_2} \leq \frac{6c_2}{3c_2}$$
$$\frac{c_1}{c_2} \leq 2.$$

Since $\frac{c_1}{c_2} \geq -\frac{1}{2}$ and $\frac{c_1}{c_2} \leq 2$, the desired range for $\frac{c_1}{c_2}$ is $\left[-\frac{1}{2}, 2\right]$, which corresponds to choice (b).

3.3   Applications of Linear Programming

1. Let x represent the number of product A made and y represent the number of product B. Each item of A uses 2 hr on the machine, so $2x$ represents the total hours required for x items of product A. Similarly, $3y$ represents the total hours used for product B. There are only 45 hr available, so

$$2x + 3y \leq 45.$$

3. Let x represent the number of green pills and y represent the number of red pills. Then $4x$ represents the number of vitamin units provided by the green pills, and y represents the vitamin units provided by the red ones. Since at least 25 units are needed per day,

$$4x + y \geq 25.$$

5. Let x represent the number of pounds of \$6 coffee and y represent the number of pounds of \$5 coffee. Since the mixture must weigh at least 50 lb,

$$x + y \geq 50.$$

(Notice that the price per pound is not used in setting up this inequality.)

7. Let $x =$ the number of engines to ship to plant I and $y =$ the number of engines to ship to plant II.

Minimize $z = 20x + 35y$

subject to:
$$x \geq 50$$
$$y \geq 27$$
$$x + y \leq 85$$
$$15x + 10y \geq 1110$$
$$x \geq 0$$
$$y \geq 0.$$

Sketch the feasible region in quadrant I, and identify the corner points.

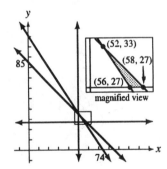

The corner points are: $(56, 27)$, which is the intersection of $y = 27$ and $15x + 10y = 1110$; $(52, 33)$, which is the intersection of $x + y = 85$ and $15x + 10y = 1110$; and $(58, 27)$, which is the intersection of $y = 27$ and $x + y = 85$. Use the corner points to find the minimum value of the objective function.

Corner Point	Value of $z = 20x + 35y$	
$(56, 27)$	$20(56) + 35(27) = 2065$	Minimum
$(52, 33)$	$20(52) + 35(33) = 2195$	
$(58, 27)$	$20(58) + 35(27) = 2105$	

The minimum cost is $2065 which occurs when 56 engines are shipped to plant I and 27 engines are shipped to plant II.

9. Let $x =$ the number of units of policy A and $y =$ the number of units of policy B.

    (a) Minimize $z = 50x + 40y$

    subject to:

    $$10,000x + 15,000y \geq 300,000$$
    $$180,000x + 120,000y \geq 3,000,000$$
    $$x \geq 0$$
    $$y \geq 0.$$

    Sketch the feasible region in quadrant I, and identify the corner points. The corner point $(6, 16)$ can be found by solving the system

    $$10,000x + 15,000y = 300,000$$
    $$180,000x + 120,000y = 3,000,000,$$

    which can be simplified as

    $$2x + 3y = 60$$
    $$3x + 2y = 50.$$

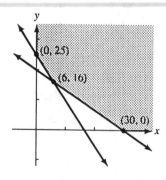

Corner Point	Value of $z = 50x + 40y$	
$(0, 25)$	$50(0) + 40(25) = 1000$	
$(6, 16)$	$50(6) + 40(16) = 940$	Minimum
$(30, 0)$	$50(30) + 40(0) = 1500$	

The minimum cost is $940 which occurs when 6 units of policy A and 16 units of policy B are purchased.

(b) The objective function changes to $z = 25x + 40y$, but the constraints remain the same. Use the same corner points as in part (a).

Corner Point	Value of $z = 25x + 40y$	
$(0, 25)$	$25(0) + 40(25) = 1000$	
$(6, 16)$	$25(6) + 40(16) = 790$	
$(30, 0)$	$25(30) + 40(0) = 750$	Minimum

The minimum cost is $750 which occurs when 30 units of policy A and no units of policy B are purchased.

11. Let $x =$ the number of type 1 bolts and $y =$ the number of type 2 bolts.

    Maximize $z = .10x + .12y$

    subject to: $.1x + .1y \leq 240$
    $$.1x + .4y \leq 720$$
    $$.1x + .02y \leq 160$$
    $$x \geq 0$$
    $$y \geq 0.$$

    Sketch the feasible region in quadrant I, and identify the corner points.

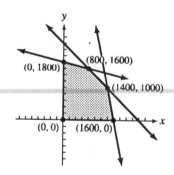

Corner Point	Value of $z = .10x + .12y$	
$(0, 0)$	0	
$(0, 1800)$	216	
$(800, 1600)$	272	Maximum
$(1400, 1000)$	260	
$(1600, 0)$	160	

The maximum revenue is $272 which occurs when 800 type 1 bolts and 1600 type 2 bolts are produced.

13. **(a)** Let $x =$ the number of kilograms of half-and-half mix

    and $y =$ the number of kilograms of the other mix.

Maximize $z = 6x + 4.8y$

subject to: $\dfrac{1}{2}x + \dfrac{1}{3}y \leq 100$

$\dfrac{1}{2}x + \dfrac{2}{3}y \leq 125$

$x \geq 0$

$y \geq 0.$

Sketch the feasible region in quadrant I, and identify the corner points.

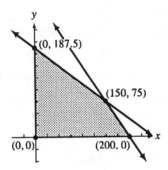

The corner point $(150, 75)$ is the point of intersection of $\frac{1}{2}x + \frac{1}{3}y = 100$ and $\frac{1}{2}x + \frac{2}{3}y = 125$.

Corner Point	Value of $z = 6x + 4.8y$	
$(0,0)$	0	
$(0, 187.5)$	900	
$(150, 75)$	1260	Maximum
$(200, 0)$	1200	

The company should prepare 150 kg of the half-and-half mix and 75 kg of the other mix for a maximum revenue of $1260.

(b) The objective function to be maximized is altered to

$$z = 8x + 4.8y,$$

but the corner points remain the same.

Corner Point	Value of $z = 8x + 4.8y$	
$(0,0)$	0	
$(0, 187.5)$	900	
$(150, 75)$	1560	
$(200, 0)$	1600	Maximum

In order to maximize the revenue under the altered conditions, the company should prepare 200 kg of the half-and-half mix and 0 kg of the other mix for a maximum revenue of $1600.

15. Let $x =$ the number of gallons from dairy I and $y =$ the number of gallons from dairy II.

Maximize $z = .037x + .032y$

subject to: $.60x + .20y \leq 36$

$x \leq 50$

$y \leq 80$

$x + y \leq 100$

$x \geq 0$

$y \geq 0.$

Sketch the feasible region, and identify the corner points.

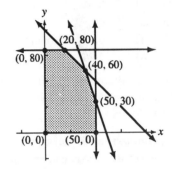

Corner Point	Value of $z = .037x + .032y$	
$(0,0)$	0	
$(0, 80)$	2.56	
$(20, 80)$	3.30	
$(40, 60)$	3.40	Maximum
$(50, 30)$	2.81	
$(50, 0)$	1.85	

The maximum amount of butterfat is 3.4 gal which occurs when 40 gal are purchased from dairy I and 60 gal are purchased from dairy II.

17. Let $x =$ amount invested in U.S. Treasury Bonds and $y =$ amount invested in mutual funds.

Maximize $z = .12x + .08y$

subject to: $x + y \leq 40$

$x \geq 20$

$y \geq 15$

$300x + 100y \leq 8400$

$x \geq 0$

$y \geq 0.$

Sketch the feasible region in quadrant I, and identify the corner points.

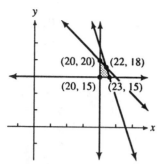

Corner Point	Value of $z = .12x + .08y$	
$(20, 15)$	3.6	
$(20, 20)$	4	
$(22, 18)$	4.08	Maximum
$(23, 15)$	3.96	

Invest \$22 million in U.S. Treasury Bonds and \$18 million in mutual funds for maximum annual interest of \$4.08 million.

19. Beta is limited to 400 units per day, so Beta ≤ 400. The correct answer is choice (a).

21. Let $x =$ the number of pill #1 and $y =$ the number of pill #2.

Minimize $z = .15x + .30y$

subject to: $8x + 2y \geq 16$
$x + y \geq 5$
$2x + 7y \geq 20$
$x \geq 0$
$y \geq 0.$

Sketch the feasible region in quadrant I.

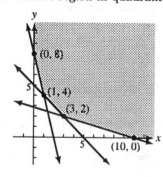

The corner points $(0, 8)$ and $(10, 0)$ can be identified from the graph. The coordinates of the corner point $(1, 4)$ can be found by solving the system

$$8x + 2y = 16$$
$$x + y = 5.$$

The coordinates of the corner point $(3, 2)$ can be found by solving the system

$$2x + 7y = 20$$
$$x + y = 5.$$

Corner Point	Value of $z = .15x + .30y$	
$(1, 4)$	1.35	
$(3, 2)$	1.05	Minimum
$(0, 8)$	2.40	
$(10, 0)$	1.50	

A minimum daily cost of \$1.05 is incurred by taking 3 #1 pills and 2 #2 pills.

23. Let $x =$ the number of servings of product A and $y =$ the number of servings of product B.

Minimize $z = .25x + .40y$

subject to: $3x + 2y \geq 15$
$2x + 4y \geq 15$
$x \geq 0$
$y \geq 0.$

Sketch the feasible region in quadrant I.

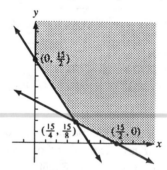

The corner points are $\left(0, \frac{15}{2}\right)$, $\left(\frac{15}{4}, \frac{15}{8}\right)$, which can be found by solving the system

$$3x + 2y = 15$$
$$2x + 4y = 15,$$

and $\left(\frac{15}{2}, 0\right)$.

Corner Point	Value of $z = .25x + .40y$	
$\left(0, \frac{15}{2}\right)$	3	
$\left(\frac{15}{4}, \frac{15}{8}\right)$	1.6875	Minimum
$\left(\frac{15}{2}, 0\right)$	1.875	

$\frac{15}{4}$ or $3\frac{3}{4}$ servings of A and $\frac{15}{8}$ or $1\frac{7}{8}$ servings of B will satisfy the requirements at a minimum cost of \$1.69 (rounded to the nearest cent).

25. Let $x =$ the number of square feet of window space and $y =$ the number of square feet of wall space.

Maximize $z = x + y$

subject to:
$$x \geq \frac{1}{6}y$$
$$10x + 20y \leq 12,000$$
$$.32x + .20y \leq 160$$
$$x \geq 0$$
$$y \geq 0.$$

Graph the feasible region on a graphing calculator and identify the corner points.

Corner Point	Value of $z = x + y$
$(0, 0)$	0
$(92.31, 553.85)$	646.16
$(181.82, 509.09)$	690.91 Maximum
$(500, 0)$	500

The maximum total area is 690.91 sq ft occurring when 181.82 sq ft is used for windows and 509.09 sq ft is used for walls.

Chapter 3 Review Exercises

For Exercises 3-13, see the answer graphs in the back of the textbook.

3. $y \geq 2x + 3$

Graph $y = 2x + 3$ as a solid line, using the intercepts $(0, 3)$ and $\left(-\frac{3}{2}, 0\right)$. Using the origin as a test point, we get $0 \geq 2(0) + 3$ or $0 \geq 3$, which is false. Shade the region that does not contain the origin, that is, the half-plane above the line.

5. $3x + 4y \leq 12$

Graph $3x + 4y = 12$ as a solid line, using the intercepts $(0, 3)$ and $(4, 0)$. Using the origin as a test point, we get $0 \leq 12$, which is true. Shade the region that contains the origin, that is, the half-plane below the line.

7. $y \geq x$

Graph $y = x$ as a solid line. Since this line contains the origin, choose a point other than $(0, 0)$ as a test point. If we use $(1, 4)$, we get $4 \geq 1$, which is true. Shade the region that contains the test point, that is, the half-plane above the line.

9. $x + y \leq 6$
$2x - y \geq 3$

$x + y \leq 6$ is the half-plane on or below the line $x + y = 6$; $2x - y \geq 3$ is the half-plane on or below the line $2x - y = 3$. Shade the overlapping part of these two half-planes, which is the region below both lines. The only corner point is the intersection of the two boundary lines, the point $(3, 3)$.

11. $-4 \leq x \leq 2$
$-1 \leq y \leq 3$
$x + y \leq 4$

$-4 \leq x \leq 2$ is the rectangular region lying on or between the two vertical lines, $x = -4$ and $x = 2$; $-1 \leq y \leq 3$ is the rectangular region lying on or between the two horizontal lines, $y = -1$ and $y = 3$; $x + y \leq 4$ is the half-plane lying on or below the line $x + y = 4$. Shade the overlapping part of these three regions. The corner points are $(-4, -1)$, $(-4, 3)$, $(1, 3)$, $(2, 2)$, and $(2, -1)$.

13. $x + 3y \geq 6$
$4x - 3y \leq 12$
$x \geq 0$
$y \geq 0$

$x + 3y \geq 6$ is the half-plane on or above the line $x + 3y = 6$; $4x - 3y \leq 12$ is the half-plane on or above the line $4x - 3y = 12$; $x \geq 0$ and $y \geq 0$ together restrict the graph to the first quadrant. Shade the portion of the first quadrant where the half-planes overlap. The corner points are $(0, 2)$ and $\left(\frac{18}{5}, \frac{4}{5}\right)$, which can be found by solving the system
$$x + 3y = 6$$
$$4x - 3y = 12.$$

15. Evaluate the objective function at $z = 2x + 4y$ at all the corner points.

Corner Point	Value of $z = 2x + 4y$
$(1, 6)$	$2(1) + 4(6) = 26$
$(6, 7)$	$2(6) + 4(7) = 40$ Maximum
$(7, 3)$	$2(7) + 4(3) = 26$
$\left(1, 2\frac{1}{2}\right)$	$2(1) + 4\left(2\frac{1}{2}\right) = 12$
$(2, 1)$	$2(2) + 4(1) = 8$ Minimum

The maximum value of 40 occurs at $(6, 7)$, and the minimum value of 8 occurs at $(2, 1)$.

17. Maximize $z = 2x + 4y$

subject to: $3x + 2y \leq 12$
$5x + y \geq 5$
$x \geq 0$
$y \geq 0.$

Sketch the feasible region in quadrant I.

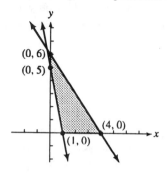

The corner points are $(0,5)$, $(0,6)$, $(4,0)$, and $(1,0)$.

Corner Point	Value of $z = 2x + 4y$
$(0,5)$	20
$(0,6)$	24 Maximum
$(4,0)$	8
$(1,0)$	2

The maximum value is 24 at $(0,6)$.

19. Minimize $z = 4x + 2y$

subject to: $x + y \leq 50$
$2x + y \geq 20$
$x + 2y \geq 30$
$x \geq 0$
$y \geq 0.$

Sketch the feasible region.

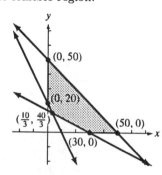

The corner points are $(0,20)$, $\left(\frac{10}{3}, \frac{40}{3}\right)$, $(30,0)$, $(50,0)$, and $(0,50)$. The corner point $\left(\frac{10}{3}, \frac{40}{3}\right)$ can be found by solving the system

$$2x + y = 20$$
$$x + 2y = 30.$$

Corner Point	Value of $z = 4x + 2y$
$(0,20)$	40 Minimum
$\left(\frac{10}{3}, \frac{40}{3}\right)$	40 Minimum
$(30,0)$	120
$(50,0)$	200
$(0,50)$	100

Thus, the minimum value is 40 and occurs at every point on the line segment joining $(0, 20)$ and $\left(\frac{10}{3}, \frac{40}{3}\right)$.

23. Maximize $z = 2x + 5y$

subject to: $3x + 2y \leq 6$
$-x + 2y \leq 4$
$x \geq 0$
$y \geq 0.$

(a) Sketch the feasible region. All corner points except one can be read from the graph. Solving the system

$$3x + 2y = 6$$
$$-x + 2y = 4$$

gives the final corner point, $\left(\frac{1}{2}, \frac{9}{4}\right)$. See the answer graph in the back of the textbook.

(b) See the answer graph in the back of the textbook.

25. Let $x = $ the number of batches of cakes and $y = $ the number of batches of cookies.

Then we have the following inequalities.

$$2x + \frac{3}{2}y \leq 15 \quad \text{(oven time)}$$

$$3x + \frac{2}{3}y \leq 13 \quad \text{(decorating)}$$

$$x \geq 0$$
$$y \geq 0$$

See the answer graph in the back of the textbook.

27. From the graph for Exercise 25, the corner points are $(0, 10)$, $(3, 6)$, $\left(\frac{13}{3}, 0\right)$, and $(0, 0)$. Since x was the number of batches of cakes and y the number of batches of cookies, the revenue function is

$$z = 30x + 20y.$$

Evaluate this objective function at each corner point.

Corner Point	Value of $z = 30x + 20y$	
$(0, 10)$	200	
$(3, 6)$	210	Maximum
$\left(\frac{13}{3}, 0\right)$	130	
$(0, 0)$	0	

Therefore, 3 batches of cakes and 6 batches of cookies should be made to produce a maximum profit of $210.

29. Let $x =$ the number of hours Charles should spend with his math tutor

and $y =$ the number of hours he should spend with his accounting tutor.

The number of points he expects to get on the two tests combined is $z = 3x + 5y$. The given information translates into the following problem.

Maximize $z = 3x + 5y$

subject to: $20x + 40y \le 220$ (finances)
$2x + y \le 16$ (aspirin)
$x + 3y \le 15$ (sleep)
$x \ge 0$
$y \ge 0.$

Sketch the feasible region.

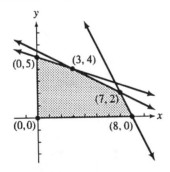

The corner points of the feasible region are $(0,0)$, $(0,5)$, $(3,4)$, $(7,2)$, and $(8,0)$. Evaluate the objective function at each corner point.

Corner Point	Value of $z = 3x + 5y$	
$(0, 0)$	0	
$(0, 5)$	25	
$(3, 4)$	29	
$(7, 2)$	31	Maximum
$(8, 0)$	24	

Therefore, Charles should spend 7 hr with the math tutor and 2 hr with the accounting tutor in order to earn a maximum of 31 points.

Chapter 3 Test

[3.1]

1. Graph the following linear inequality.

$$2x + y \leq 4$$

[3.1–3.2]

2. Graph the feasible region for the following system of inequalities. Find all corner points.

$$\begin{aligned} x + y &\leq 4 \\ 2x + y &\geq 6 \\ x &\geq 0 \\ y &\geq 0 \end{aligned}$$

[3.2]

3. Find the maximum and minimum values of the objective function $z = 3x + 2y$ for the region sketched below.

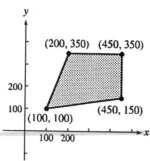

Use the graphical method to solve the following linear programming problems.

4.   Maximize $z = 4x + 3y$

subject to: $\begin{aligned} 3x + y &\leq 12 \\ x + y &\geq 3 \\ x &\geq 0 \\ y &\geq 0. \end{aligned}$

5.   Minimize $z = x + 10y$

subject to: $\begin{aligned} 5x + 2y &\geq 20 \\ x + y &\geq 7 \\ x + 6y &\geq 27 \\ y &\geq 3 \\ x &\geq 0. \end{aligned}$

[3.3]

6. The Gigantic Zipper Company manufactures two kinds of zippers. Type I zippers require 2 minutes on machine A and 3 minutes on machine B. Type II zippers require 1 minute on machine A and 4 minutes on machine B. Machine A is available for 20 minutes, while machine B is available for 12 minutes. The profit on each type I zipper is $.30 and on each type II zipper is $.20. How many of each type of zipper should be manufactured to ensure the maximum profit?

Chapter 3 Test Answers

1.
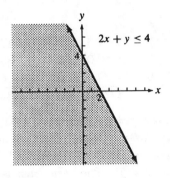

$$2x + y \le 4$$

2.
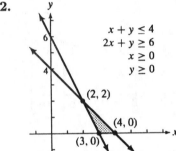

$$x + y \le 4$$
$$2x + y \ge 6$$
$$x \ge 0$$
$$y \ge 0$$

Corner points: $(3,0)$, $(4,0)$, $(2,2)$

3. Maximum of 2050 at $(450, 350)$; minimum of 500 at $(100, 100)$

4. Maximum of 36 at $(0, 12)$

5. Minimum of 39 at $(9, 3)$

6. 4 type I zippers and 0 type II zippers

LINEAR PROGRAMMING: THE SIMPLEX METHOD

4.1   Slack Variables and the Pivot

1. $x_1 + 2x_2 \leq 6$

Add s_1 to the given inequality to obtain

$$x_1 + 2x_2 + s_1 = 6.$$

3. $2x_1 + 4x_2 + 3x_3 \leq 100$

Add s_1 to the given inequality to obtain

$$2x_1 + 4x_2 + 3x_3 + s_1 = 100.$$

5. (a) Since there are three constraints to be converted into equations, we need three slack variables.

(b) We use s_1, s_2, and s_3 for the slack variables.

(c) The equations are

$$
\begin{aligned}
4x_1 + 2x_2 + s_1 &= 20 \\
5x_1 + x_2 + s_2 &= 50 \\
2x_1 + 3x_2 + s_3 &= 25.
\end{aligned}
$$

7. (a) There are two constraints to be converted into equations, so we must introduce two slack variables.

(b) Call the slack variables s_1 and s_2.

(c) The equations are

$$
\begin{aligned}
7x_1 + 6x_2 + 8x_3 + s_1 &= 118 \\
4x_1 + 5x_2 + 10x_3 + s_2 &= 220.
\end{aligned}
$$

9.

$$
\begin{bmatrix}
x_1 & x_2 & x_3 & s_1 & s_2 & z & \\
2 & 2 & 0 & 3 & 1 & 0 & 15 \\
3 & 4 & 1 & 6 & 0 & 0 & 20 \\
\hline
-2 & -1 & 0 & 1 & 0 & 1 & 10
\end{bmatrix}
$$

The variables x_3 and s_2 are basic variables, because the columns for these variables have all zeros except for one nonzero entry. If the remaining variables x_1, x_2, and s_1 are zero, then $x_3 = 20$ and $s_2 = 15$. From the bottom row, $z = 10$. The basic feasible solution is $x_1 = 0, x_2 = 0, x_3 = 20, s_1 = 0, s_2 = 15$, and $z = 10$.

11.

$$
\begin{bmatrix}
x_1 & x_2 & x_3 & s_1 & s_2 & s_3 & z & \\
6 & 2 & 2 & 3 & 0 & 0 & 0 & 16 \\
2 & 2 & 0 & 1 & 0 & 5 & 0 & 35 \\
2 & 1 & 0 & 3 & 1 & 0 & 0 & 6 \\
\hline
-3 & -2 & 0 & 2 & 0 & 0 & 3 & 36
\end{bmatrix}
$$

The basic variables are x_3, s_2, and s_3. If x_1, x_2, and s_1 are zero, then $2x_3 = 16$, so $x_3 = 8$. Similarly, $s_2 = 6$ and $5s_3 = 35$, so $s_3 = 7$. From the bottom row, $3z = 36$, so $z = 12$. The basic feasible solution is $x_1 = 0, x_2 = 0, x_3 = 8, s_1 = 0, s_2 = 6, s_3 = 7$, and $z = 12$.

13.

$$
\begin{bmatrix}
x_1 & x_2 & x_3 & s_1 & s_2 & z & \\
1 & 2 & 4 & 1 & 0 & 0 & 56 \\
2 & \boxed{2} & 1 & 0 & 1 & 0 & 40 \\
\hline
-1 & -3 & -2 & 0 & 0 & 1 & 0
\end{bmatrix}
$$

Clear the x_2 column.

$$
-R_2 + R_1 \to R_1 \quad
\begin{bmatrix}
x_1 & x_2 & x_3 & s_1 & s_2 & z & \\
-1 & 0 & 3 & 1 & -1 & 0 & 16 \\
2 & \boxed{2} & 1 & 0 & 1 & 0 & 40 \\
\hline
-1 & -3 & -2 & 0 & 0 & 1 & 0
\end{bmatrix}
$$

$$
3R_2 + 2R_3 \to R_3 \quad
\begin{bmatrix}
x_1 & x_2 & x_3 & s_1 & s_2 & z & \\
-1 & 0 & 3 & 1 & -1 & 0 & 16 \\
2 & 2 & 1 & 0 & 1 & 0 & 40 \\
\hline
4 & 0 & -1 & 0 & 3 & 2 & 120
\end{bmatrix}
$$

x_2 and s_1 are now basic. The solution is $x_1 = 0, x_2 = 20, x_3 = 0, s_1 = 16, s_2 = 0$, and $z = 60$.

15.

$$
\begin{bmatrix}
x_1 & x_2 & x_3 & s_1 & s_2 & s_3 & z & \\
2 & 2 & \boxed{1} & 1 & 0 & 0 & 0 & 12 \\
1 & 2 & 3 & 0 & 1 & 0 & 0 & 45 \\
3 & 1 & 1 & 0 & 0 & 1 & 0 & 20 \\
\hline
-2 & -1 & -3 & 0 & 0 & 0 & 1 & 0
\end{bmatrix}
$$

Clear the x_3 column.

$$
\begin{array}{l}
\\
-3R_1 + R_2 \to R_2 \\
-R_1 + R_3 \to R_3 \\
3R_1 + R_4 \to R_4
\end{array}
\begin{bmatrix}
x_1 & x_2 & x_3 & s_1 & s_2 & s_3 & z & \\
2 & 2 & 1 & 1 & 0 & 0 & 0 & 12 \\
-5 & -4 & 0 & -3 & 1 & 0 & 0 & 9 \\
1 & -1 & 0 & -1 & 0 & 1 & 0 & 8 \\
4 & 5 & 0 & 3 & 0 & 0 & 1 & 36
\end{bmatrix}
$$

x_3, s_2, and s_3 are now basic. The solution is $x_1 = 0, x_2 = 0, x_3 = 12, s_1 = 0, s_2 = 9, s_3 = 8$, and $z = 36$.

17.

$$\begin{array}{ccccccc|c} x_1 & x_2 & x_3 & s_1 & s_2 & s_3 & z & \\ 1 & 1 & 1 & 1 & 0 & 0 & 0 & 60 \\ 3 & 1 & \boxed{2} & 0 & 1 & 0 & 0 & 100 \\ 1 & 2 & 3 & 0 & 0 & 1 & 0 & 200 \\ \hline -1 & -1 & -2 & 0 & 0 & 0 & 1 & 0 \end{array}$$

Clear the x_3 column.

$$\begin{array}{l} \\ -R_2 + 2R_1 \rightarrow R_1 \\ \\ -3R_2 + 2R_3 \rightarrow R_3 \\ R_2 + \ R_4 \rightarrow R_4 \end{array} \begin{array}{ccccccc|c} x_1 & x_2 & x_3 & s_1 & s_2 & s_3 & z & \\ -1 & 1 & 0 & 2 & -1 & 0 & 0 & 20 \\ 3 & 1 & 2 & 0 & 1 & 0 & 0 & 100 \\ -7 & 1 & 0 & 0 & -3 & 2 & 0 & 100 \\ \hline 2 & 0 & 0 & 0 & 1 & 0 & 1 & 100 \end{array}$$

x_3, s_1, and s_3 are now basic. The solution is $x_1 = 0, x_2 = 0, x_3 = 50, s_1 = 10, s_2 = 0, s_3 = 50$, and $z = 100$.

19. Find $x_1 \geq 0$ and $x_2 \geq 0$ such that

$$2x_1 + 3x_2 \leq 6$$
$$4x_1 + \ x_2 \leq 6$$

and $z = 5x_1 + x_2$ is maximized.

We need two slack variables, s_1 and s_2. Then the problem can be restated as:

Find $x_1 \geq 0, x_2 \geq 0, s_1 \geq 0$, and $s_2 \geq 0$ such that

$$2x_1 + 3x_2 + s_1 \ \ \ \ = 6$$
$$4x_1 + \ x_2 \ \ \ \ + s_2 = 6$$

and $z = 5x_1 + x_2$ is maximized.

Rewrite the objective function as

$$-5x_1 - x_2 + z = 0.$$

The initial simplex tableau is

$$\begin{array}{ccccc|c} x_1 & x_2 & s_1 & s_2 & z & \\ 2 & 3 & 1 & 0 & 0 & 6 \\ 4 & 1 & 0 & 1 & 0 & 6 \\ \hline -5 & -1 & 0 & 0 & 1 & 0 \end{array}.$$

21. Find $x_1 \geq 0$ and $x_2 \geq 0$ such that

$$x_1 + \ x_2 \leq 10$$
$$5x_1 + 2x_2 \leq 20$$
$$x_1 + 2x_2 \leq 36$$

and $z = x_1 + 3x_2$ is maximized.

Using slack variables s_1, s_2, and s_3, the problem can be restated as:

Find $x_1 \geq 0, x_2 \geq 0, s_1 \geq 0, s_2 \geq 0$, and $s_3 \geq 0$ such that

$$x_1 + \ x_2 + s_1 \ \ \ \ \ \ \ \ = 10$$
$$5x_1 + 2x_2 \ \ \ \ + s_2 \ \ \ \ = 20$$
$$x_1 + 2x_2 \ \ \ \ \ \ \ \ + s_3 = 36$$

and $z = x_1 + 3x_2$ is maximized.

Rewrite the objective function as

$$-x_1 - 3x_2 + z = 0.$$

The initial simplex tableau is

$$\begin{array}{cccccc|c} x_1 & x_2 & s_1 & s_2 & s_3 & z & \\ 1 & 1 & 1 & 0 & 0 & 0 & 10 \\ 5 & 2 & 0 & 1 & 0 & 0 & 20 \\ 1 & 2 & 0 & 0 & 1 & 0 & 36 \\ \hline -1 & -3 & 0 & 0 & 0 & 1 & 0 \end{array}.$$

23. Find $x_1 \geq 0$ and $x_2 \geq 0$ such that

$$3x_1 + x_2 \leq 12$$
$$x_1 + x_2 \leq 15$$

and $z = 2x_1 + x_2$ is maximized.

Using slack variables s_1 and s_2, the problem can be restated as:

Find $x_1 \geq 0, x_2 \geq 0, s_1 \geq 0$, and $s_2 \geq 0$ such that

$$3x_1 + x_2 + s_1 \ \ \ \ = 12$$
$$x_1 + x_2 \ \ \ \ + s_2 = 15$$

and $z = 2x_1 + x_2$ is maximized.

Rewrite the objective function as

$$-2x_1 - x_2 + z = 0.$$

The initial simplex tableau is

$$\begin{array}{ccccc|c} x_1 & x_2 & s_1 & s_2 & z & \\ 3 & 1 & 1 & 0 & 0 & 12 \\ 1 & 1 & 0 & 1 & 0 & 15 \\ \hline -2 & -1 & 0 & 0 & 1 & 0 \end{array}.$$

25. Let x_1 represent the number of simple figures, x_2 the number of figures with additions, and x_3 the number of computer-drawn sketches. Organize the information in a table.

	Simple Figures	Figures with Additions	Computer-Drawn Sketches	Maximum Allowed
Cost	20	35	60	2200
Royalties	95	200	325	

The cost constraint is

$$20x_1 + 35x_2 + 60x_3 \leq 2200.$$

The limit of 400 figures leads to the constraint

$$x_1 + x_2 + x_3 \leq 400.$$

The other stated constraints are

$$x_3 \leq x_1 + x_2 \text{ and } x_1 \geq 2x_2,$$

and these can be rewritten in standard form as

$$-x_1 - x_2 + x_3 \leq 0 \text{ and } -x_1 + 2x_2 \leq 0$$

respectively. The problem may be stated as:

Find $x_1 \geq 0, x_2 \geq 0$, and $x_3 \geq 0$ such that

$$\begin{aligned}
20x_1 + 35x_2 + 60x_3 &\leq 2200 \\
x_1 + x_2 + x_3 &\leq 400 \\
-x_1 - x_2 + x_3 &\leq 0 \\
-x_1 + 2x_2 \quad\;\; &\leq 0
\end{aligned}$$

and $z = 95x_1 + 200x_2 + 325x_3$ is maximized.

Introduce slack variables s_1, s_2, s_3, and s_4, and the problem can be restated as:

Find $x_1 \geq 0, x_2 \geq 0, x_3 \geq 0, s_1 \geq 0, s_2 \geq 0, s_3 \geq 0$, and $s_4 \geq 0$ such that

$$\begin{aligned}
20x_1 + 35x_2 + 60x_3 + s_1 \qquad\qquad\qquad &= 2200 \\
x_1 + x_2 + x_3 \qquad + s_2 \qquad\qquad &= 400 \\
-x_1 - x_2 + x_3 \qquad\qquad + s_3 \qquad &= 0 \\
-x_1 + 2x_2 \qquad\qquad\qquad\qquad + s_4 &= 0
\end{aligned}$$

and $z = 95x_1 + 200x_2 + 325x_3$ is maximized.

Rewrite the objective function as

$$-95x_1 - 200x_2 - 325x_3 + z = 0.$$

The initial simplex tableau is

$$\left[\begin{array}{ccccccc|c}
x_1 & x_2 & x_3 & s_1 \; s_2 \; s_3 \; s_4 \; z & & & & \\
20 & 35 & 60 & 1 \;\; 0 \;\; 0 \;\; 0 \;\; 0 & & & & 2200 \\
1 & 1 & 1 & 0 \;\; 1 \;\; 0 \;\; 0 \;\; 0 & & & & 400 \\
-1 & -1 & 1 & 0 \;\; 0 \;\; 1 \;\; 0 \;\; 0 & & & & 0 \\
-1 & 2 & 0 & 0 \;\; 0 \;\; 0 \;\; 1 \;\; 0 & & & & 0 \\
\hline
-95 & -200 & -325 & 0 \;\; 0 \;\; 0 \;\; 0 \;\; 1 & & & & 0
\end{array}\right].$$

27. Let x_1 represent the number of redwood tables, x_2 the number of stained Douglas fir tables, and x_3 the number of stained white spruce tables. Organize the information in a table.

	Redwood	Douglas Fir	White Spruce	Maximum Available
Assembly Time	8	7	8	90 8-hr days = 720 hr
Staining Time	0	2	2	60 8-hr days = 480 hr
Cost	$159	$138.85	$129.35	$15,000

The limit of 720 hr for carpenters leads to the constraint

$$8x_1 + 7x_2 + 8x_3 \leq 720.$$

The limit of 480 hr for staining leads to the constraint

$$2x_2 + 2x_3 \leq 480.$$

The cost constraint is

$$159x_1 + 138.85x_2 + 129.35x_3 \leq 15,000.$$

The problem may be stated as:

Find $x_1 \geq 0, x_2 \geq 0$, and $x_3 \geq 0$ such that

$$\begin{aligned}
8x_1 + 7x_2 + 8x_3 &\leq 720 \\
2x_2 + 2x_3 &\leq 480 \\
159x_1 + 138.85x_2 + 129.35x_3 &\leq 15,000
\end{aligned}$$

and $z = x_1 + x_2 + x_3$ is maximized.

Introduce slack variables s_1, s_2, and s_3, and the problem can be restated as:

Find $x_1 \geq 0, x_2 \geq 0, x_3 \geq 0, s_1 \geq 0, s_2 \geq 0$, and $s_3 \geq 0$ such that

$$\begin{aligned}
8x_1 + 7x_2 + 8x_3 + s_1 \qquad\qquad &= 720 \\
2x_2 + 2x_3 \qquad + s_2 \qquad &= 480 \\
159x_1 + 138.85x_2 + 129.35x_3 \qquad\qquad + s_3 &= 15,000
\end{aligned}$$

and $z = x_1 + x_2 + x_3$ is maximized.

Rewrite the objective function as

$$-x_1 - x_2 - x_3 + z = 0.$$

The initial simplex tableau is

$$\left[\begin{array}{cccccccc|c}
x_1 & x_2 & x_3 & s_1 & s_2 & s_3 & z & & \\
8 & 7 & 8 & 1 & 0 & 0 & 0 & & 720 \\
0 & 2 & 2 & 0 & 1 & 0 & 0 & & 480 \\
159 & 138.85 & 129.35 & 0 & 0 & 1 & 0 & & 15,000 \\
\hline
-1 & -1 & -1 & 0 & 0 & 0 & 1 & & 0
\end{array}\right].$$

29. Let x_1 represent the number of newspaper ads, x_2 the number of radio ads, and x_3 the number of TV ads. Organize the information in a table.

	Newspaper Ads	Radio Ads	TV Ads
Cost per Ad	400	200	1200
Maximum Number	20	30	6
Women Seeing/ Hearing Ads	2000	1200	10,000

The cost constraint is

$$400x_1 + 200x_2 + 1200x_3 \le 8000.$$

The constraints on number of ads are

$$x_1 \le 20$$
$$x_2 \le 30$$
$$x_3 \le 6.$$

The problem may be stated as:

Find $x_1 \ge 0, x_2 \ge 0$, and $x_3 \ge 0$ such that

$$400x_1 + 200x_2 + 1200x_3 \le 8000$$
$$x_1 \qquad\qquad\qquad \le 20$$
$$\qquad x_2 \qquad\qquad \le 30$$
$$\qquad\qquad x_3 \le 6$$

and $z = 2000x_1 + 1200x_2 + 10{,}000x_3$ is maximized.

Introduce slack variables s_1, s_2, s_3, and s_4, and the problem can be restated as:

Find $x_1 \ge 0, x_2 \ge 0, x_3 \ge 0, s_1 \ge 0, s_2 \ge 0, s_3 \ge 0$, and $s_4 \ge 0$ such that

$$400x_1 + 200x_2 + 1200x_3 + s_1 \qquad\qquad\qquad = 8000$$
$$x_1 \qquad\qquad\qquad + s_2 \qquad\qquad = 20$$
$$\qquad x_2 \qquad\qquad\qquad + s_3 \qquad = 30$$
$$\qquad\qquad x_3 \qquad\qquad\qquad + s_4 = 6$$

and $z = 2000x_1 + 1200x_2 + 10{,}000x_3$ is maximized.

Rewrite the objective function as

$$-2000x_1 - 1200x_2 - 10{,}000x_3 + z = 0.$$

The initial simplex tableau is

$$
\begin{bmatrix}
x_1 & x_2 & x_3 & s_1 & s_2 & s_3 & s_4 & z & \\
400 & 200 & 1200 & 1 & 0 & 0 & 0 & 0 & 8000 \\
1 & 0 & 0 & 0 & 1 & 0 & 0 & 0 & 20 \\
0 & 1 & 0 & 0 & 0 & 1 & 0 & 0 & 30 \\
0 & 0 & 1 & 0 & 0 & 0 & 1 & 0 & 6 \\
\hline
-2000 & -1200 & -10{,}000 & 0 & 0 & 0 & 0 & 1 & 0
\end{bmatrix}
$$

4.2   Maximization Problems

1.
$$
\begin{bmatrix}
x_1 & x_2 & x_3 & s_1 & s_2 & z & \\
1 & 2 & 4 & 1 & 0 & 0 & 8 \\
2 & 2 & 1 & 0 & 1 & 0 & 10 \\
\hline
-2 & -5 & -1 & 0 & 0 & 1 & 0
\end{bmatrix}
$$

The most negative indicator is -5, in the second column. Find the quotients $\frac{8}{2} = 4$ and $\frac{10}{2} = 5$; since 4 is the smallest quotient, 2 in row 1, column 2 is the pivot.

$$
\begin{array}{c}
\frac{8}{2} = 4 \\[4pt]
\frac{10}{2} = 5
\end{array}
\begin{bmatrix}
x_1 & x_2 & x_3 & s_1 & s_2 & z & \\
1 & \boxed{2} & 4 & 1 & 0 & 0 & 8 \\
2 & 2 & 1 & 0 & 1 & 0 & 10 \\
\hline
-2 & -5 & -1 & 0 & 0 & 1 & 0
\end{bmatrix}
$$

Performing row transformations, we get the following tableau.

$$
\begin{array}{c}
\\
-R_1 + R_2 \to R_2 \\
5R_1 + 2R_3 \to R_3
\end{array}
\begin{bmatrix}
x_1 & x_2 & x_3 & s_1 & s_2 & z & \\
1 & 2 & 4 & 1 & 0 & 0 & 8 \\
1 & 0 & -3 & -1 & 1 & 0 & 2 \\
1 & 0 & 18 & 5 & 0 & 2 & 40
\end{bmatrix}
$$

All of the numbers in the last row are nonnegative, so we are finished pivoting. Create a 1 in the columns corresponding to x_2 and z.

$$
\begin{array}{c}
\frac{1}{2}R_1 \to R_1 \\[4pt]
\\
\frac{1}{2}R_3 \to R_3
\end{array}
\begin{bmatrix}
x_1 & x_2 & x_3 & s_1 & s_2 & z & \\
\frac{1}{2} & 1 & 2 & \frac{1}{2} & 0 & 0 & 4 \\
1 & 0 & -3 & -1 & 1 & 0 & 2 \\
\frac{1}{2} & 0 & 9 & \frac{5}{2} & 0 & 1 & 20
\end{bmatrix}
$$

The maximum value is 20 and occurs when $x_1 = 0, x_2 = 4, x_3 = 0, s_1 = 0$, and $s_2 = 2$.

3.
$$
\begin{bmatrix}
x_1 & x_2 & s_1 & s_2 & s_3 & z & \\
1 & 3 & 1 & 0 & 0 & 0 & 12 \\
2 & 1 & 0 & 1 & 0 & 0 & 10 \\
1 & 1 & 0 & 0 & 1 & 0 & 4 \\
\hline
-2 & -1 & 0 & 0 & 0 & 1 & 0
\end{bmatrix}
$$

The most negative indicator is -2, in the first column. Find the quotients $\frac{12}{1} = 12, \frac{10}{2} = 5$, and $\frac{4}{1} = 4$; since 4 is the smallest quotient, 1 in row 3, column 1 is the pivot.

$$
\begin{bmatrix}
x_1 & x_2 & s_1 & s_2 & s_3 & z & \\
1 & 3 & 1 & 0 & 0 & 0 & 12 \\
2 & 1 & 0 & 1 & 0 & 0 & 10 \\
\boxed{1} & 1 & 0 & 0 & 1 & 0 & 4 \\
\hline
-2 & -1 & 0 & 0 & 0 & 1 & 0
\end{bmatrix}
$$

$$-R_3 + R_1 \rightarrow R_1$$
$$-2R_3 + R_2 \rightarrow R_2$$
$$2R_3 + R_4 \rightarrow R_4$$

x_1	x_2	s_1	s_2	s_3	z	
0	2	1	0	-1	0	8
0	-1	0	1	-2	0	2
1	1	0	0	1	0	4
0	1	0	0	2	1	8

This is a final tableau since all of the numbers in the last row are nonnegative. The maximum value is 8 when $x_1 = 4, x_2 = 0, s_1 = 8, s_2 = 2$, and $s_3 = 0$.

5.

x_1	x_2	x_3	s_1	s_2	s_3	z	
2	2	8	1	0	0	0	40
4	-5	6	0	1	0	0	60
2	-2	6	0	0	1	0	24
-14	-10	-12	0	0	0	1	0

The most negative indicator is -14, in the first column. Find the quotients $\frac{40}{2} = 20, \frac{60}{4} = 15$, and $\frac{24}{2} = 12$; since 12 is the smallest quotient, 2 in row 3, column 1 is the pivot.

x_1	x_2	x_3	s_1	s_2	s_3	z	
2	2	8	1	0	0	0	40
4	-5	6	0	1	0	0	60
[2]	-2	6	0	0	1	0	24
-14	-10	-12	0	0	0	1	0

Performing row transformations, we get the following tableau.

$$-R_3 + R_1 \rightarrow R_1$$
$$-2R_3 + R_2 \rightarrow R_2$$
$$7R_3 + R_4 \rightarrow R_4$$

x_1	x_2	x_3	s_1	s_2	s_3	z	
0	[4]	2	1	0	-1	0	16
0	-1	-6	0	1	-2	0	12
2	-2	6	0	0	1	0	24
0	-24	30	0	0	7	1	168

Since there is still a negative indicator, we must repeat the process. The second pivot is the 4 in column 2, since $\frac{16}{4}$ is the only nonnegative quotient in the only column with a negative indicator. Performing row transformations again, we get the following tableau.

$$R_1 + 4R_2 \rightarrow R_2$$
$$R_1 + 2R_3 \rightarrow R_3$$
$$6R_1 + R_4 \rightarrow R_4$$

x_1	x_2	x_3	s_1	s_2	s_3	z	
0	4	2	1	0	-1	0	16
0	0	-22	1	4	-9	0	64
4	0	14	1	0	1	0	64
0	0	42	6	0	1	1	264

All of the numbers in the last row are nonnegative, so we are finished pivoting. Create a 1 in the columns corresponding to x_1, x_2, and s_2.

$$\tfrac{1}{4}R_1 \rightarrow R_1$$
$$\tfrac{1}{4}R_2 \rightarrow R_2$$
$$\tfrac{1}{4}R_3 \rightarrow R_3$$

x_1	x_2	x_3	s_1	s_2	s_3	z	
0	1	$\frac{1}{2}$	$\frac{1}{4}$	0	$-\frac{1}{4}$	0	4
0	0	$-\frac{11}{2}$	$\frac{1}{4}$	1	$-\frac{9}{4}$	0	16
1	0	$\frac{7}{2}$	$\frac{1}{4}$	0	$\frac{1}{4}$	0	16
0	0	42	6	0	1	1	264

The maximum value is 264 and occurs when $x_1 = 16, x_2 = 4, x_3 = 0, s_1 = 0, s_2 = 16$, and $s_3 = 0$.

7. Maximize $z = 4x_1 + 3x_2$
subject to: $2x_1 + 3x_2 \leq 11$
     $x_1 + 2x_2 \leq 6$
with $x_1 \geq 0, x_2 \geq 0$.

Two slack variables, s_1 and s_2, need to be introduced. The problem can be restated as:

Maximize $z = 4x_1 + 3x_2$
subject to: $2x_1 + 3x_2 + s_1 \quad\;\; = 11$
     $x_1 + 2x_2 \quad\;\; + s_2 = 6$
with $x_1 \geq 0, x_2 \geq 0, s_1 \geq 0, s_2 \geq 0$.

Rewrite the objective function as

$$-4x_1 - 3x_2 + z = 0.$$

The initial simplex tableau follows.

x_1	x_2	s_1	s_2	z	
2	3	1	0	0	11
1	2	0	1	0	6
-4	-3	0	0	1	0

The most negative indicator is -4, in column 1; to select the pivot from column 1, find the quotients $\frac{11}{2}$ and $\frac{6}{1}$. The smallest is $\frac{11}{2}$, so 2 is the pivot.

x_1	x_2	s_1	s_2	z	
[2]	3	1	0	0	11
1	2	0	1	0	6
-4	-3	0	0	1	0

$$-R_1 + 2R_2 \rightarrow R_2$$
$$2R_1 + R_3 \rightarrow R_3$$

x_1	x_2	s_1	s_2	z	
2	3	1	0	0	11
0	1	-1	2	0	1
0	3	2	0	1	22

All of the indicators are nonnegative. Create a 1 in the columns corresponding to x_1 and s_2.

$$
\begin{array}{c}
\frac{1}{2}R_1 \to R_1 \\
\frac{1}{2}R_2 \to R_2
\end{array}
\begin{array}{c}
\begin{array}{cccccc}
x_1 & x_2 & s_1 & s_2 & z &
\end{array} \\
\left[
\begin{array}{ccccc|c}
1 & \frac{3}{2} & \frac{1}{2} & 0 & 0 & \frac{11}{2} \\
0 & \frac{1}{2} & -\frac{1}{2} & 1 & 0 & \frac{1}{2} \\
\hline
0 & 3 & 2 & 0 & 1 & 22
\end{array}
\right]
\end{array}
$$

The maximum value is 22 when $x_1 = 5.5, x_2 = 0, s_1 = 0$, and $s_2 = .5$.

9. Maximize $z = 10x_1 + 12x_2$

subject to: $4x_1 + 2x_2 \le 20$
$$5x_1 + x_2 \le 50$$
$$2x_1 + 2x_2 \le 24$$

with $x_1 \ge 0, x_2 \ge 0.$

Three slack variables, s_1, s_2, and s_3, need to be introduced. The initial tableau is as follows.

$$
\begin{array}{c}
\begin{array}{cccccc}
x_1 & x_2 & s_1 & s_2 & s_3 & z
\end{array} \\
\left[
\begin{array}{cccccc|c}
4 & 2 & 1 & 0 & 0 & 0 & 20 \\
5 & 1 & 0 & 1 & 0 & 0 & 50 \\
2 & 2 & 0 & 0 & 1 & 0 & 24 \\
\hline
-10 & -12 & 0 & 0 & 0 & 1 & 0
\end{array}
\right]
\end{array}
$$

The most negative indicator is -12, in column 2. The quotients are $\frac{20}{2} = 10, \frac{50}{1} = 50$, and $\frac{24}{2} = 12$; the smallest is 10, so 2 in row 1, column 2 is the pivot.

$$
\begin{array}{c}
\begin{array}{cccccc}
x_1 & x_2 & s_1 & s_2 & s_3 & z
\end{array} \\
\left[
\begin{array}{cccccc|c}
4 & \boxed{2} & 1 & 0 & 0 & 0 & 20 \\
5 & 1 & 0 & 1 & 0 & 0 & 50 \\
2 & 2 & 0 & 0 & 1 & 0 & 24 \\
\hline
-10 & -12 & 0 & 0 & 0 & 1 & 0
\end{array}
\right]
\end{array}
$$

$$
\begin{array}{c}

\end{array}
\begin{array}{c}
\begin{array}{cccccc}
x_1 & x_2 & s_1 & s_2 & s_3 & z
\end{array} \\
\begin{array}{l}
 \\
-R_1 + 2R_2 \to R_2 \\
-R_1 + R_3 \to R_3 \\
6R_1 + R_4 \to R_4
\end{array}
\left[
\begin{array}{cccccc|c}
4 & 2 & 1 & 0 & 0 & 0 & 20 \\
6 & 0 & -1 & 2 & 0 & 0 & 80 \\
-2 & 0 & -1 & 0 & 1 & 0 & 4 \\
14 & 0 & 6 & 0 & 0 & 1 & 120
\end{array}
\right]
\end{array}
$$

All of the indicators are nonnegative, so we are finished pivoting. Create a 1 in the columns corresponding to x_2 and s_2.

$$
\begin{array}{c}
\frac{1}{2}R_1 \to R_1 \\
\frac{1}{2}R_2 \to R_2
\end{array}
\begin{array}{c}
\begin{array}{cccccc}
x_1 & x_2 & s_1 & s_2 & s_3 & z
\end{array} \\
\left[
\begin{array}{cccccc|c}
2 & 1 & \frac{1}{2} & 0 & 0 & 0 & 10 \\
3 & 0 & -\frac{1}{2} & 1 & 0 & 0 & 40 \\
\hline
-2 & 0 & -1 & 0 & 1 & 0 & 4 \\
14 & 0 & 6 & 0 & 0 & 1 & 120
\end{array}
\right]
\end{array}
$$

The maximum value is 120 when $x_1 = 0, x_2 = 10, s_1 = 0, s_2 = 40$, and $s_3 = 4$.

11. Maximize $z = 8x_1 + 3x_2 + x_3$

subject to: $x_1 + 6x_2 + 8x_3 \le 118$
$$x_1 + 5x_2 + 10x_3 \le 220$$

with $x_1 \ge 0, x_2 \ge 0, x_3 \ge 0.$

Two slack variables, s_1 and s_2, need to be introduced. The initial simplex tableau is as follows.

$$
\begin{array}{c}
\begin{array}{cccccc}
x_1 & x_2 & x_3 & s_1 & s_2 & z
\end{array} \\
\left[
\begin{array}{cccccc|c}
\boxed{1} & 6 & 8 & 1 & 0 & 0 & 118 \\
1 & 5 & 10 & 0 & 1 & 0 & 220 \\
\hline
-8 & -3 & -1 & 0 & 0 & 1 & 0
\end{array}
\right]
\end{array}
$$

The most negative indicator is -8, in the first column. The quotients are $\frac{118}{1} = 118$ and $\frac{220}{1} = 220$; since 118 is the smallest, 1 in row 1, column 1 is the pivot. Performing row transformations, we get the following tableau.

$$
\begin{array}{c}
 \\
-R_1 + R_2 \to R_2 \\
8R_1 + R_3 \to R_3
\end{array}
\begin{array}{c}
\begin{array}{cccccc}
x_1 & x_2 & x_3 & s_1 & s_2 & z
\end{array} \\
\left[
\begin{array}{cccccc|c}
1 & 6 & 8 & 1 & 0 & 0 & 118 \\
0 & -1 & 2 & -1 & 1 & 0 & 102 \\
0 & 45 & 63 & 8 & 0 & 1 & 944
\end{array}
\right]
\end{array}
$$

All of the indicators are nonnegative, so we are finished pivoting. The maximum value is 944 when $x_1 = 118, x_2 = 0, x_3 = 0, s_1 = 0$, and $s_2 = 102$.

13. Maximize $z = x_1 + 2x_2 + x_3 + 5x_4$

subject to: $x_1 + 2x_2 + x_3 + x_4 \le 50$
$$3x_1 + x_2 + 2x_3 + x_4 \le 100$$

with $x_1 \ge 0, x_2 \ge 0, x_3 \ge 0, x_4 \ge 0.$

Two slack variables, s_1 and s_2, need to be introduced. The initial simplex tableau is as follows.

$$
\begin{array}{c}
\begin{array}{ccccccc}
x_1 & x_2 & x_3 & x_4 & s_1 & s_2 & z
\end{array} \\
\left[
\begin{array}{ccccccc|c}
1 & 2 & 1 & \boxed{1} & 1 & 0 & 0 & 50 \\
3 & 1 & 2 & 1 & 0 & 1 & 0 & 100 \\
\hline
-1 & -2 & -1 & -5 & 0 & 0 & 1 & 0
\end{array}
\right]
\end{array}
$$

In the column with the most negative indicator, -5, the quotients are $\frac{50}{1} = 50$ and $\frac{100}{1} = 100$; the smallest is 50, so 1 in row 1, column 4 is the pivot.

$$
\begin{array}{c}
 \\
-R_1 + R_2 \to R_2 \\
5R_1 + R_3 \to R_3
\end{array}
\begin{array}{c}
\begin{array}{ccccccc}
x_1 & x_2 & x_3 & x_4 & s_1 & s_2 & z
\end{array} \\
\left[
\begin{array}{ccccccc|c}
1 & 2 & 1 & 1 & 1 & 0 & 0 & 50 \\
2 & -1 & 1 & 0 & -1 & 1 & 0 & 50 \\
4 & 8 & 4 & 0 & 5 & 0 & 1 & 250
\end{array}
\right]
\end{array}
$$

This is a final tableau, since all of the indicators are nonnegative. The maximum value is 250 when $x_1 = 0, x_2 = 0, x_3 = 0, x_4 = 50, s_1 = 0$, and $s_2 = 50$.

17. Organize the information in a table.

	Church Group	Labor Union	Maximum Time Available
Letter Writing	2	2	16
Follow-up	1	3	12
Money Raised	$100	$200	

Let x_1 and x_2 be the number of church groups and labor unions contacted respectively. We need two slack variables, s_1 and s_2.

$$\text{Maximize} \quad z = 100x_1 + 200x_2$$
$$\text{subject to:} \quad 2x_1 + 2x_2 + s_1 \qquad = 16$$
$$x_1 + 3x_2 \qquad + s_2 = 12$$
$$\text{with} \quad x_1 \geq 0, x_2 \geq 0, s_1 \geq 0, s_2 \geq 0.$$

The initial simplex tableau is as follows.

$$\begin{array}{ccccc}
x_1 & x_2 & s_1 & s_2 & z \\
\end{array}$$
$$\left[\begin{array}{ccccc|c}
2 & 2 & 1 & 0 & 0 & 16 \\
1 & \boxed{3} & 0 & 1 & 0 & 12 \\
\hline
-100 & -200 & 0 & 0 & 1 & 0
\end{array}\right]$$

Pivot on the 3 in row 2, column 2.

$$\begin{array}{ccccc}
 & x_1 & x_2 & s_1 & s_2 & z \\
\end{array}$$

$-2R_2 + 3R_1 \to R_1$
$$\left[\begin{array}{ccccc|c}
\boxed{4} & 0 & 3 & -2 & 0 & 24 \\
1 & 3 & 0 & 1 & 0 & 12 \\
\hline
\end{array}\right.$$
$200R_2 + 3R_3 \to R_3$
$$\left.\begin{array}{ccccc|c}
-100 & 0 & 0 & 200 & 3 & 2400
\end{array}\right]$$

Pivot on the 4 in row 1, column 1.

$$\begin{array}{ccccc}
x_1 & x_2 & s_1 & s_2 & z \\
\end{array}$$
$$\left[\begin{array}{ccccc|c}
4 & 0 & 3 & -2 & 0 & 24 \\
\end{array}\right.$$
$-R_1 + 4R_2 \to R_2$
$$\begin{array}{ccccc|c}
0 & 12 & -3 & 6 & 0 & 24 \\
\end{array}$$
$25R_1 + R_3 \to R_3$
$$\left.\begin{array}{ccccc|c}
0 & 0 & 75 & 150 & 3 & 3000
\end{array}\right]$$

This is a final tableau, since all of the indicators are nonnegative. Create a 1 in the columns corresponding to $x_1, x_2,$ and z.

$$\begin{array}{ccccc}
x_1 & x_2 & s_1 & s_2 & z \\
\end{array}$$
$\frac{1}{4}R_1 \to R_1$
$$\left[\begin{array}{ccccc|c}
1 & 0 & \frac{3}{4} & -\frac{1}{2} & 0 & 6 \\
\end{array}\right.$$
$\frac{1}{12}R_2 \to R_2$
$$\begin{array}{ccccc|c}
0 & 1 & -\frac{1}{4} & \frac{1}{2} & 0 & 2 \\
\end{array}$$
$\frac{1}{3}R_3 \to R_3$
$$\left.\begin{array}{ccccc|c}
0 & 0 & 25 & 50 & 1 & 1000
\end{array}\right]$$

The maximum amount of money raised is $1000/mo when $x_1 = 6$ and $x_2 = 2$, that is, when 6 churches and 2 labor unions are contacted.

19. Organize the information in a table.

	Recording	Mixing	Editing	Profit
Jazz	4	2	6	$ .80
Blues	4	8	2	$ .60
Reggae	10	4	6	$1.20
Maximum Available	80	52	54	

Let x_1, x_2, and x_3 represent the number of jazz, blues, and reggae albums respectively. We need three slack variables, s_1, s_2, and s_3.

$$\text{Maximize} \quad z = .8x_1 + .6x_2 + 1.2x_3$$
$$\text{subject to:} \quad 4x_1 + 4x_2 + 10x_3 + s_1 \qquad\qquad = 80$$
$$2x_1 + 8x_2 + 4x_3 \qquad + s_2 \qquad = 52$$
$$6x_1 + 2x_2 + 6x_3 \qquad\qquad + s_3 = 54$$
$$\text{with} \quad x_1 \geq 0, x_2 \geq 0, x_3 \geq 0, s_1 \geq 0, s_2 \geq 0, \text{ and } s_3 \geq 0.$$

The initial simplex tableau is as follows.

	x_1	x_2	x_3	s_1	s_2	s_3	z	
	4	4	10	1	0	0	0	80
	2	8	4	0	1	0	0	52
	6	2	6	0	0	1	0	54
	$-.8$	$-.6$	-1.2	0	0	0	1	0

Pivot on the 10 in row 1, column 3.

	x_1	x_2	x_3	s_1	s_2	s_3	z	
	4	4	10	1	0	0	0	80
$-2R_1 + 5R_2 \rightarrow R_2$	2	32	0	-2	5	0	0	100
$-3R_1 + 5R_3 \rightarrow R_3$	18	-2	0	-3	0	5	0	30
$3R_1 + 25R_4 \rightarrow R_4$	-8	-3	0	3	0	0	25	240

Pivot on the 18 in row 3, column 1.

	x_1	x_2	x_3	s_1	s_2	s_3	z	
$-2R_3 + 9R_1 \rightarrow R_1$	0	40	90	15	0	-10	0	660
$-R_3 + 9R_2 \rightarrow R_2$	0	290	0	-15	45	-5	0	870
	18	-2	0	-3	0	5	0	30
$4R_3 + 9R_4 \rightarrow R_4$	0	-35	0	15	0	20	225	2280

Pivot on the 290 in row 2, column 2.

	x_1	x_2	x_3	s_1	s_2	s_3	z	
$-4R_2 + 29R_1 \rightarrow R_1$	0	0	2610	495	-180	-270	0	15,660
	0	290	0	-15	45	-5	0	870
$R_2 + 145R_3 \rightarrow R_3$	2610	0	0	-450	45	720	0	5220
$7R_2 + 58R_4 \rightarrow R_4$	0	0	0	765	315	1125	13,050	138,330

This is a final tableau, since all of the indicators are nonnegative. Create a 1 in the columns corresponding to x_1, x_2, x_3, and z.

	x_1	x_2	x_3	s_1	s_2	s_3	z	
$\frac{1}{2610}R_1 \rightarrow R_1$	0	0	1	$\frac{11}{58}$	$-\frac{2}{29}$	$-\frac{3}{29}$	0	6
$\frac{1}{290}R_2 \rightarrow R_2$	0	1	0	$-\frac{3}{58}$	$\frac{9}{58}$	$-\frac{1}{58}$	0	3
$\frac{1}{2610}R_3 \rightarrow R_3$	1	0	0	$-\frac{5}{29}$	$\frac{1}{58}$	$\frac{8}{29}$	0	2
$\frac{1}{13,050}R_4 \rightarrow R_4$	0	0	0	$\frac{17}{290}$	$\frac{7}{290}$	$\frac{5}{58}$	1	10.6

The maximum weekly profit is \$10.60 when $x_1 = 2, x_2 = 3$, and $x_3 = 6$, that is, when 2 jazz albums, 3 blues albums, and 6 reggae albums are produced.

21. (a) Let x_1 represent the number of one-speed bicycles, x_2 the number of three-speed bicycles, and x_3 the number of ten-speed bicycles.

From Exercise 26 in Section 4.1, the initial simplex tableau is as follows.

$$
\begin{array}{c}
\begin{array}{cccccc} x_1 & x_2 & x_3 & s_1 & s_2 & z \end{array} \\
\left[\begin{array}{cccccc|c}
17 & 27 & \boxed{34} & 1 & 0 & 0 & 91{,}800 \\
12 & 21 & 15 & 0 & 1 & 0 & 42{,}000 \\
\hline
-8 & -12 & -22 & 0 & 0 & 1 & 0
\end{array} \right]
\end{array}
$$

Pivot on the 34 in row 1, column 3.

$$
\begin{array}{r}
\\
-15R_1 + 34R_2 \rightarrow R_2 \\
11R_1 + 17R_3 \rightarrow R_3
\end{array}
\begin{array}{c}
\begin{array}{cccccc} x_1 & x_2 & x_3 & s_1 & s_2 & z \end{array} \\
\left[\begin{array}{cccccc|c}
17 & 27 & 34 & 1 & 0 & 0 & 91{,}800 \\
153 & 309 & 0 & -15 & 34 & 0 & 51{,}000 \\
51 & 93 & 0 & 11 & 0 & 17 & 1{,}009{,}800
\end{array} \right]
\end{array}
$$

This is a final tableau, since all of the indicators are nonnegative. Create a 1 in the columns corresponding to $x_3, s_2,$ and z.

$$
\begin{array}{r}
\frac{1}{34}R_1 \rightarrow R_1 \\
\frac{1}{34}R_2 \rightarrow R_2 \\
\frac{1}{17}R_3 \rightarrow R_3
\end{array}
\begin{array}{c}
\begin{array}{cccccc} x_1 & x_2 & x_3 & s_1 & s_2 & z \end{array} \\
\left[\begin{array}{cccccc|c}
\frac{1}{2} & \frac{27}{34} & 1 & \frac{1}{34} & 0 & 0 & 2700 \\
\frac{9}{2} & \frac{309}{34} & 0 & -\frac{15}{34} & 1 & 0 & 1500 \\
3 & \frac{93}{17} & 0 & \frac{11}{17} & 0 & 1 & 59{,}400
\end{array} \right]
\end{array}
$$

From the tableau, $x_1 = 0, x_2 = 0,$ and $x_3 = 2700.$ The company should make no one-speed or three-speed bicycles and 2700 ten-speed bicycles.

(b) From the third row of the final tableau, the maximum profit is $59,400.

---

23. (a) Let x_1 represent the number of newspaper ads, x_2 the number of radio ads, and x_3 the number of TV ads.

From Exercise 29 in Section 4.1, the initial simplex tableau is as follows.

$$
\begin{array}{c}
\begin{array}{cccccccc} x_1 & x_2 & x_3 & s_1 & s_2 & s_3 & s_4 & z \end{array} \\
\left[\begin{array}{cccccccc|c}
400 & 200 & 1200 & 1 & 0 & 0 & 0 & 0 & 8000 \\
1 & 0 & 0 & 0 & 1 & 0 & 0 & 0 & 20 \\
0 & 1 & 0 & 0 & 0 & 1 & 0 & 0 & 30 \\
0 & 0 & \boxed{1} & 0 & 0 & 0 & 1 & 0 & 6 \\
\hline
-2000 & -1200 & -10{,}000 & 0 & 0 & 0 & 0 & 1 & 0
\end{array} \right]
\end{array}
$$

Pivot on the 1 in row 4, column 3.

$$
\begin{array}{r}
-1200R_3 + R_1 \rightarrow R_1 \\
\\
\\
\\
10{,}000R_4 + R_5 \rightarrow R_5
\end{array}
\begin{array}{c}
\begin{array}{ccccccccc} x_1 & x_2 & x_3 & s_1 & s_2 & s_3 & s_4 & z \end{array} \\
\left[\begin{array}{cccccccc|c}
\boxed{400} & 200 & 0 & 1 & 0 & 0 & -1200 & 0 & 800 \\
1 & 0 & 0 & 0 & 1 & 0 & 0 & 0 & 20 \\
0 & 1 & 0 & 0 & 0 & 1 & 0 & 0 & 30 \\
0 & 0 & 1 & 0 & 0 & 0 & 1 & 0 & 6 \\
\hline
-2000 & -1200 & 0 & 0 & 0 & 0 & 10{,}000 & 1 & 60{,}000
\end{array} \right]
\end{array}
$$

Now pivot on the 400 in row 1, column 1.

$$\begin{array}{c} \\ \\ -R_1 + 400R_2 \rightarrow R_2 \\ \\ \\ 5R_1 + R_5 \rightarrow R_5 \end{array}
\begin{array}{cccccccc} x_1 & x_2 & x_3 & s_1 & s_2 & s_3 & s_4 & z \end{array}$$

x_1	x_2	x_3	s_1	s_2	s_3	s_4	z	
400	[200]	0	1	0	0	−1200	0	800
0	−200	0	−1	400	0	1200	0	7200
0	1	0	0	0	1	0	0	30
0	0	1	0	0	0	1	0	6
0	−200	0	5	0	0	4000	1	64,000

with $-R_1 + 400R_2 \rightarrow R_2$ and $5R_1 + R_5 \rightarrow R_5$.

Pivot on the 200 in row 1, column 2.

x_1	x_2	x_3	s_1	s_2	s_3	s_4	z	
400	200	0	1	0	0	−1200	0	800
400	0	0	0	400	0	0	0	15,200
400	0	0	0	0	−200	−1200	0	−5200
0	0	1	0	0	0	1	0	6
400	0	0	6	0	0	2800	1	64,800

with $R_1 + R_2 \rightarrow R_2$, $R_1 - 200R_3 \rightarrow R_3$, and $R_1 + R_5 \rightarrow R_5$.

Create a 1 in the columns corresponding to $x_2, s_2,$ and s_3.

x_1	x_2	x_3	s_1	s_2	s_3	s_4	z	
2	1	0	$\frac{1}{200}$	0	0	−6	0	4
1	0	0	0	1	0	0	0	38
−2	0	0	0	0	1	6	0	26
0	0	1	0	0	0	1	0	6
400	0	0	6	0	0	2800	1	64,800

with $\frac{1}{200}R_1 \rightarrow R_1$, $\frac{1}{400}R_2 \rightarrow R_2$, and $-\frac{1}{200}R_3 \rightarrow R_3$.

This is the final tableau. The maximum exposure is 64,800 women when $x_1 = 0, x_2 = 4,$ and $x_3 = 6,$ that is, when no newspaper ads, 4 radio ads, and 6 TV ads are run.

(b) Newspaper ads are the most costly. By dividing the cost of the ad by the number of women who see the ad, we get

$$\text{Newspaper} = \tfrac{400}{2000} = .20 \text{ or } 20\cancel{c} \text{ per woman per ad;}$$

$$\text{Radio} = \tfrac{200}{1200} = .17 \text{ or } 17\cancel{c} \text{ per woman per ad;}$$

$$\text{TV} = \tfrac{1200}{10,000} = .12 \text{ or } 12\cancel{c} \text{ per woman per ad.}$$

Therefore, with radio and TV you are getting a better value for your money.

25. (a) Look at the first table, which has to do with the profits. The profit-maximization formula is

$$\$2A + \$5B + \$4C = X,$$

so the answer is choice (1).

(b) Look at the "Painting" row of the second chart. The "Painting" constraint is

$$1A + 2B + 2C \leq 38,000,$$

so the answer is choice (3).

27. Let x_1 represent the number of minutes for the senator, x_2 the number of minutes for the congresswoman, and x_3 the number of minutes for the governor.

Of the half-hour show's time, only $30 - 3 = 27$ min are available to be allotted to the politicians. The given information leads to the equation

$$x_1 + x_2 + x_3 = 27$$

and the inequalities

$$x_1 \geq 2x_3 \quad \text{and} \quad x_1 + x_3 \geq 2x_2,$$

and we are to maximize the objective function

$$z = 40x_1 + 60x_2 + 50x_3.$$

Rewrite the equation as

$$x_3 = 27 - x_1 - x_2$$

and the inequalities as

$$x_1 - 2x_3 \geq 0 \quad \text{and} \quad x_1 - 2x_2 + x_3 \geq 0.$$

Substitute $27 - x_1 - x_2$ for x_3 in the objective function and the inequalities, and the problem is as follows.

Maximize $z = 40x_1 + 60x_2 + 50(27 - x_1 - x_2)$
subject to: $x_1 \quad\; - 2(27 - x_1 - x_2) \geq 0$
             $x_1 - 2x_2 + (27 - x_1 - x_2) \geq 0$
with $x_1 \geq 0, x_2 \geq 0$, and $x_3 = 27 - x_1 - x_2 \geq 0$.

Simplify to obtain the following.

Maximize $z = 1350 - 10x_1 + 10x_2$
subject to: $3x_1 + 2x_2 \geq 54$
               $x_2 \leq 9$
with $x_1 \geq 0, x_2 \geq 0$, and $x_3 = 27 - x_1 - x_2 \geq 0$.

This linear programming problem is not in standard maximum form, so we will attempt to use reason rather than the simplex algorithm. (The graphical method of solution examined in Chapter 3 would also be acceptable.) Decreasing x_2 would increase x_1 (since $x_1 + x_2 + x_3$ must remain equal to 27) and would decrease z (since $z = 1350 - 10x_1 + 10x_2$). Therefore, choose x_2 to be as large as possible, which means $x_2 = 9$. The constraint

$$3x_1 + 2x_2 \geq 54$$

can be rewritten as

$$x_1 \geq \frac{54 - 2x_2}{3}.$$

Since we have chosen $x_2 = 9$, this means

$$x_1 \geq \frac{54 - 2(9)}{3} \text{ or } x_1 \geq 12.$$

Increasing x_1 would decrease z, so choose x_1 to be as small as possible. This means $x_1 = 12$. With $x_1 = 12$ and $x_2 = 9$, we have

$$x_3 = 27 - 12 - 9 = 6,$$

and

$$z = 1350 - 10(12) + 10(9) = 1320.$$

For a maximum of 1,320,000 viewers, the time allotments should be 12 min for the senator, 9 min for the congresswoman, and 6 min for the governor.

29. (a) Let x_1 represent the number of toy trucks and x_2 the number of toy fire engines.

Maximize $z = 8.50x_1 + 12.10x_2$
subject to: $-2x_1 + 3x_2 \leq \qquad 0$
            $x_1 \qquad\quad \leq \quad 6700$
               $x_2 \leq \quad 5500$
           $x_1 + \quad x_2 \leq 12,000$
with $x_1 \geq 0, x_2 \geq 0.$

This exercise should be solved by graphing calculator or computer methods. The answer is to produce 6700 trucks and 4467 fire engines for a maximum profit of $110,997.

(b) Many solutions are possible.

(c) Many solutions are possible.

4.3   Minimization Problems; Duality

1. To form the transpose of a matrix, the rows of the original matrix are written as the columns of the transpose. The transpose of

$$\begin{bmatrix} 1 & 2 & 3 \\ 3 & 2 & 1 \\ 1 & 10 & 0 \end{bmatrix}$$

is

$$\begin{bmatrix} 1 & 3 & 1 \\ 2 & 2 & 10 \\ 3 & 1 & 0 \end{bmatrix}.$$

3. The transpose of

$$\begin{bmatrix} -1 & 4 & 6 & 12 \\ 13 & 25 & 0 & 4 \\ -2 & -1 & 11 & 3 \end{bmatrix}$$

is

$$\begin{bmatrix} -1 & 13 & -2 \\ 4 & 25 & -1 \\ 6 & 0 & 11 \\ 12 & 4 & 3 \end{bmatrix}.$$

5. Maximize $z = 4x_1 + 3x_2 + 2x_3$
subject to: $x_1 + x_2 + x_3 \le 5$
$x_1 + x_2 \qquad \le 4$
$2x_1 + x_2 + 3x_3 \le 15$
with $x_1 \ge 0, x_2 \ge 0, x_3 \ge 0.$

To form the dual, first write the augmented matrix for the given problem.

$$\begin{bmatrix} 1 & 1 & 1 & 5 \\ 1 & 1 & 0 & 4 \\ 2 & 1 & 3 & 15 \\ \hline 4 & 3 & 2 & 0 \end{bmatrix}$$

Then form the transpose of this matrix.

$$\begin{bmatrix} 1 & 1 & 2 & 4 \\ 1 & 1 & 1 & 3 \\ 1 & 0 & 3 & 2 \\ \hline 5 & 4 & 15 & 0 \end{bmatrix}$$

The dual problem is stated from this second matrix (using y instead of x).

Minimize $w = 5y_1 + 4y_2 + 15y_3$
subject to: $y_1 + y_2 + 2y_3 \ge 4$
$y_1 + y_2 + y_3 \ge 3$
$y_1 \qquad + 3y_3 \ge 2$
with $y_1 \ge 0, y_2 \ge 0, y_3 \ge 0.$

7. Minimize $w = y_1 + 2y_2 + y_3 + 5y_4$
subject to: $y_1 + y_2 + y_3 + y_4 \ge 50$
$3y_1 + y_2 + 2y_3 + y_4 \ge 100$
with $y_1 \ge 0, y_2 \ge 0, y_3 \ge 0, y_4 \ge 0.$

To find the dual, first write the augmented matrix for the given problem.

$$\begin{bmatrix} 1 & 1 & 1 & 1 & 50 \\ 3 & 1 & 2 & 1 & 100 \\ \hline 1 & 2 & 1 & 5 & 0 \end{bmatrix}$$

Then form the transpose of this matrix.

$$\begin{bmatrix} 1 & 3 & 1 \\ 1 & 1 & 2 \\ 1 & 2 & 1 \\ 1 & 1 & 5 \\ \hline 50 & 100 & 0 \end{bmatrix}$$

The dual problem is stated from this second matrix (using x instead of y).

Maximize $z = 50x_1 + 100x_2$
subject to: $x_1 + 3x_2 \le 1$
$x_1 + x_2 \le 2$
$x_1 + 2x_2 \le 1$
$x_1 + x_2 \le 5$
with $x_1 \ge 0, x_2 \ge 0.$

9. Find $y_1 \ge 0$ and $y_2 \ge 0$ such that

$$2y_1 + 3y_2 \ge 6$$
$$2y_1 + y_2 \ge 7$$

and $w = 5y_1 + 2y_2$ is minimized.

Write the augmented matrix for this problem.

$$\begin{bmatrix} 2 & 3 & 6 \\ 2 & 1 & 7 \\ \hline 5 & 2 & 0 \end{bmatrix}$$

Form the transpose of this matrix.

$$\begin{bmatrix} 2 & 2 & 5 \\ 3 & 1 & 2 \\ \hline 6 & 7 & 0 \end{bmatrix}$$

Use this matrix to write the dual problem.

Find $x_1 \ge 0$ and $x_2 \ge 0$ such that

$$2x_1 + 2x_2 \le 5$$
$$3x_1 + x_2 \le 2$$

and $z = 6x_1 + 7x_2$ is maximized.

Introduce slack variables s_1 and s_2. The initial tableau is as follows.

$$\begin{array}{ccccc} x_1 & x_2 & s_1 & s_2 & z \\ \end{array}$$
$$\begin{bmatrix} 2 & 2 & 1 & 0 & 0 & 5 \\ 3 & \boxed{1} & 0 & 1 & 0 & 2 \\ \hline -6 & -7 & 0 & 0 & 1 & 0 \end{bmatrix}$$

Pivot on the 1 in row 2, column 2, since that column has the most negative indicator and that row has the smallest nonnegative quotient.

$$\begin{array}{ccccc} x_1 & x_2 & s_1 & s_2 & z \\ \end{array}$$

$-2R_2 + R_1 \to R_1$
$7R_2 + R_3 \to R_3$
$$\begin{bmatrix} -4 & 0 & 1 & -2 & 0 & 1 \\ 3 & 1 & 0 & 1 & 0 & 2 \\ \hline 15 & 0 & 0 & 7 & 1 & 14 \end{bmatrix}$$

The minimum value of w is the same as the maximum value of z. The minimum value of w is 14 when $y_1 = 0$ and $y_2 = 7$. (Note that the values of y_1 and y_2 are given by the entries in the bottom row of the columns corresponding to the slack variables in the final tableau.)

11. Find $y_1 \geq 0$ and $y_2 \geq 0$ such that

$$10y_1 + 5y_2 \geq 100$$
$$20y_1 + 10y_2 \geq 150$$

and $w = 4y_1 + 5y_2$ is minimized.

Write the augmented matrix for this problem.

$$\begin{bmatrix} 10 & 5 & | & 100 \\ 20 & 10 & | & 150 \\ \hline 4 & 5 & | & 0 \end{bmatrix}$$

Form the transpose of this matrix.

$$\begin{bmatrix} 10 & 20 & | & 4 \\ 5 & 10 & | & 5 \\ \hline 100 & 150 & | & 0 \end{bmatrix}$$

Write the dual problem from this matrix.

Find $x_1 \geq 0$ and $x_2 \geq 0$ such that

$$10x_1 + 20x_2 \leq 4$$
$$5x_1 + 10x_2 \leq 5$$

and $z = 100x_1 + 150x_2$ is maximized.

The initial simplex tableau is as follows.

$$\begin{array}{ccccc} x_1 & x_2 & s_1 & s_2 & z \\ \begin{bmatrix} 10 & \boxed{20} & 1 & 0 & 0 & | & 4 \\ 5 & 10 & 0 & 1 & 0 & | & 5 \\ \hline -100 & -150 & 0 & 0 & 1 & | & 0 \end{bmatrix} \end{array}$$

Pivot on the 20 in row 1, column 2.

$$\begin{array}{ccccc} & x_1 & x_2 & s_1 & s_2 & z \\ & \begin{bmatrix} \boxed{10} & 20 & 1 & 0 & 0 & | & 4 \\ 0 & 0 & -1 & 2 & 0 & | & 6 \\ \hline -50 & 0 & 15 & 0 & 2 & | & 60 \end{bmatrix} \end{array}$$

$-R_1 + 2R_1 \rightarrow R_2$
$15R_1 + 2R_3 \rightarrow R_3$

Pivot on the 10 in row 1, column 1.

$$\begin{array}{ccccc} & x_1 & x_2 & s_1 & s_2 & z \\ & \begin{bmatrix} 10 & 20 & 1 & 0 & 0 & | & 4 \\ 0 & 0 & -1 & 2 & 0 & | & 6 \\ \hline 0 & 100 & 20 & 0 & 2 & | & 80 \end{bmatrix} \end{array}$$

$5R_1 + R_3 \rightarrow R_3$

Create a 1 in the columns corresponding to x_1, s_2, and z.

$$\begin{array}{ccccc} & x_1 & x_2 & s_1 & s_2 & z \\ \frac{1}{10}R_1 \rightarrow R_1 & \begin{bmatrix} 1 & 2 & \frac{1}{10} & 0 & 0 & | & \frac{2}{5} \\ \frac{1}{2}R_2 \rightarrow R_2 & 0 & 0 & -\frac{1}{2} & 1 & 0 & | & 3 \\ \frac{1}{2}R_3 \rightarrow R_3 & 0 & 50 & 10 & 0 & 1 & | & 40 \end{bmatrix} \end{array}$$

The minimum value of w is 40 when $y_1 = 10$ and $y_2 = 0$. (These values of y_1 and y_2 are read from the last row of the columns corresponding to s_1 and s_2 in the final tableau.)

13. Minimize $w = 2y_1 + y_2 + 3y_3$
subject to: $y_1 + y_2 + y_3 \geq 100$
                $2y_1 + y_2 \qquad \geq 50$
with $y_1 \geq 0, y_2 \geq 0, y_3 \geq 0.$

Write the augmented matrix.

$$\begin{bmatrix} 1 & 1 & 1 & | & 100 \\ 2 & 1 & 0 & | & 50 \\ \hline 2 & 1 & 3 & | & 0 \end{bmatrix}$$

Form the transpose of this matrix.

$$\begin{bmatrix} 1 & 2 & | & 2 \\ 1 & 1 & | & 1 \\ 1 & 0 & | & 3 \\ \hline 100 & 50 & | & 0 \end{bmatrix}$$

The dual problem is as follows.

Maximize $z = 100x_1 + 50x_2$
subject to: $x_1 + 2x_2 \leq 2$
              $x_1 + x_2 \leq 1$
              $x_1 \qquad \leq 3$
with $x_1 \geq 0, x_2 \geq 0.$

The initial simplex tableau is as follows.

$$\begin{array}{cccccc} x_1 & x_2 & s_1 & s_2 & s_3 & z \\ \begin{bmatrix} 1 & 2 & 1 & 0 & 0 & 0 & | & 2 \\ \boxed{1} & 1 & 0 & 1 & 0 & 0 & | & 1 \\ 1 & 0 & 0 & 0 & 1 & 0 & | & 3 \\ \hline -100 & -50 & 0 & 0 & 0 & 1 & | & 0 \end{bmatrix} \end{array}$$

Pivot on the 1 in row 2, column 1.

$$\begin{array}{cccccc} & x_1 & x_2 & s_1 & s_2 & s_3 & z \\ -R_2 + R_1 \rightarrow R_1 & \begin{bmatrix} 0 & 1 & 1 & -1 & 0 & 0 & | & 1 \\ 1 & 1 & 0 & 1 & 0 & 0 & | & 1 \\ -R_2 + R_3 \rightarrow R_3 & 0 & -1 & 0 & -1 & 1 & 0 & | & 2 \\ 100R_2 + R_4 \rightarrow R_4 & 0 & 50 & 0 & 100 & 0 & 1 & | & 100 \end{bmatrix} \end{array}$$

The minimum value of w is 100 when $y_1 = 0, y_2 = 100$, and $y_3 = 0$.

15. Minimize $z = x_1 + 2x_2$
  subject to: $-2x_1 + x_2 \geq 1$
          $x_1 - 2x_2 \geq 1$
  with $x_1 \geq 0, x_2 \geq 0$.

A quick sketch of the constraints $-2x_1 + x_2 \geq 1$ and $x_1 - 2x_2 \geq 1$ will verify that the two corresponding half planes do not overlap in the first quadrant of the x_1x_2-plane. Therefore, this problem (P) has no feasible solution. The dual of the given problem is as follows:

Maximize $w = y_1 + y_2$
  subject to: $-2y_1 + y_2 \leq 1$
          $y_1 - 2y_2 \leq 2$
  with $y_1 \geq 0, y_2 \geq 0$.

A quick sketch here will verify that there is a feasible region in the y_1y_2-plane, and it is unbounded. Therefore, there is no maximum value of w in this problem (D).

(P) has no feasible solution and the objective function of (D) is unbounded; this is choice (a).

17. Maximize $x_1 + 1.5x_2 = z$
  subject to: $x_1 + 2x_2 \leq 200$
          $4x_1 + 3x_2 \leq 600$
          $0 \leq x_2 \leq 90$
  with $x_1 \geq 0$.

The final simplex tableau of this problem is as follows.

$$
\begin{array}{cccccc|c}
x_1 & x_2 & s_1 & s_2 & s_3 & z & \\
\hline
0 & 1 & .8 & -.2 & 0 & 0 & 40 \\
1 & 0 & -.6 & .4 & 0 & 0 & 120 \\
0 & 0 & -.8 & .2 & 1 & 0 & 50 \\
\hline
0 & 0 & .6 & .1 & 0 & 1 & 180
\end{array}
$$

(a) The corresponding dual problem is as follows:

Minimize $w = 200y_1 + 600y_2 + 90y_3$
  subject to: $y_1 + 4y_2 \geq 1$
          $2y_1 + 3y_2 + y_3 \geq 1.5$
  with $y_1 \geq 0, y_2 \geq 0, y_3 \geq 0$.

(b) From the given final tableau, the optimal solution to the dual problem is $y_1 = .6, y_2 = .1, y_3 = 0$, and $w = 180$.

(c) The shadow value for felt is .6; an increase in supply of 10 units of felt will increase profit to

$$\$180 + .6(10) = \$186.$$

(d) The shadow values are .1 for stuffing and 0 for trim. If stuffing and trim are each decreased by 10 units, the profit will be

$$\$180 - .1(10) - 0(10) = \$179.$$

19. Let $y_1 =$ the number of small test tubes and $y_2 =$ the number of large test tubes.

Minimize $w = 15y_1 + 12y_2$
  subject to: $y_1 + y_2 \geq 2100$
          $y_1 \geq 2y_2$
          $y_1 \geq 800$
          $y_2 \geq 500$
  with $y_1 \geq 0, y_2 \geq 0$.

The second constraint can be rewritten as

$$y_1 - 2y_2 \geq 0.$$

Write the augmented matrix for this problem.

$$
\begin{bmatrix}
1 & 1 & 2100 \\
1 & -2 & 0 \\
1 & 0 & 800 \\
0 & 1 & 500 \\
\hline
15 & 12 & 0
\end{bmatrix}
$$

Transpose to get the matrix for the dual problem.

$$
\begin{bmatrix}
1 & 1 & 1 & 0 & 15 \\
1 & -2 & 0 & 1 & 12 \\
\hline
2100 & 0 & 800 & 500 & 0
\end{bmatrix}
$$

Write the dual problem:

Maximize $z = 2100x_1 + 800x_3 + 500x_4$
  subject to: $x_1 + x_2 + x_3 \leq 15$
          $x_1 - 2x_2 + x_4 \leq 12$
  with $x_1 \geq 0, x_2 \geq 0, x_3 \geq 0, x_4 \geq 0$.

Write the initial tableau.

$$
\begin{array}{ccccccc|c}
x_1 & x_2 & x_3 & x_4 & s_1 & s_2 & z & \\
\hline
1 & 1 & 1 & 0 & 1 & 0 & 0 & 15 \\
\boxed{1} & -2 & 0 & 1 & 0 & 1 & 0 & 12 \\
\hline
-2100 & 0 & -800 & -500 & 0 & 0 & 1 & 0
\end{array}
$$

Pivot on the 1 in row 2, column 1.

$$
\begin{array}{c}
 \\
-R_2 + R_1 \rightarrow R_1 \\
 \\
2100R_2 + R_3 \rightarrow R_3
\end{array}
\begin{array}{cccccccc}
x_1 & x_2 & x_3 & x_4 & s_1 & s_2 & z & \\
\end{array}
\left[
\begin{array}{ccccccc|c}
0 & \boxed{3} & 1 & -1 & 1 & -1 & 0 & 3 \\
1 & -2 & 0 & 1 & 0 & 1 & 0 & 12 \\
\hline
0 & -4200 & -800 & 1600 & 0 & 2100 & 1 & 25,200
\end{array}
\right]
$$

Pivot on the 3 in row 1, column 2.

$$
\begin{array}{c}
 \\
2R_2 + 3R_3 \rightarrow R_2 \\
1400R_1 + R_3 \rightarrow R_3
\end{array}
\begin{array}{ccccccc}
x_1 & x_2 & x_3 & x_4 & s_1 & s_2 & z \\
\end{array}
\left[
\begin{array}{ccccccc|c}
0 & 3 & 1 & -1 & 1 & -1 & 0 & 3 \\
3 & 0 & 2 & 1 & 2 & 1 & 0 & 42 \\
\hline
0 & 0 & 600 & 200 & 1400 & 700 & 1 & 29,400
\end{array}
\right]
$$

Create a 1 in the columns corresponding to x_1 and x_2.

$$
\begin{array}{c}
\frac{1}{3}R_1 \rightarrow R_1 \\
\frac{1}{3}R_2 \rightarrow R_2
\end{array}
\begin{array}{ccccccc}
x_1 & x_2 & x_3 & x_4 & s_1 & s_2 & z \\
\end{array}
\left[
\begin{array}{ccccccc|c}
0 & 1 & \frac{1}{3} & -\frac{1}{3} & \frac{1}{3} & -\frac{1}{3} & 0 & 1 \\
1 & 0 & \frac{2}{3} & \frac{1}{3} & \frac{2}{3} & \frac{1}{3} & 0 & 14 \\
\hline
0 & 0 & 600 & 200 & 1400 & 700 & 1 & 29,400
\end{array}
\right]
$$

The minimum cost is 29,400¢ or \$294 when $y_1 = 1400$ and $y_2 = 700$, that is, when 1400 small test tubes and 700 large test tubes are ordered.

21. Let $y_1 =$ the number of political interviews conducted

and $y_2 =$ the number of market interviews conducted.

The problem is:

Minimize $\quad w = 45y_1 + 55y_2$
subject to: $\quad y_1 + y_2 \geq 8$
$\quad 8y_1 + 10y_2 \geq 60$
$\quad 6y_1 + 5y_2 \geq 40$
with $\quad\quad\quad y_1 \geq 0, y_2 \geq 0.$

Write the augmented matrix.

$$
\left[
\begin{array}{cc|c}
1 & 1 & 8 \\
8 & 10 & 60 \\
6 & 5 & 40 \\
\hline
45 & 55 & 0
\end{array}
\right]
$$

Transpose to get the matrix for the dual problem.

$$
\left[
\begin{array}{ccc|c}
1 & 8 & 6 & 45 \\
1 & 10 & 5 & 55 \\
\hline
8 & 60 & 40 & 0
\end{array}
\right]
$$

Write the dual problem:

Maximize $\quad z = 8x_1 + 60x_2 + 40x_3$
subject to: $\quad x_1 + 8x_2 + 6x_3 \leq 45$
$\quad x_1 + 10x_2 + 5x_3 \leq 55$
with $\quad\quad\quad x_1 \geq 0, x_2 \geq 0, x_3 \geq 0.$

Write the initial tableau.

$$
\begin{array}{cccccc}
x_1 & x_2 & x_3 & s_1 & s_2 & z \\
\end{array}
\left[
\begin{array}{cccccc|c}
1 & 8 & 6 & 1 & 0 & 0 & 45 \\
1 & \boxed{10} & 5 & 0 & 1 & 0 & 55 \\
\hline
-8 & -60 & -40 & 0 & 0 & 1 & 0
\end{array}
\right]
$$

Pivot on the 10 in row 2, column 2.

$$
\begin{array}{c}
-4R_2 + 5R_1 \rightarrow R_1 \\
 \\
6R_2 + R_3 \rightarrow R_3
\end{array}
\begin{array}{cccccc}
x_1 & x_2 & x_3 & s_1 & s_2 & z \\
\end{array}
\left[
\begin{array}{cccccc|c}
1 & 0 & \boxed{10} & 5 & -4 & 0 & 5 \\
1 & 10 & 5 & 0 & 1 & 0 & 55 \\
\hline
-2 & 0 & -10 & 0 & 6 & 1 & 330
\end{array}
\right]
$$

Pivot on the 10 in row 1, column 3.

$$
\begin{array}{c}
 \\
-R_1 + 2R_2 \rightarrow R_2 \\
R_1 + R_3 \rightarrow R_3
\end{array}
\begin{array}{cccccc}
x_1 & x_2 & x_3 & s_1 & s_2 & z \\
\end{array}
\left[
\begin{array}{cccccc|c}
\boxed{1} & 0 & 10 & 5 & -4 & 0 & 5 \\
1 & 20 & 0 & -5 & 6 & 0 & 105 \\
\hline
-1 & 0 & 0 & 5 & 2 & 1 & 335
\end{array}
\right]
$$

Pivot on the 1 in row 1, column 1.

$$
\begin{array}{c}
 \\
-R_1 + R_2 \rightarrow R_2 \\
R_1 + R_3 \rightarrow R_3
\end{array}
\begin{array}{cccccc}
x_1 & x_2 & x_3 & s_1 & s_2 & z \\
\end{array}
\left[
\begin{array}{cccccc|c}
1 & 0 & 10 & 5 & -4 & 0 & 5 \\
0 & 20 & -10 & -10 & \boxed{10} & 0 & 100 \\
\hline
0 & 0 & 10 & 10 & -2 & 1 & 340
\end{array}
\right]
$$

Pivot on the 10 in row 2, column 5.

$$
\begin{array}{c}
\\
2R_2 + 5R_1 \rightarrow R_1 \\
\\
R_2 + 5R_3 \rightarrow R_3
\end{array}
\begin{array}{cccccc|c}
x_1 & x_2 & x_3 & s_1 & s_2 & z & \\
5 & 40 & 30 & 5 & 0 & 0 & 225 \\
0 & 20 & -10 & -10 & 10 & 0 & 100 \\
0 & 20 & 40 & 40 & 0 & 5 & 1800
\end{array}
$$

Create a 1 in the columns corresponding to x_1, s_2, and z.

$$
\begin{array}{c}
\frac{1}{5}R_1 \rightarrow R_1 \\
\frac{1}{10}R_2 \rightarrow R_2 \\
\frac{1}{5}R_3 \rightarrow R_3
\end{array}
\begin{array}{cccccc|c}
x_1 & x_2 & x_3 & s_1 & s_2 & z & \\
1 & 8 & 6 & 1 & 0 & 0 & 45 \\
0 & 2 & -1 & -1 & 1 & 0 & 10 \\
0 & 4 & 8 & 8 & 0 & 1 & 360
\end{array}
$$

The minimum time spent is 360 min when $y_1 = 8$ and $y_2 = 0$, that is, when 8 political interviews and no market interviews are done.

23. Let $y_1 =$ the number of gallons of ingredient 1,
    $y_2 =$ the number of gallons of ingredient 2,
    $y_3 =$ the number of gallons of ingredient 3,
    $y_4 =$ the number of gallons of ingredient 4,
    $y_5 =$ the number of gallons of ingredient 5,
and $y_6 =$ the number of gallons of ingredient 6.

Note that $10\%(15{,}000) = 1500$, and $.01(15{,}000) = 150$.

The problem becomes:

Minimize

$$w = .48y_1 + .32y_2 + .53y_3 + .28y_4 + .43y_5 + .04y_6$$

subject to:

$$
\begin{array}{rl}
.28y_1 + .19y_2 + .43y_3 + .57y_4 + .22y_5 & \le 1500 \\
y_3 + y_4 & \ge 150 \\
y_2 \qquad\qquad + y_5 & \ge 150 \\
y_1 \qquad\qquad + y_4 & \ge 150 \\
y_1 + y_2 + y_3 + y_4 + y_5 + y_6 & = 15{,}000
\end{array}
$$

with $y_1 \ge 0, y_2 \ge 0, y_3 \ge 0, y_4 \ge 0, y_5 \ge 0, y_6 \ge 0$.

This exercise should be solved by graphing calculator or computer methods. The answer is to use 0 gal of ingredient 1, 150 gal of 2, 0 gal of 3, 150 gal of 4, 0 gal of 5, and 14,700 gal of water for a minimum cost of $678.

25. Let $y_1 =$ amount of product A and $y_2 =$ amount of product B.

The problem is:

Minimize $w = 25y_1 + 40y_2$
subject to: $3y_1 + 2y_2 \ge 15$
$2y_1 + 4y_2 \ge 15$
with $y_1 \ge 0, y_2 \ge 0$.

Write the augmented matrix.

$$
\left[
\begin{array}{cc|c}
3 & 2 & 15 \\
2 & 4 & 15 \\
\hline
25 & 40 & 0
\end{array}
\right]
$$

Transpose to get the matrix for the dual problem.

$$
\left[
\begin{array}{cc|c}
3 & 2 & 25 \\
2 & 4 & 40 \\
\hline
15 & 15 & 0
\end{array}
\right].
$$

Write the dual problem:

Maximize $z = 15x_1 + 15x_2$
subject to: $3x_1 + 2x_2 \le 25$
$2x_1 + 4x_2 \le 40$
with $x_1 \ge 0, x_2 \ge 0$.

Write the initial tableau.

$$
\left[
\begin{array}{ccccc|c}
x_1 & x_2 & s_1 & s_2 & z & \\
\boxed{3} & 2 & 1 & 0 & 0 & 25 \\
2 & 4 & 0 & 1 & 0 & 40 \\
\hline
-15 & -15 & 0 & 0 & 1 & 0
\end{array}
\right]
$$

Pivot on the 3 in row 1, column 1.

$$
\begin{array}{c}
\\
-2R_1 + 3R_2 \rightarrow R_2 \\
5R_1 + R_3 \rightarrow R_3
\end{array}
\left[
\begin{array}{ccccc|c}
x_1 & x_2 & s_1 & s_2 & z & \\
3 & 2 & 1 & 0 & 0 & 25 \\
0 & \boxed{8} & -2 & 3 & 0 & 70 \\
\hline
0 & -5 & 5 & 0 & 1 & 125
\end{array}
\right]
$$

Pivot on the 8 in row 2, column 2.

$$
\begin{array}{c}
-R_2 + 4R_1 \rightarrow R_1 \\
\\
5R_2 + 8R_3 \rightarrow R_3
\end{array}
\left[
\begin{array}{ccccc|c}
x_1 & x_2 & s_1 & s_2 & z & \\
12 & 0 & 6 & -3 & 0 & 30 \\
0 & 8 & -2 & 3 & 0 & 70 \\
\hline
0 & 0 & 30 & 15 & 8 & 1350
\end{array}
\right]
$$

Create a 1 in the columns corresponding to x_1, x_2, and z.

$$
\begin{array}{c}
\frac{1}{12}R_1 \rightarrow R_1 \\
\frac{1}{8}R_2 \rightarrow R_2 \\
\\
\frac{1}{8}R_3 \rightarrow R_3
\end{array}
\left[
\begin{array}{ccccc|c}
x_1 & x_2 & s_1 & s_2 & z & \\
1 & 0 & \frac{1}{2} & -\frac{1}{4} & 0 & \frac{5}{2} \\
0 & 1 & -\frac{1}{4} & \frac{3}{8} & 0 & \frac{35}{4} \\
\hline
0 & 0 & \frac{15}{4} & \frac{15}{8} & 1 & \frac{675}{4}
\end{array}
\right]
$$

The minimum cost is $\frac{675}{4}$, which is 169¢ or $1.69, when $y_1 = \frac{15}{4} = 3\frac{3}{4}$ and $y_2 = \frac{15}{8} = 1\frac{7}{8}$, that is, when $3\frac{3}{4}$ servings of A and $1\frac{7}{8}$ servings of B are used.

4.4   Nonstandard Problems

1. $2x_1 + 3x_2 \leq 8$
$x_1 + 4x_2 \geq 7$

Introduce the slack variable s_1 and the surplus variable s_2 to obtain the following equations:

$$2x_1 + 3x_2 + s_1 \quad\quad = 8$$
$$x_1 + 4x_2 \quad\quad - s_2 = 7.$$

3. $x_1 + x_2 + x_3 \leq 100$
$x_1 + x_2 + x_3 \geq \ 75$
$x_1 + x_2 \quad\quad\ \geq \ 27$

Introduce the slack variable s_1 and the surplus variables s_2 and s_3 to obtain the following equations:

$$x_1 + x_2 + x_3 + s_1 \quad\quad\quad = 100$$
$$x_1 + x_2 + x_3 \quad - s_2 \quad\quad = \ 75$$
$$x_1 + x_2 \quad\quad\quad\quad - s_3 = \ 27.$$

5. Minimize $\quad w = 4y_1 + 3y_2 + 2y_3$
subject to: $\quad y_1 + y_2 + \ y_3 \geq \ 5$
$\quad\quad\quad\quad\ y_1 + y_2 \quad\quad\ \geq \ 4$
$\quad\quad\quad\quad\ 2y_1 + y_2 + 3y_3 \geq 15$
with $\quad\quad\quad y_1 \geq 0, y_2 \geq 0, y_3 \geq 0.$

Change this to a maximization problem by letting $z = -w$. The problem can now be stated equivalently as follows:

Maximize $\quad z = -4y_1 - 3y_2 - 2y_3$
subject to: $\quad y_1 + y_2 + \ y_3 \geq \ 5$
$\quad\quad\quad\quad\ y_1 + y_2 \quad\quad\ \geq \ 4$
$\quad\quad\quad\quad\ 2y_1 + y_2 + 3y_3 \geq 15$
with $\quad\quad\quad y_1 \geq 0, y_2 \geq 0, y_3 \geq 0.$

7. Minimize $\quad w = y_1 + 2y_2 + y_3 + 5y_4$
subject to: $\quad y_1 + y_2 + \ y_3 + y_4 \geq \ 50$
$\quad\quad\quad\quad\ 3y_1 + y_2 + 2y_3 + y_4 \geq 100$
with $\quad\quad\quad y_1 \geq 0, y_2 \geq 0, y_3 \geq 0, y_4 \geq 0.$

Change this to a maximization problem by letting $z = -w$. The problem can now be stated equivalently as follows:

Maximize $\quad z = -y_1 - 2y_2 - y_3 - 5y_4$
subject to: $\quad y_1 + y_2 + \ y_3 + y_4 \geq \ 50$
$\quad\quad\quad\quad\ 3y_1 + y_2 + 2y_3 + y_4 \geq 100$
with $\quad\quad\quad y_1 \geq 0, y_2 \geq 0, y_3 \geq 0, y_4 \geq 0.$

9. Find $x_1 \geq 0$ and $x_2 \geq 0$ such that

$$x_1 + 2x_2 \geq 24$$
$$x_1 + \ x_2 \leq 40$$

and $z = 12x_1 + 10x_2$ is maximized.

Subtracting the surplus variable s_1 and adding the slack variable s_2 leads to the equations

$$x_1 + 2x_2 - s_1 \quad\quad = 24$$
$$x_1 + \ x_2 \quad\quad + s_2 = 40.$$

The initial simplex tableau is as follows.

$$
\begin{array}{ccccc|c}
x_1 & x_2 & s_1 & s_2 & z & \\
\hline
\boxed{1} & 2 & -1 & 0 & 0 & 24 \\
1 & 1 & 0 & 1 & 0 & 40 \\
\hline
-12 & -10 & 0 & 0 & 1 & 0
\end{array}
$$

The initial basic solution is not feasible since $s_1 = -24$ is negative, so row transformations must be used. Pivot on the 1 in row 1, column 1, since it is the positive entry that is farthest to the left in the first row (the row containing the -1) and since, in the first column, $\frac{24}{1} = 24$ is a smaller quotient than $\frac{40}{1} = 40$. After row transformations, we obtain the following tableau.

$$
\begin{array}{r}
\\
-R_1 + R_2 \to R_2 \\
\\
12R_1 + R_3 \to R_3
\end{array}
\begin{array}{ccccc|c}
x_1 & x_2 & s_1 & s_2 & z & \\
\hline
1 & 2 & -1 & 0 & 0 & 24 \\
0 & -1 & \boxed{1} & 1 & 0 & 16 \\
\hline
0 & 14 & -12 & 0 & 1 & 288
\end{array}
$$

The basic solution is now feasible, but the problem is not yet finished since there is a negative indicator. Continue in the usual way. The 1 in column 3 is the next pivot. After row transformations, we get the following tableau.

$$
\begin{array}{r}
R_1 + R_2 \to R_1 \\
\\
12R_2 + R_3 \to R_3
\end{array}
\begin{array}{ccccc|c}
x_1 & x_2 & s_1 & s_2 & z & \\
\hline
1 & 1 & 0 & 1 & 0 & 40 \\
0 & -1 & 1 & 1 & 0 & 16 \\
\hline
0 & 2 & 0 & 12 & 1 & 480
\end{array}
$$

This is a final tableau since the entries in the last row are all nonnegative. The maximum value is 480 when $x_1 = 40$ and $x_2 = 0$.

11. Find $x_1 \geq 0, x_2 \geq 0$, and $x_3 \geq 0$ such that

$$x_1 + x_2 + x_3 \leq 150$$
$$x_1 + x_2 + x_3 \geq 100$$

and $z = 2x_1 + 5x_2 + 3x_3$ is maximized.

The initial tableau is as follows.

$$\begin{array}{cccccc|c} x_1 & x_2 & x_3 & s_1 & s_2 & z & \\ 1 & 1 & 1 & 1 & 0 & 0 & 150 \\ 1 & \boxed{1} & 1 & 0 & -1 & 0 & 100 \\ \hline -2 & -5 & -3 & 0 & 0 & 1 & 0 \end{array}$$

Note that s_1 is a slack variable, while s_2 is a surplus variable. The initial basic solution is not feasible, since $s_2 = -100$ is negative. Pivot on the 1 in row 2, column 2. (Note that any column with a nonnegative entry in the second row could be used as the pivot column.)

$$\begin{array}{c} -R_2+R_1 \to R_1 \\ \\ 5R_2+R_3 \to R_3 \end{array}\begin{array}{cccccc|c} x_1 & x_2 & x_3 & s_1 & s_2 & z & \\ 0 & 0 & 0 & 1 & \boxed{1} & 0 & 50 \\ 1 & 1 & 1 & 0 & -1 & 0 & 100 \\ \hline 3 & 0 & 2 & 0 & -5 & 1 & 500 \end{array}$$

The basic solution is now feasible but there is a negative indicator. Pivot on the 1 in row 1, column 5.

$$\begin{array}{c} \\ R_1+R_2 \to R_2 \\ 5R_1+R_3 \to R_3 \end{array}\begin{array}{cccccc|c} x_1 & x_2 & x_3 & s_1 & s_2 & z & \\ 0 & 0 & 0 & 1 & 1 & 0 & 50 \\ 1 & 1 & 1 & 1 & 0 & 0 & 150 \\ \hline 3 & 0 & 2 & 5 & 0 & 1 & 750 \end{array}$$

This is a final tableau. The maximum value is 750 when $x_1 = 0$, $x_2 = 150$, and $x_3 = 0$.

13. Find $x_1 \geq 0$ and $x_2 \geq 0$ such that

$$x_1 + x_2 \leq 100$$
$$x_1 + x_2 \geq 50$$
$$2x_1 + x_2 \leq 110$$

and $z = -2x_1 + 3x_2$ is maximized.

The initial tableau is as follows.

$$\begin{array}{cccccc|c} x_1 & x_2 & s_1 & s_2 & s_3 & z & \\ 1 & 1 & 1 & 0 & 0 & 0 & 100 \\ 1 & \boxed{1} & 0 & -1 & 0 & 0 & 50 \\ 2 & 1 & 0 & 0 & 1 & 0 & 110 \\ \hline 2 & -3 & 0 & 0 & 0 & 1 & 0 \end{array}$$

The initial basic solution is not feasible since $s_2 = -50$. Pivot on the 1 in row 2, column 2.

$$\begin{array}{c} -R_2+R_1 \to R_1 \\ \\ -R_2+R_3 \to R_3 \\ 3R_2+R_4 \to R_4 \end{array}\begin{array}{cccccc|c} x_1 & x_2 & s_1 & s_2 & s_3 & z & \\ 0 & 0 & 1 & \boxed{1} & 0 & 0 & 50 \\ 1 & 1 & 0 & -1 & 0 & 0 & 50 \\ 1 & 0 & 0 & 1 & 1 & 0 & 60 \\ 5 & 0 & 0 & -3 & 0 & 1 & 150 \end{array}$$

The basic solution is now feasible, but there is a negative indicator. Pivot on the 1 in row 1, column 4. (It would also be acceptable to pivot on the 1 in row 2, column 1.)

$$\begin{array}{c} \\ -R_1+R_2 \to R_2 \\ -R_1+R_3 \to R_3 \\ 3R_1+R_4 \to R_4 \end{array}\begin{array}{cccccc|c} x_1 & x_2 & s_1 & s_2 & s_3 & z & \\ 0 & 0 & 1 & 1 & 0 & 0 & 50 \\ 1 & 1 & 1 & 0 & 0 & 0 & 100 \\ 1 & 0 & -1 & 0 & 1 & 0 & 10 \\ 5 & 0 & 3 & 0 & 0 & 1 & 300 \end{array}$$

This is a final tableau. The maximum value is 300 when $x_1 = 0$ and $x_2 = 100$.

15. Find $y_1 \geq 0$ and $y_2 \geq 0$ such that

$$10y_1 + 15y_2 \leq 150$$
$$20y_1 + 5y_2 \geq 100$$

and $w = 4y_1 + 5y_2$ is minimized.

Let $z = -w = -4y_1 - 5y_2$. Maximize z. The initial tableau is as follows.

$$\begin{array}{ccccc|c} y_1 & y_2 & s_1 & s_2 & z & \\ 10 & 15 & 1 & 0 & 0 & 150 \\ \boxed{20} & 5 & 0 & -1 & 0 & 100 \\ \hline 4 & 5 & 0 & 0 & 1 & 0 \end{array}$$

The solution is not feasible since $s_2 = -100$. Pivot on the 20 in row 2, column 1.

$$\begin{array}{c} -R_2+2R_1 \to R_1 \\ \\ -R_2+5R_3 \to R_3 \end{array}\begin{array}{ccccc|c} y_1 & y_2 & s_1 & s_2 & z & \\ 0 & \boxed{25} & 2 & 1 & 0 & 200 \\ 20 & 5 & 0 & -1 & 0 & 100 \\ 0 & 20 & 0 & 1 & 5 & -100 \end{array}$$

Create a 1 in the columns corresponding to y_1, s_1, and z.

$$\begin{array}{c} \frac{1}{2}R_1 \to R_1 \\ \frac{1}{20}R_2 \to R_2 \\ \\ \frac{1}{5}R_3 \to R_3 \end{array}\begin{array}{ccccc|c} y_1 & y_2 & s_1 & s_2 & z & \\ 0 & \frac{25}{2} & 1 & \frac{1}{2} & 0 & 100 \\ 1 & \frac{1}{4} & 0 & -\frac{1}{20} & 0 & 5 \\ 0 & 4 & 0 & \frac{1}{5} & 1 & -20 \end{array}$$

This is a final tableau. The minimum value is 20 when $y_1 = 5$ and $y_2 = 0$.

17. Maximize $z = 3x_1 + 2x_2$
subject to: $x_1 + x_2 = 50$
$4x_1 + 2x_2 \geq 120$
$5x_1 + 2x_2 \leq 200$
with $x_1 \geq 0, x_2 \geq 0$.

The artificial variable a_1 is used to rewrite $x_1 + x_2 = 50$ as $x_1 + x_2 + a_1 = 50$; note that a_1 must equal 0 for this equation to be a true statement. Also the surplus variable s_1 and the slack variable s_2 are needed. The initial tableau is as follows.

$$
\begin{array}{cccccc|c}
x_1 & x_2 & a_1 & s_1 & s_2 & z & \\
1 & 1 & 1 & 0 & 0 & 0 & 50 \\
\boxed{4} & 2 & 0 & -1 & 0 & 0 & 120 \\
5 & 2 & 0 & 0 & 1 & 0 & 200 \\
\hline
-3 & -2 & 0 & 0 & 0 & 1 & 0
\end{array}
$$

The initial basic solution is not feasible. Pivot on the 4 in row 2, column 1.

$$
\begin{array}{c}
-R_2 + 4R_1 \to R_1 \\
\\
-5R_2 + 4R_3 \to R_3 \\
\\
3R_2 + 4R_4 \to R_4
\end{array}
\begin{array}{cccccc|c}
x_1 & x_2 & a_1 & s_1 & s_2 & z & \\
0 & 2 & 4 & 1 & 0 & 0 & 80 \\
4 & 2 & 0 & -1 & 0 & 0 & 120 \\
0 & -2 & 0 & \boxed{5} & 4 & 0 & 200 \\
\hline
0 & -2 & 0 & -3 & 0 & 4 & 360
\end{array}
$$

The basic solution is now feasible, but there are negative indicators. Pivot on the 5 in row 3, column 4 (which is the column with the most negative indicator and the row with the smallest non-negative quotient).

$$
\begin{array}{c}
-R_3 + 5R_1 \to R_1 \\
R_3 + 5R_2 \to R_2 \\
\\
3R_3 + 5R_4 \to R_4
\end{array}
\begin{array}{cccccc|c}
x_1 & x_2 & a_1 & s_1 & s_2 & z & \\
0 & \boxed{12} & 20 & 0 & -4 & 0 & 200 \\
20 & 8 & 0 & 0 & 4 & 0 & 800 \\
0 & -2 & 0 & 5 & 4 & 0 & 200 \\
\hline
0 & -16 & 0 & 0 & 12 & 20 & 2400
\end{array}
$$

Pivot on the 12 in row 1, column 2.

$$
\begin{array}{c}
\\
-2R_1 + 3R_2 \to R_2 \\
R_1 + 6R_3 \to R_3 \\
4R_1 + 3R_4 \to R_4
\end{array}
\begin{array}{cccccc|c}
x_1 & x_2 & a_1 & s_1 & s_2 & z & \\
0 & 12 & 20 & 0 & -4 & 0 & 200 \\
60 & 0 & -40 & 0 & 20 & 0 & 2000 \\
0 & 0 & 20 & 30 & 20 & 0 & 1400 \\
\hline
0 & 0 & 80 & 0 & 20 & 60 & 8000
\end{array}
$$

We now have $a_1 = 0$, so drop the a_1 column.

$$
\begin{array}{ccccc|c}
x_1 & x_2 & s_1 & s_2 & z & \\
0 & 12 & 0 & -4 & 0 & 200 \\
60 & 0 & 0 & 20 & 0 & 2000 \\
0 & 0 & 30 & 20 & 0 & 1400 \\
\hline
0 & 0 & 0 & 20 & 60 & 8000
\end{array}
$$

We are finished pivoting. Create a 1 in the columns corresponding to $x_1, x_2, s_1,$ and z.

$$
\begin{array}{c}
\frac{1}{12}R_1 \to R_1 \\
\frac{1}{60}R_2 \to R_2 \\
\frac{1}{30}R_3 \to R_3 \\
\frac{1}{60}R_4 \to R_4
\end{array}
\begin{array}{ccccc|c}
x_1 & x_2 & s_1 & s_2 & z & \\
0 & 1 & 0 & -\frac{1}{3} & 0 & 16\frac{2}{3} \\
1 & 0 & 0 & \frac{1}{3} & 0 & 33\frac{1}{3} \\
0 & 0 & 1 & \frac{2}{3} & 0 & 46\frac{2}{3} \\
\hline
0 & 0 & 0 & \frac{1}{3} & 1 & 133\frac{1}{3}
\end{array}
$$

The maximum value is $133\frac{1}{3}$ when $x_1 = 33\frac{1}{3}$ and $x_2 = 16\frac{2}{3}$.

19. Minimize $w = 32y_1 + 40y_2$
subject to: $20y_1 + 10y_2 = 200$
$25y_1 + 40y_2 \leq 500$
$18y_1 + 24y_2 \geq 300$
with $y_1 \geq 0, y_2 \geq 0$.

With artificial, slack, and surplus variables, this problem becomes

Maximize $z = -32y_1 - 40y_2$
subject to: $20y_1 + 10y_2 + a_1 \qquad\qquad = 200$
$25y_1 + 40y_2 \qquad + s_1 \qquad = 500$
$18y_1 + 24y_2 \qquad\qquad - s_2 = 300.$

The initial tableau is as follows.

$$
\begin{array}{cccccc|c}
y_1 & y_2 & a_1 & s_1 & s_2 & z & \\
\boxed{20} & 10 & 1 & 0 & 0 & 0 & 200 \\
25 & 40 & 0 & 1 & 0 & 0 & 500 \\
18 & 24 & 0 & 0 & -1 & 0 & 300 \\
\hline
32 & 40 & 0 & 0 & 0 & 1 & 0
\end{array}
$$

The initial basic solution is not feasible. Pivot on the 20 in row 1, column 1.

$$
\begin{array}{c}
\\
-5R_1 + 4R_2 \to R_2 \\
-9R_1 + 10R_3 \to R_3 \\
-8R_1 + 5R_4 \to R_4
\end{array}
\begin{array}{cccccc|c}
y_1 & y_2 & a_1 & s_1 & s_2 & z & \\
20 & 10 & 1 & 0 & 0 & 0 & 200 \\
0 & 110 & -5 & 4 & 0 & 0 & 1000 \\
0 & 150 & -9 & 0 & -10 & 0 & 1200 \\
\hline
0 & 120 & -8 & 0 & 0 & 5 & -1600
\end{array}
$$

Eliminate the a_1 column.

$$
\begin{array}{ccccc|c}
y_1 & y_2 & s_1 & s_2 & z & \\
20 & 10 & 0 & 0 & 0 & 200 \\
0 & 110 & 4 & 0 & 0 & 1000 \\
0 & \boxed{150} & 0 & -10 & 0 & 1200 \\
\hline
0 & 120 & 0 & 0 & 5 & -1600
\end{array}
$$

Since $s_2 = -120$ is not feasible, pivot on the 150 in row 3, column 2.

$$
\begin{array}{c}
\\
-R_3 + 15R_1 \to R_1 \\
-11R_3 + 15R_2 \to R_2 \\
\\
-4R_3 + 5R_4 \to R_4
\end{array}
\begin{array}{c}
\begin{array}{ccccc}
y_1 & y_2 & s_1 & s_2 & z
\end{array} \\
\left[
\begin{array}{ccccc|c}
300 & 0 & 0 & 10 & 0 & 1800 \\
0 & 0 & 60 & 110 & 0 & 1800 \\
0 & 150 & 0 & -10 & 0 & 1200 \\
\hline
0 & 0 & 0 & 40 & 25 & -12,800
\end{array}
\right]
\end{array}
$$

$$
\begin{array}{c}
\\
\frac{1}{300}R_1 \to R_1 \\
\frac{1}{60}R_2 \to R_2 \\
\frac{1}{150}R_3 \to R_3 \\
\frac{1}{25}R_4 \to R_4
\end{array}
\begin{array}{c}
\begin{array}{ccccc}
y_1 & y_2 & s_1 & s_2 & z
\end{array} \\
\left[
\begin{array}{ccccc|c}
1 & 0 & 0 & \frac{1}{30} & 0 & 6 \\
0 & 0 & 1 & \frac{11}{6} & 0 & 30 \\
0 & 1 & 0 & -\frac{1}{15} & 0 & 8 \\
\hline
0 & 0 & 0 & \frac{8}{5} & 1 & -512
\end{array}
\right]
\end{array}
$$

The minimum value is 512 when $y_1 = 6$ and $y_2 = 8$.

23. Let $y_1 =$ amount shipped from S_1 to D_1,
  $y_2 =$ amount shipped from S_1 to D_2,
  $y_3 =$ amount shipped from S_2 to D_1,
and $y_4 =$ amount shipped from S_2 to D_2.

Minimize $w = 30y_1 + 20y_2 + 25y_3 + 22y_4$

subject to:
$$
\begin{aligned}
y_1 + y_3 &\geq 3000 \\
y_2 + y_4 &\geq 5000 \\
y_1 + y_2 &\leq 5000 \\
y_3 + y_4 &\leq 5000 \\
2y_1 + 6y_2 + 5y_3 + 4y_4 &\leq 40,000
\end{aligned}
$$
with $y_1 \geq 0, y_2 \geq 0, y_3 \geq 0, y_4 \geq 0.$

Maximize $z = -w = -30y_1 - 20y_2 - 25y_3 - 22y_4.$

$$
\begin{array}{c}
\begin{array}{ccccccccccc}
y_1 & y_2 & y_3 & y_4 & s_1 & s_2 & s_3 & s_4 & s_5 & z &
\end{array} \\
\left[
\begin{array}{cccccccccc|c}
\boxed{1} & 0 & 1 & 0 & -1 & 0 & 0 & 0 & 0 & 0 & 3000 \\
0 & 1 & 0 & 1 & 0 & -1 & 0 & 0 & 0 & 0 & 5000 \\
1 & 1 & 0 & 0 & 0 & 0 & 1 & 0 & 0 & 0 & 5000 \\
0 & 0 & 1 & 1 & 0 & 0 & 0 & 1 & 0 & 0 & 5000 \\
2 & 6 & 5 & 4 & 0 & 0 & 0 & 0 & 1 & 0 & 40,000 \\
\hline
30 & 20 & 25 & 22 & 0 & 0 & 0 & 0 & 0 & 1 & 0
\end{array}
\right]
\end{array}
$$

Pivot on the 1 in row 1, column 1 since the feasible solution has a negative value, $s_1 = -3000$.

$$
\begin{array}{c}
\\
\\
-R_1 + R_3 \to R_3 \\
\\
-2R_1 + R_5 \to R_5 \\
-30R_1 + R_6 \to R_6
\end{array}
\begin{array}{c}
\begin{array}{ccccccccccc}
y_1 & y_2 & y_3 & y_4 & s_1 & s_2 & s_3 & s_4 & s_5 & z &
\end{array} \\
\left[
\begin{array}{cccccccccc|c}
1 & 0 & 1 & 0 & -1 & 0 & 0 & 0 & 0 & 0 & 3000 \\
0 & \boxed{1} & 0 & 1 & 0 & -1 & 0 & 0 & 0 & 0 & 5000 \\
0 & 1 & -1 & 0 & 1 & 0 & 1 & 0 & 0 & 0 & 2000 \\
0 & 0 & 1 & 1 & 0 & 0 & 0 & 1 & 0 & 0 & 5000 \\
0 & 6 & 3 & 4 & 2 & 0 & 0 & 0 & 1 & 0 & 34,000 \\
\hline
0 & 20 & -5 & 22 & 30 & 0 & 0 & 0 & 0 & 1 & -90,000
\end{array}
\right]
\end{array}
$$

Since the feasible solution has a negative value ($s_2 = -5000$), pivot on the 1 in row 2, column 2.

	y_1	y_2	y_3	y_4	s_1	s_2	s_3	s_4	s_5	z	
	1	0	1	0	-1	0	0	0	0	0	3000
	0	1	0	1	0	-1	0	0	0	0	5000
$R_2 - R_3 \to R_3$	0	0	[1]	1	-1	-1	-1	0	0	0	3000
	0	0	1	1	0	0	0	1	0	0	5000
$-6R_2 + R_5 \to R_5$	0	0	3	-2	2	6	0	0	1	0	4000
$-20R_2 + R_6 \to R_6$	0	0	-5	2	30	20	0	0	0	1	-190,000

Since the feasible solution has a negative value ($s_3 = -3000$), pivot on the 1 in row 3, column 3.

	y_1	y_2	y_3	y_4	s_1	s_2	s_3	s_4	s_5	z	
$-R_3 + R_1 \to R_1$	1	0	0	-1	0	1	1	0	0	0	0
	0	1	0	1	0	-1	0	0	0	0	5000
	0	0	1	1	-1	-1	-1	0	0	0	3000
$-R_3 + R_4 \to R_4$	0	0	0	0	1	1	1	1	0	0	2000
$-3R_3 + R_5 \to R_5$	0	0	0	[-5]	5	9	3	0	1	0	-5000
$5R_3 + R_6 \to R_6$	0	0	0	7	25	15	-5	0	0	1	-175,000

Since the feasible solution has a negative value ($s_5 = -5000$), pivot on the -5 in row 5, column 4.

	y_1	y_2	y_3	y_4	s_1	s_2	s_3	s_4	s_5	z	
$R_5 + (-5R_1) \to R_1$	-5	0	0	0	5	4	-2	0	1	0	-5000
$R_5 + 5R_2 \to R_2$	0	5	0	0	5	4	3	0	1	0	20,000
$R_5 + 5R_3 \to R_3$	0	0	5	0	0	4	-2	0	1	0	10,000
	0	0	0	0	1	1	[1]	1	0	0	2000
	0	0	0	-5	5	9	3	0	1	0	-5000
$7R_5 + 5R_6 \to R_6$	0	0	0	0	160	138	-4	0	7	5	-910,000

Now that the basic feasible solution has all nonnegative values, pivot on the 1 in row 4, column 7.

	y_1	y_2	y_3	y_4	s_1	s_2	s_3	s_4	s_5	z	
$2R_4 + R_1 \to R_1$	-5	0	0	0	7	6	0	2	1	0	-1000
$-3R_4 + R_2 \to R_2$	0	5	0	0	2	1	0	-3	1	0	14,000
$2R_4 + R_3 \to R_3$	0	0	5	0	2	6	0	2	1	0	14,000
	0	0	0	0	1	1	1	1	0	0	2000
$-3R_4 + R_5 \to R_5$	0	0	0	-5	2	6	0	-3	1	0	-11,000
$4R_4 + R_6 \to R_6$	0	0	0	0	164	142	0	4	7	5	-902,000

Create a 1 in the columns corresponding to y_1, y_2, y_3, y_4, and z.

	y_1	y_2	y_3	y_4	s_1	s_2	s_3	s_4	s_5	z	
$-\frac{1}{5}R_1 \to R_1$	1	0	0	0	$-\frac{7}{5}$	$-\frac{6}{5}$	0	$-\frac{2}{5}$	$-\frac{1}{5}$	0	200
$\frac{1}{5}R_2 \to R_2$	0	1	0	0	$\frac{2}{5}$	$\frac{1}{5}$	0	$-\frac{3}{5}$	$\frac{1}{5}$	0	2800
$\frac{1}{5}R_3 \to R_3$	0	0	1	0	$\frac{2}{5}$	$\frac{6}{5}$	0	$\frac{2}{5}$	$\frac{1}{5}$	0	2800
	0	0	0	0	1	1	1	1	0	0	2000
$-\frac{1}{5}R_5 \to R_5$	0	0	0	1	$-\frac{2}{5}$	$-\frac{6}{5}$	0	$\frac{3}{5}$	$-\frac{1}{5}$	0	2200
$\frac{1}{5}R_6 \to R_6$	0	0	0	0	$\frac{164}{5}$	$\frac{142}{5}$	0	$\frac{4}{5}$	$\frac{7}{5}$	1	$-180{,}400$

Here, $y_1 = 200, y_2 = 2800, y_3 = 2800, y_4 = 2200$, and $z = -w = 180{,}400$. So, ship 200 barrels of oil from supplier S_1 to distributor D_1. Ship 2800 barrels of oil from supplier S_1 to distributor D_2. Ship 2800 barrels of oil from supplier S_2 to distributor D_1. Ship 2200 barrels of oil from supplier S_2 to distributor D_2. The minimum cost is $180{,}400.

25. Let $x_1 =$ the number of million dollars for home loans
and $x_2 =$ the number of million dollars for commercial loans.

$$\text{Maximize} \quad z = .12x_1 + .10x_2$$
$$\text{subject to:} \quad x_1 \qquad\;\; \ge 4x_2 \text{ or } x_1 - 4x_2 \ge 0$$
$$x_1 + \;\; x_2 \ge 10$$
$$3x_1 + 2x_2 \le 72$$
$$x_1 + \;\; x_2 \le 25$$
$$\text{with} \qquad x_1 \ge 0, x_2 \ge 0.$$

x_1	x_2	s_1	s_2	s_3	s_4	z	
1	-4	-1	0	0	0	0	0
1	1	0	-1	0	0	0	10
3	2	0	0	1	0	0	72
1	1	0	0	0	1	0	25
$-.12$	$-.10$	0	0	0	0	1	0

Since the basic feasible solution has a negative value ($s_2 = -10$), pivot on the 1 in row 2, column 1.

	x_1	x_2	s_1	s_2	s_3	s_4	z	
$R_2 - R_1 \to R_1$	0	5	1	-1	0	0	0	10
	1	1	0	-1	0	0	0	10
$-3R_2 + R_3 \to R_3$	0	-1	0	3	1	0	0	42
$-R_2 + R_4 \to R_4$	0	0	0	1	0	1	0	15
$.12R_2 + R_5 \to R_5$	0	.02	0	$-.12$	0	0	1	1.2

Since we now have a basic feasible solution with all nonnegative values, we pivot as usual on the 3 in row 3, column 4.

	x_1	x_2	s_1	s_2	s_3	s_4	z	
$R_3 + 3R_1 \to R_1$	0	14	3	0	1	0	0	72
$R_3 + 3R_2 \to R_2$	3	2	0	0	1	0	0	72
	0	-1	0	3	1	0	0	42
$-3R_4 + R_3 \to R_4$	0	-1	0	0	1	-3	0	-3
$.12R_3 + 3R_5 \to R_5$	0	$-.06$	0	0	.12	0	3	8.64

Pivot on the -1 in row 4, column 2.

$$
\begin{array}{l}
 \\
14R_4 + R_1 \to R_1 \\
2R_4 + R_2 \to R_2 \\
-R_4 + R_3 \to R_3 \\
 \\
-.06R_4 + R_5 \to R_5
\end{array}
\begin{array}{c}
\begin{array}{ccccccc}
x_1 & x_2 & s_1 & s_2 & s_3 & s_4 & z
\end{array} \\
\left[
\begin{array}{ccccccc|c}
0 & 0 & 3 & 0 & 15 & -42 & 0 & 30 \\
3 & 0 & 0 & 0 & 3 & -6 & 0 & 66 \\
0 & 0 & 0 & 3 & 0 & 3 & 0 & 45 \\
0 & -1 & 0 & 0 & 1 & -3 & 0 & -3 \\
\hline
0 & 0 & 0 & 0 & .06 & .18 & 3 & 8.82
\end{array}
\right]
\end{array}
$$

Create a 1 in the columns corresponding to $x_1, x_2, s_1, s_2,$ and z.

$$
\begin{array}{l}
\frac{1}{3}R_1 \to R_1 \\
\frac{1}{3}R_2 \to R_2 \\
\frac{1}{3}R_3 \to R_3 \\
-R_4 \to R_4 \\
\frac{1}{3}R_5 \to R_5
\end{array}
\begin{array}{c}
\begin{array}{ccccccc}
x_1 & x_2 & s_1 & s_2 & s_3 & s_4 & z
\end{array} \\
\left[
\begin{array}{ccccccc|c}
0 & 0 & 1 & 0 & 5 & -14 & 0 & 10 \\
1 & 0 & 0 & 0 & 1 & -2 & 0 & 22 \\
0 & 0 & 0 & 1 & 0 & 1 & 0 & 15 \\
0 & 1 & 0 & 0 & -1 & 3 & 0 & 3 \\
\hline
0 & 0 & 0 & 0 & .02 & .06 & 1 & 2.94
\end{array}
\right]
\end{array}
$$

Here, $x_1 = 22, x_2 = 3$, and $z = 2.94$. Make \$22 million (\$22,000,000) in home loans and \$3 million (\$3,000,000) in commercial loans for a maximum return of \$2.94 million, or \$2,940,000.

27. Let $x_1 =$ the number of pounds of bluegrass seed,
     $x_2 =$ the number of pounds of rye seed,
and $x_3 =$ the number of pounds of Bermuda seed.

Minimize $w = 12x_1 + 15x_2 + 5x_3$

subject to: $x_1 + x_2 + x_3 = 5000$

$$x_3 \le \tfrac{2}{3}x_2 \text{ or } -\tfrac{2}{3}x_2 + x_3 \le 0$$

$$x_1 \ge .2(x_1 + x_2 + x_3) \text{ or } .8x_1 - .2x_2 - .2x_3 \ge 0$$

with $x_1 \ge 0, x_2 \ge 0, x_3 \ge 0.$

Maximize $z = -w = -12x_1 - 15x_2 - 5x_3.$

$$
\begin{array}{c}
\begin{array}{ccccccc}
x_1 & x_2 & x_3 & a_1 & s_1 & s_2 & z
\end{array} \\
\left[
\begin{array}{ccccccc|c}
\boxed{1} & 1 & 1 & 1 & 0 & 0 & 0 & 5000 \\
0 & -\frac{2}{3} & 1 & 0 & 1 & 0 & 0 & 0 \\
.8 & -.2 & -.2 & 0 & 0 & -1 & 0 & 0 \\
\hline
12 & 15 & 5 & 0 & 0 & 0 & 1 & 0
\end{array}
\right]
\end{array}
$$

Pivot on the 1 in row 1, column 1.

$$
\begin{array}{l}
 \\
 \\
-.8R_1 + R_3 \to R_3 \\
-12R_1 + R_4 \to R_4
\end{array}
\begin{array}{c}
\begin{array}{ccccccc}
x_1 & x_2 & x_3 & a_1 & s_1 & s_2 & z
\end{array} \\
\left[
\begin{array}{ccccccc|c}
1 & 1 & 1 & 1 & 0 & 0 & 0 & 5000 \\
0 & -\frac{2}{3} & 1 & 0 & 1 & 0 & 0 & 0 \\
0 & -1 & -1 & -.8 & 0 & -1 & 0 & -4000 \\
\hline
0 & 3 & -7 & -12 & 0 & 0 & 1 & -60,000
\end{array}
\right]
\end{array}
$$

Since $a_1 = 0$, we can drop the a_1 column.

$$
\begin{array}{c}
\begin{array}{cccccc}
x_1 & x_2 & x_3 & s_1 & s_2 & z
\end{array} \\
\left[
\begin{array}{cccccc|c}
1 & 1 & 1 & 0 & 0 & 0 & 5000 \\
0 & -\frac{2}{3} & 1 & 1 & 0 & 0 & 0 \\
0 & -1 & \boxed{-1} & 0 & -1 & 0 & -4000 \\
\hline
0 & 3 & -7 & 0 & 0 & 1 & -60,000
\end{array}
\right]
\end{array}
$$

Pivot on the -1 in row 3, column 3.

$$
\begin{array}{c}
\\
R_3 + R_1 \to R_1 \\
R_3 + R_2 \to R_2 \\
\\
\\
-7R_3 + R_4 \to R_4
\end{array}
\begin{array}{ccccccc}
x_1 & x_2 & x_3 & s_1 & s_2 & z & \\
\end{array}
\left[
\begin{array}{cccccc|c}
1 & 0 & 0 & 0 & -1 & 0 & 1000 \\
0 & \boxed{-\frac{5}{3}} & 0 & 1 & -1 & 0 & -4000 \\
0 & -1 & -1 & 0 & -1 & 0 & -4000 \\
0 & 10 & 0 & 0 & 7 & 1 & -32{,}000
\end{array}
\right]
$$

Since the basic solution is not feasible ($s_1 = -4000$), pivot on the $-\frac{5}{3}$ in row 2, column 2.

$$
\begin{array}{c}
\\
\\
R_2 + \left(-\frac{5}{3}R_3\right) \to R_3 \\
\\
10R_2 + \frac{5}{3}R_4 \to R_4
\end{array}
\begin{array}{ccccccc}
x_1 & x_2 & x_3 & s_1 & s_2 & z & \\
\end{array}
\left[
\begin{array}{cccccc|c}
1 & 0 & 0 & 0 & -1 & 0 & 1000 \\
0 & -\frac{5}{3} & 0 & 1 & -1 & 0 & -4000 \\
0 & 0 & \frac{5}{3} & 1 & \frac{2}{3} & 0 & \frac{8000}{3} \\
\hline
0 & 0 & 0 & 10 & \frac{5}{3} & \frac{5}{3} & -\frac{280{,}000}{3}
\end{array}
\right]
$$

Create a 1 in the columns corresponding to x_2, x_3, and z.

$$
\begin{array}{c}
\\
-\frac{3}{5}R_2 \to R_2 \\
\frac{3}{5}R_3 \to R_3 \\
\frac{3}{5}R_4 \to R_4
\end{array}
\begin{array}{cccccc}
x_1 & x_2 & x_3 & s_1 & s_2 & z \\
\end{array}
\left[
\begin{array}{cccccc|c}
1 & 0 & 0 & 0 & -1 & 0 & 1000 \\
0 & 1 & 0 & -\frac{3}{5} & \frac{3}{5} & 0 & 2400 \\
0 & 0 & 1 & \frac{3}{5} & \frac{2}{5} & 0 & 1600 \\
\hline
0 & 0 & 0 & 6 & 1 & 1 & -56{,}000
\end{array}
\right]
$$

Here, $x_1 = 1000, x_2 = 2400, x_3 = 1600$, and $z = -w = 56{,}000$. Therefore, use 1000 lb of bluegrass, 2400 lb of rye, and 1600 lb of Bermuda for a minimum cost of 56,000¢, that is, \$560.

29. Let $x_1 =$ the number of computers shipped from W_1 to D_1,

$x_2 =$ the number of computers shipped from W_1 to D_2,

$x_3 =$ the number of computers shipped from W_2 to D_1,

and $x_4 =$ the number of computers shipped from W_2 to D_2.

Minimize $w = 14x_1 + 12x_2 + 12x_3 + 10x_4$

subject to: $x_1 + x_3 \geq 32$

$x_2 + x_4 \geq 20$

$x_1 + x_2 \leq 25$

$x_3 + x_4 \leq 30$

with $x_1 \geq 0, x_2 \geq 0, x_3 \geq 0, x_4 \geq 0.$

Maximize $z = -w = -14x_1 - 12x_2 - 12x_3 - 10x_4.$

$$
\begin{array}{ccccccccc}
x_1 & x_2 & x_3 & x_4 & s_1 & s_2 & s_3 & s_4 & z \\
\end{array}
\left[
\begin{array}{ccccccccc|c}
\boxed{1} & 0 & 1 & 0 & -1 & 0 & 0 & 0 & 0 & 32 \\
0 & 1 & 0 & 1 & 0 & -1 & 0 & 0 & 0 & 20 \\
1 & 1 & 0 & 0 & 0 & 0 & 1 & 0 & 0 & 25 \\
0 & 0 & 1 & 1 & 0 & 0 & 0 & 1 & 0 & 30 \\
\hline
14 & 12 & 12 & 10 & 0 & 0 & 0 & 0 & 1 & 0
\end{array}
\right]
$$

Since the basic solution is not feasible ($s_1 = -32$), pivot on the 1 in row 1, column 1.

$$R_1 - R_3 \rightarrow R_3$$

$$-14R_1 + R_5 \rightarrow R_5$$

	x_1	x_2	x_3	x_4	s_1	s_2	s_3	s_4	z	
	1	0	1	0	-1	0	0	0	0	32
	0	1	0	1	0	-1	0	0	0	20
	0	-1	1	0	-1	0	-1	0	0	7
	0	0	1	1	0	0	0	1	0	30
	0	12	-2	10	14	0	0	0	1	-448

Pivot on the 1 in row 2, column 2 since $s_2 = -20$.

$$R_2 + R_3 \rightarrow R_3$$

$$-12R_2 + R_5 \rightarrow R_5$$

	x_1	x_2	x_3	x_4	s_1	s_2	s_3	s_4	z	
	1	0	1	0	-1	0	0	0	0	32
	0	1	0	1	0	-1	0	0	0	20
	0	0	1	1	-1	-1	-1	0	0	27
	0	0	1	1	0	0	0	1	0	30
	0	0	-2	-2	14	12	0	0	1	-688

Pivot on the 1 in row 3, column 3 since $s_3 = -27$.

$$-R_3 + R_1 \rightarrow R_1$$

$$-R_3 + R_4 \rightarrow R_4$$

$$2R_3 + R_5 \rightarrow R_5$$

	x_1	x_2	x_3	x_4	s_1	s_2	s_3	s_4	z	
	1	0	0	-1	0	1	1	0	0	5
	0	1	0	1	0	-1	0	0	0	20
	0	0	1	1	-1	-1	-1	0	0	27
	0	0	0	0	1	1	1	1	0	3
	0	0	0	0	12	10	-2	0	1	-634

Pivot on the 1 in row 4, column 7.

$$-R_4 + R_1 \rightarrow R_1$$

$$R_4 + R_3 \rightarrow R_3$$

$$2R_4 + R_5 \rightarrow R_5$$

	x_1	x_2	x_3	x_4	s_1	s_2	s_3	s_4	z	
	1	0	0	-1	-1	0	0	-1	0	2
	0	1	0	1	0	-1	0	0	0	20
	0	0	1	1	0	0	0	1	0	30
	0	0	0	0	1	1	1	1	0	3
	0	0	0	0	14	12	0	2	1	-628

Here, $x_1 = 2, x_2 = 20, x_3 = 30, x_4 = 0$, and $z = -w = 628$. Therefore, ship 2 computers from W_1 to D_1, 20 computers from W_1 to D_2, 30 computers from W_2 to D_1, and 0 computers from W_2 to D_2 for a minimum cost of \$628.

31. Let $y_1 = $ the number of ounces of ingredient I,

   $y_2 = $ the number of ounces of ingredient II,

and $y_3 = $ the number of ounces of ingredient III.

The problem is:

$$\begin{aligned}
\text{Minimize} \quad & w = .30y_1 + .09y_2 + .27y_3 \\
\text{subject to:} \quad & y_1 + y_2 + y_3 \geq 10 \\
& y_1 + y_2 + y_3 \leq 15 \\
& y_1 \qquad\qquad \geq \tfrac{1}{4}y_2 \\
& \qquad\qquad y_3 \geq y_1 \\
\text{with} \quad & y_1 \geq 0, y_2 \geq 0, y_3 \geq 0.
\end{aligned}$$

Use a graphing calculator or computer to find that the minimum is $w = 1.55$ when $y_1 = 1\frac{2}{3}, y_2 = 6\frac{2}{3}$, and $y_3 = 1\frac{2}{3}$. Therefore, the additive should consist of $1\frac{2}{3}$ oz of ingredient I, $6\frac{2}{3}$ oz of ingredient II, and $1\frac{2}{3}$ oz of ingredient III, for a minimum cost of \$1.55/gal. The amount of additive that should be used per gallon of gasoline is $1\frac{2}{3} + 6\frac{2}{3} + 1\frac{2}{3} = 10$ oz.

Chapter 4 Review Exercises

1. The simplex method should be used for problems with more than two variables or problems with two variables and many constants.

3. Maximize $\quad z = 5x_1 + 3x_2$

   subject to: $\quad 2x_1 + 5x_2 \le 50$

   $\qquad\qquad\quad x_1 + 3x_2 \le 25$

   $\qquad\qquad\, 4x_1 + \;\; x_2 \le 18$

   $\qquad\qquad\quad x_1 + \;\; x_2 \le 12$

   with $\qquad\quad x_1 \ge 0, x_2 \ge 0.$

   (a) Adding the slack variables s_1, s_2, s_3, and s_4, we obtain the following equations:

   $$2x_1 + 5x_2 + s_1 \qquad\qquad\quad = 50$$
   $$x_1 + 3x_2 \quad + s_2 \qquad\qquad = 25$$
   $$4x_1 + \;\, x_2 \qquad + s_3 \qquad = 18$$
   $$x_1 + \;\, x_2 \qquad\qquad + s_4 = 12.$$

   (b) The initial tableau is

   $$\begin{bmatrix}
   x_1 & x_2 & s_1 & s_2 & s_3 & s_4 & z & \\
   2 & 5 & 1 & 0 & 0 & 0 & 0 & 50 \\
   1 & 3 & 0 & 1 & 0 & 0 & 0 & 25 \\
   4 & 1 & 0 & 0 & 1 & 0 & 0 & 18 \\
   1 & 1 & 0 & 0 & 0 & 1 & 0 & 12 \\
   \hline
   -5 & -3 & 0 & 0 & 0 & 0 & 1 & 0
   \end{bmatrix}.$$

5. Maximize $\quad z = 5x_1 + 8x_2 + 6x_3$

   subject to: $\quad x_1 + \;\, x_2 + x_3 \le \;\; 90$

   $\qquad\qquad 2x_1 + 5x_2 + x_3 \le 120$

   $\qquad\qquad\; x_1 + 3x_2 \qquad\;\; \ge \;\; 80$

   with $\qquad x_1 \ge 0, x_2 \ge 0, x_3 \ge 0.$

   (a) Adding the slack variables s_1 and s_2 and subtracting the surplus variable s_3, we obtain the following equations:

   $$x_1 + \;\, x_2 + x_3 + s_1 \qquad\qquad = \;\; 90$$
   $$2x_1 + 5x_2 + x_3 \qquad + s_2 \qquad = 120$$
   $$x_1 + 3x_2 \qquad\qquad\qquad - s_3 = \;\; 80.$$

(b) The initial tableau

$$\begin{bmatrix}
x_1 & x_2 & x_3 & s_1 & s_2 & s_3 & z & \\
1 & 1 & 1 & 1 & 0 & 0 & 0 & 90 \\
2 & 5 & 1 & 0 & 1 & 0 & 0 & 120 \\
1 & 3 & 0 & 0 & 0 & -1 & 0 & 80 \\
\hline
-5 & -8 & -6 & 0 & 0 & 0 & 1 & 0
\end{bmatrix}.$$

7. $$\begin{bmatrix}
x_1 & x_2 & x_3 & s_1 & s_2 & z & \\
1 & 2 & 3 & 1 & 0 & 0 & 28 \\
\boxed{2} & 4 & 1 & 0 & 1 & 0 & 32 \\
\hline
-5 & -2 & -3 & 0 & 0 & 1 & 0
\end{bmatrix}$$

The most negative entry in the last row is -5, and the smaller of the two quotients is $\frac{32}{2} = 16$. Hence, the 2 in row 2, column 1, is the first pivot. Performing row transformations leads to the following tableau.

$$\begin{array}{c}
-R_2 + 2R_1 \to R_1 \\
\\
5R_2 + 2R_3 \to R_3
\end{array}
\begin{bmatrix}
x_1 & x_2 & x_3 & s_1 & s_2 & z & \\
0 & 0 & \boxed{5} & 2 & -1 & 0 & 24 \\
2 & 4 & 1 & 0 & 1 & 0 & 32 \\
\hline
0 & 16 & -1 & 0 & 5 & 2 & 160
\end{bmatrix}$$

Pivot on the 5 in row 1, column 3.

$$\begin{array}{c}
\\
-R_1 + 5R_2 \to R_2 \\
R_1 + 5R_3 \to R_3
\end{array}
\begin{bmatrix}
x_1 & x_2 & x_3 & s_1 & s_2 & z & \\
0 & 0 & 5 & 2 & -1 & 0 & 24 \\
10 & 20 & 0 & -2 & 6 & 0 & 136 \\
0 & 80 & 0 & 2 & 24 & 10 & 824
\end{bmatrix}$$

Create a 1 in the columns corresponding to x_1, x_3, and z.

$$\begin{array}{c}
\frac{1}{5}R_1 \to R_1 \\
\frac{1}{10}R_2 \to R_2 \\
\frac{1}{10}R_3 \to R_3
\end{array}
\begin{bmatrix}
x_1 & x_2 & x_3 & s_1 & s_2 & z & \\
0 & 0 & 1 & .4 & -.2 & 0 & 4.8 \\
1 & 2 & 0 & -.2 & .6 & 0 & 13.6 \\
0 & 8 & 0 & .2 & 2.4 & 1 & 82.4
\end{bmatrix}$$

The maximum value is 82.4 when $x_1 = 13.6, x_2 = 0, x_3 = 4.8, s_1 = 0$, and $s_2 = 0$.

9. $$\begin{bmatrix}
x_1 & x_2 & x_3 & s_1 & s_2 & s_3 & z & \\
1 & 2 & 2 & 1 & 0 & 0 & 0 & 50 \\
\boxed{3} & 1 & 0 & 0 & 1 & 0 & 0 & 20 \\
1 & 0 & 2 & 0 & 0 & -1 & 0 & 15 \\
\hline
-5 & -3 & -2 & 0 & 0 & 0 & 1 & 0
\end{bmatrix}$$

The initial basic solution is not feasible since $s_3 = -15$. In the third row where the negative coefficient appears, the nonnegative entry that appears farthest to the left is the 1 in the first column. In the first column, the smallest nonnegative quotient is $\frac{20}{3}$. Pivot on the 3 in row 2, column 1.

$$
\begin{array}{c}
 \\
-R_2 + 3R_1 \to R_1 \\
 \\
 \\
-R_2 + 3R_3 \to R_3 \\
5R_2 + 3R_4 \to R_4
\end{array}
\begin{array}{ccccccc|c}
x_1 & x_2 & x_3 & s_1 & s_2 & s_3 & z & \\
0 & 5 & 6 & 3 & -1 & 0 & 0 & 130 \\
3 & 1 & 0 & 0 & 1 & 0 & 0 & 20 \\
0 & -1 & \boxed{6} & 0 & -1 & -3 & 0 & 25 \\
\hline
0 & -4 & -6 & 0 & 5 & 0 & 3 & 100
\end{array}
$$

Continue by pivoting on each circled entry.

$$
\begin{array}{c}
 \\
-R_3 + R_2 \to R_1 \\
 \\
 \\
R_3 + R_4 \to R_4
\end{array}
\begin{array}{ccccccc|c}
x_1 & x_2 & x_3 & s_1 & s_2 & s_3 & z & \\
0 & \boxed{6} & 0 & 3 & 0 & 3 & 0 & 105 \\
3 & 1 & 0 & 0 & 1 & 0 & 0 & 20 \\
0 & -1 & 6 & 0 & -1 & -3 & 0 & 25 \\
\hline
0 & -5 & 0 & 0 & 4 & -3 & 3 & 125
\end{array}
$$

The basic solution is now feasible, but there are negative indicators.
Continue pivoting.

$$
\begin{array}{c}
 \\
-R_1 + 6R_2 \to R_2 \\
R_1 + 6R_3 \to R_3 \\
5R_1 + 6R_4 \to R_4
\end{array}
\begin{array}{ccccccc|c}
x_1 & x_2 & x_3 & s_1 & s_2 & s_3 & z & \\
0 & 6 & 0 & 3 & 0 & \boxed{3} & 0 & 105 \\
18 & 0 & 0 & -3 & 6 & -3 & 0 & 15 \\
0 & 0 & 36 & 3 & 0 & -15 & 0 & 255 \\
\hline
0 & 0 & 0 & 15 & 24 & -3 & 18 & 1275
\end{array}
$$

$$
\begin{array}{c}
 \\
R_1 + R_2 \to R_2 \\
5R_1 + R_3 \to R_3 \\
R_1 + R_4 \to R_4
\end{array}
\begin{array}{ccccccc|c}
x_1 & x_2 & x_3 & s_1 & s_2 & s_3 & z & \\
0 & 6 & 0 & 3 & 0 & 3 & 0 & 105 \\
18 & 6 & 0 & 0 & 6 & 0 & 0 & 120 \\
0 & 30 & 36 & 18 & 0 & 0 & 0 & 780 \\
\hline
0 & 6 & 0 & 18 & 24 & 0 & 18 & 1380
\end{array}
$$

Create a 1 in the columns corresponding to x_1, x_3, s_3, and z.

$$
\begin{array}{c}
\frac{1}{3}R_1 \to R_1 \\
\frac{1}{18}R_2 \to R_2 \\
\frac{1}{36}R_3 \to R_3 \\
\frac{1}{18}R_4 \to R_4
\end{array}
\begin{array}{ccccccc|c}
x_1 & x_2 & x_3 & s_1 & s_2 & s_3 & z & \\
0 & 2 & 0 & 1 & 0 & 1 & 0 & 35 \\
1 & .33 & 0 & 0 & .33 & 0 & 0 & 6.67 \\
0 & .83 & 1 & .5 & 0 & 0 & 0 & 21.67 \\
\hline
0 & .33 & 0 & 1 & 1.33 & 0 & 1 & 76.67
\end{array}
$$

The maximum value is about 76.67 when $x_1 \approx 6.67, x_2 = 0, x_3 \approx 21.67, s_1 = 0, s_2 = 0$, and $s_3 = 35$.

11. Minimize $\quad w = 10y_1 + 15y_2$

subject to: $\quad y_1 + y_2 \geq 17$

$5y_1 + 8y_2 \geq 42$

with $\qquad y_1 \geq 0, y_2 \geq 0.$

Let $z = -w$, and maximize $z = -10y_1 - 15y_2$ subject to the same constraints.

To form the dual, write the augmented matrix for the given problem.

$$
\begin{bmatrix}
1 & 1 & 17 \\
5 & 8 & 42 \\
\hline
10 & 15 & 0
\end{bmatrix}
$$

Form the transpose of this matrix.

$$\left[\begin{array}{cc|c} 1 & 5 & 10 \\ 1 & 8 & 15 \\ \hline 17 & 42 & 0 \end{array}\right]$$

Write the dual problem.

Maximize $\quad z = 17x_1 + 42x_2$

subject to: $\quad x_1 + 5x_2 \le 10$

$\qquad\qquad x_1 + 8x_2 \le 15$

with $\qquad\quad x_1 \ge 0, x_2 \ge 0.$

13. Minimize $\quad w = 7y_1 + 2y_2 + 3y_3$

subject to: $\quad y_1 + y_2 + 2y_3 \ge 48$

$\qquad\qquad y_1 + y_2 \qquad\ \ \ge 12$

$\qquad\qquad\qquad\quad\ \ y_3 \ge 10$

$\qquad\quad\ \ 3y_1 \qquad + y_3 \ge 30$

with $\qquad\ y_1 \ge 0, _2 \ge 0, y_3 \ge 0.$

Let $z = -w$, and maximize $z = -7y_1 - 2y_2 - 3y_3$ subject to the same constraints.

To form the dual, write the augmented matrix for the given problem.

$$\left[\begin{array}{ccc|c} 1 & 1 & 2 & 48 \\ 1 & 1 & 0 & 12 \\ 0 & 0 & 1 & 10 \\ 3 & 0 & 1 & 30 \\ \hline 7 & 2 & 3 & 0 \end{array}\right]$$

Form the transpose of this matrix.

$$\left[\begin{array}{cccc|c} 1 & 1 & 0 & 3 & 7 \\ 1 & 1 & 0 & 0 & 2 \\ 2 & 0 & 1 & 1 & 3 \\ \hline 48 & 12 & 10 & 30 & 0 \end{array}\right]$$

Write the dual problem.

Maximize $\quad z = 48x_1 + 12x_2 + 10x_3 + 30x_4$

subject to: $\quad x_1 + x_2 \qquad\quad + 3x_4 \le 7$

$\qquad\qquad x_1 + x_2 \qquad\qquad\quad\ \le 2$

$\qquad\qquad 2x_1 \qquad + x_3 + \ x_4 \le 3$

with $\qquad\ x_1 \ge 0, x_2 \ge 0, x_3 \ge 0, x_4 \ge 0.$

15.

$$\left[\begin{array}{ccccccc|c} y_1 & y_2 & s_1 & s_2 & s_3 & s_4 & z & \\ 0 & 0 & 3 & 0 & 1 & 1 & 0 & 2 \\ 1 & 0 & -2 & 0 & 2 & 0 & 0 & 8 \\ 0 & 1 & 7 & 0 & 0 & 0 & 0 & 12 \\ 0 & 0 & 1 & 1 & -4 & 0 & 0 & 1 \\ \hline 0 & 0 & 5 & 0 & 8 & 0 & 1 & -62 \end{array}\right]$$

From this final tableau, read that the maximum value of $z = -w$ is -62 when $y_1 = 8, y_2 = 12, s_1 = 0, s_2 = 1, s_3 = 0,$ and $s_4 = 2$. Therefore, the minimum value of w is 62 when $y_1 = 8, y_2 = 12, s_1 = 0, s_2 = 1, s_3 = 0,$ and $s_4 = 2$.

17. Any maximizing or minimizing problems can be solved using slack, surplus, and artificial variables. Slack variables are used in problems involving "$\le$" constraints. Surplus variables are used in problems involving "$\ge$" constraints. Artificial variables are used in problems involving "$=$" constraints.

19.

$$\left[\begin{array}{cccccc|c} 4 & 2 & 3 & 1 & 0 & 0 & 9 \\ 5 & 4 & 1 & 0 & 1 & 0 & 10 \\ \hline -6 & -7 & -5 & 0 & 0 & 1 & 0 \end{array}\right]$$

(a) The 1 in column 4 and the 1 in column 5 indicate that the constraints involve $\le$. The problem being solved with this tableau is:

Maximize $\quad z = 6x_1 + 7x_2 + 5x_3$

subject to: $\quad 4x_1 + 2x_2 + 3x_3 \le\ \ 9$

$\qquad\qquad 5x_1 + 4x_2 + \ x_3 \le 10$

with $\qquad\ x_1 \ge 0, x_2 \ge 0, x_3 \ge 0.$

(b) If the 1 in row 1, column 4 was -1 rather than 1, then the first constraint would have a surplus variable rather than a slack variable, which means the first constraint would be $4x_1 + 2x_2 + 3x_3 \ge 9$ instead of $4x_1 + 2x_2 + 3x_3 \le 9$.

(c)

$$\left[\begin{array}{cccccc|c} x_1 & x_2 & x_3 & s_1 & s_2 & z & \\ 3 & 0 & 5 & 2 & -1 & 0 & 8 \\ 11 & 10 & 0 & -1 & 3 & 0 & 21 \\ \hline 47 & 0 & 0 & 13 & 11 & 10 & 227 \end{array}\right]$$

From this tableau, the solution is $x_1 = 0, x_2 = \frac{21}{10} = 2.1, x_3 = \frac{8}{5} = 1.6,$ and $z = \frac{227}{10} = 22.7.$

(d) The dual of the original problem is as follows:

Minimize $\quad w = 9y_1 + 10y_2$

subject to: $\quad 4y_1 + 5y_2 \ge 6$

$\qquad\qquad 2y_1 + 4y_2 \ge 7$

$\qquad\qquad 3y_1 + \ y_2 \ge 5$

with $\qquad\ y_1 \ge 0, y_2 \ge 0.$

(e) From the tableau in part (c), the solution of the dual in part (d) is $y_1 = \frac{13}{10} = 1.3, y_2 = \frac{11}{10} = 1.1,$ and $w = \frac{227}{10} = 22.7.$

21. The information is contained in the following table.

	A	B	C	Maximum Available
Buy	5	3	6	1200
Sell	1	2	2	800
Deliver	2	1	5	500
Profit	4	3	3	

(a) Let $x_1 =$ number of item A,
   $x_2 =$ number of item B,
and $x_3 =$ number of item C.

(b) The objective function is

$$z = 4x_1 + 3x_2 + 3x_3.$$

(c) The constraints are

$$5x_1 + 3x_2 + 6x_3 \leq 1200$$
$$x_1 + 2x_2 + 2x_3 \leq \ 800$$
$$2x_1 + \ x_2 + 5x_3 \leq \ 500.$$

23. (a) Let $x_1 =$ number of gallons of fruity wine
and $x_2 =$ number of gallons of crystal wine.

(b) The profit function is

$$z = 12x_1 + 15x_2.$$

(c) The ingredients available are the limitations; the constraints are

$$2x_1 + \ x_2 \leq 110$$
$$2x_1 + 3x_2 \leq 125$$
$$2x_1 + \ x_2 \leq \ 90.$$

25. Based on the information given in Exercise 21, the initial tableau is

$$\begin{bmatrix} x_1 & x_2 & x_3 & s_1 & s_2 & s_3 & z & \\ \boxed{5} & 3 & 6 & 1 & 0 & 0 & 0 & 1200 \\ 1 & 2 & 2 & 0 & 1 & 0 & 0 & 800 \\ 2 & 1 & 5 & 0 & 0 & 1 & 0 & 500 \\ \hline -4 & -3 & -3 & 0 & 0 & 0 & 1 & 0 \end{bmatrix}.$$

Pivot on the 5 in row 1, column 1.

$$\begin{array}{c} \\ -R_1 + 5R_2 \rightarrow R_2 \\ -2R_1 + 5R_3 \rightarrow R_3 \\ 4R_1 + 5R_4 \rightarrow R_4 \end{array} \begin{bmatrix} x_1 & x_2 & x_3 & s_1 & s_2 & s_3 & z & \\ 5 & \boxed{3} & 6 & 1 & 0 & 0 & 0 & 1200 \\ 0 & 7 & 4 & -1 & 5 & 0 & 0 & 2800 \\ 0 & -1 & 13 & -2 & 0 & 5 & 0 & 100 \\ 0 & -3 & 9 & 4 & 0 & 0 & 5 & 4800 \end{bmatrix}$$

Pivot on the 3 in row 1, column 2.

$$\begin{array}{c} \\ -7R_1 + 3R_2 \rightarrow R_2 \\ R_1 + 3R_3 \rightarrow R_3 \\ R_1 + \ R_4 \rightarrow R_4 \end{array} \begin{bmatrix} x_1 & x_2 & x_3 & s_1 & s_2 & s_3 & z & \\ 5 & 3 & 6 & 1 & 0 & 0 & 0 & 1200 \\ -35 & 0 & -30 & -10 & 15 & 0 & 0 & 0 \\ 5 & 0 & 45 & -5 & 0 & 15 & 0 & 1500 \\ 5 & 0 & 15 & 5 & 0 & 0 & 5 & 6000 \end{bmatrix}$$

$$
\begin{array}{c}
\frac{1}{3}R_1 \to R_1 \\
\frac{1}{15}R_2 \to R_2 \\
\frac{1}{15}R_3 \to R_3 \\
\\
\frac{1}{5}R_4 \to R_4
\end{array}
\begin{array}{c}
x_1 \quad x_2 \quad x_3 \quad s_1 \quad s_2 \quad s_3 \quad z \\
\left[
\begin{array}{ccccccc|c}
\frac{5}{3} & 1 & 2 & \frac{1}{3} & 0 & 0 & 0 & 400 \\
-\frac{7}{3} & 0 & -2 & -\frac{2}{3} & 1 & 0 & 0 & 0 \\
\frac{1}{3} & 0 & 3 & -\frac{1}{3} & 0 & 1 & 0 & 100 \\
\hline
1 & 0 & 3 & 1 & 0 & 0 & 1 & 1200
\end{array}
\right]
\end{array}
$$

This is a final tableau. She should buy none of item A, 400 of item B, and none of item C for a maximum profit of $1200.

27. Based on Exercise 23, the initial tableau is

$$
\begin{array}{c}
x_1 \quad x_2 \quad s_1 \quad s_2 \quad s_3 \quad z \\
\left[
\begin{array}{cccccc|c}
2 & 1 & 1 & 0 & 0 & 0 & 110 \\
2 & \boxed{3} & 0 & 1 & 0 & 0 & 125 \\
2 & 1 & 0 & 0 & 1 & 0 & 90 \\
\hline
-12 & -15 & 0 & 0 & 0 & 1 & 0
\end{array}
\right]
\end{array}
.
$$

Locating the first pivot in the usual way, it is found to be the 3 in row 2, column 2. After row transformations, we get the next tableau.

$$
\begin{array}{c}
-R_2 + 3R_1 \to R_1 \\
\\
-R_2 + 3R_3 \to R_3 \\
\\
5R_2 + R_4 \to R_4
\end{array}
\begin{array}{c}
x_1 \quad x_2 \quad s_1 \quad s_2 \quad s_3 \quad z \\
\left[
\begin{array}{cccccc|c}
4 & 0 & 3 & -1 & 0 & 0 & 205 \\
2 & 3 & 0 & 1 & 0 & 0 & 125 \\
\boxed{4} & 0 & 0 & -1 & 3 & 0 & 145 \\
\hline
-2 & 0 & 0 & 5 & 0 & 1 & 625
\end{array}
\right]
\end{array}
$$

Pivot on the 4 in row 3, column 1.

$$
\begin{array}{c}
-R_3 + R_1 \to R_1 \\
-R_3 + 2R_2 \to R_2 \\
\\
R_3 + 2R_4 \to R_4
\end{array}
\begin{array}{c}
x_1 \quad x_2 \quad s_1 \quad s_2 \quad s_3 \quad z \\
\left[
\begin{array}{cccccc|c}
0 & 0 & 3 & 0 & -3 & 0 & 60 \\
0 & 6 & 0 & 3 & -3 & 0 & 105 \\
4 & 0 & 0 & -1 & 3 & 0 & 145 \\
\hline
0 & 0 & 0 & 9 & 3 & 2 & 1395
\end{array}
\right]
\end{array}
$$

$$
\begin{array}{c}
\frac{1}{3}R_1 \to R_1 \\
\frac{1}{6}R_2 \to R_2 \\
\frac{1}{4}R_3 \to R_3 \\
\\
\frac{1}{2}R_4 \to R_4
\end{array}
\begin{array}{c}
x_1 \quad x_2 \quad s_1 \quad s_2 \quad s_3 \quad z \\
\left[
\begin{array}{cccccc|c}
0 & 0 & 1 & 0 & -1 & 0 & 20 \\
0 & 1 & 0 & \frac{1}{2} & -\frac{1}{2} & 0 & \frac{35}{2} \\
1 & 0 & 0 & -\frac{1}{4} & \frac{3}{4} & 0 & \frac{145}{4} \\
\hline
0 & 0 & 0 & \frac{9}{2} & \frac{3}{2} & 1 & \frac{1395}{2}
\end{array}
\right]
\end{array}
$$

The final tableau gives the solution $x_1 = \frac{145}{4}, x_2 = \frac{35}{2}$, and $z = \frac{1395}{2} = 697.5$. 36.25 gal of fruity wine and 17.5 gal of crystal wine should be produced for a maximum profit of $697.50.

Extended Application: A Minimum Cost Balanced Organic Fertilizer Mix

1. There are 6 parts blood meal, 11 parts bone meal, and 6 parts sul-po-mag. Since this is a total of 23 parts, $\frac{6}{23}$ of the total is blood meal, $\frac{11}{23}$ is bone meal, and $\frac{6}{23}$ is sul-po-mag. The N-P-K percent for blood meal is 11-0-0, for bone meal is 6-12-0, and for sul-po-mag is 0-0-22.

   Amount of nitrogen:

   $$\frac{6}{23}(.11) + \frac{11}{23}(.06) + \frac{6}{23}(0) = \frac{1.32}{23};$$

   Amount of phosphorus:

   $$\frac{6}{23}(0) + \frac{11}{23}(.12) + \frac{6}{23}(0) = \frac{1.32}{23};$$

   Amount of potassium:

   $$\frac{6}{23}(0) + \frac{11}{23}(0) + \frac{6}{23}(.22) = \frac{1.32}{23}.$$

   Look at the percentage of each:

   Blood Meal: $\dfrac{4.5454}{17.4242} \approx .260867;$

   Bone Meal: $\dfrac{8.3333}{17.4242} \approx .4782601;$

   Sul-Po-Mag: $\dfrac{4.5454}{17.4242} \approx .260867.$

   With proportions of 6-11-6, we obtain

   Blood Meal: $\dfrac{6}{23} \approx .260870;$

   Bone Meal: $\dfrac{11}{23} \approx .4782609;$

   Sul-Po-Mag: $\dfrac{6}{23} \approx .260870.$

   Since the decimal representations are approximately the same, the minimum cost mixture has proportions of 6-11-6.

2. Looking at Exercise 1, the percentage of each nutrient is $\frac{1.32}{23} \approx .0574$ or .06. This would be a 6-6-6 balanced fertilizer.

Chapter 4 Test

1. For the following maximization problem,

    (a) determine the number of slack variables needed;

    (b) convert each constraint into a linear equation.

    $$\text{Maximize } z = 50x_1 + 80x_2$$
    $$\text{subject to:} \quad x_1 + 2x_2 \leq 32$$
    $$3x_1 + 4x_2 \leq 84$$
    $$x_2 \leq 12$$
    $$\text{with} \quad x_1 \geq 0, \; x_2 \geq 0.$$

2. For the following maximization problem,

    (a) set up the initial simplex tableau;

    (b) determine the initial basic solution;

    (c) find the first pivot element and justify your choice.

    $$\text{Maximize } z = 10x_1 + 5x_2$$
    $$\text{subject to:} \quad 6x_1 + 2x_2 \leq 36$$
    $$2x_1 + 4x_2 \leq 32$$
    $$\text{with} \quad x_1 \geq 0, \; x_2 \geq 0.$$

3. Determine the basic feasible solution from the following tableau.

$$
\begin{array}{ccccccc}
x_1 & x_2 & x_3 & s_1 & s_2 & s_3 & z \\
\end{array}
$$

$$
\left[
\begin{array}{ccccccc|c}
1 & 0 & -2 & 1 & 0 & 2 & 0 & 6 \\
0 & 1 & 4 & 1 & 0 & -1 & 0 & 14 \\
0 & 0 & 3 & 0 & 1 & 3 & 0 & 12 \\
\hline
0 & 0 & 5 & 7 & 0 & 6 & 1 & 40 \\
\end{array}
\right]
$$

4. Solve the problem with given initial tableau.

$$
\begin{array}{ccccccc}
x_1 & x_2 & x_3 & s_1 & s_2 & s_3 & z \\
\end{array}
$$

$$
\left[
\begin{array}{ccccccc|c}
1 & 1 & 1 & 1 & 0 & 0 & 0 & 1000 \\
40 & 20 & 30 & 0 & 1 & 0 & 0 & 3200 \\
1 & 2 & 1 & 0 & 0 & 1 & 0 & 160 \\
\hline
-100 & -300 & -200 & 0 & 0 & 0 & 1 & 0 \\
\end{array}
\right]
$$

Use the simplex method to solve each linear programming problem.

5.   Maximize $z = 6x_1 + 9x_2 + 6x_3$

    subject to: $2x_1 + 3x_2 + 3x_3 \leq 30$

                   $2x_1 + 2x_2 + x_3 \leq 20$

                   $2x_1 + 5x_2 + x_3 \leq 40$

    with $x_1 \geq 0,\ x_2 \geq 0,\ x_3 \geq 0.$

6. Mammoth Micros markets computers with single-sided and double-sided disk drives. They obtain these drives from Large Disks, Inc. and Double Drives Are Us. Large Disk charges \$250 for a single-sided and \$350 for a double-sided disk. Double drives charges \$290 and \$320 for single-sided and double-sided disks, respectively. Each month Large Disks can supply at most 1000 drives in all. Double Drives can supply at most 2000. Mammoth needs at least 1200 single and 1600 double drives. How many of each type should they buy from each company to minimize their total costs? What is the minimum cost?

[4.3]

7. Find the transpose of the following matrix.

$$\begin{bmatrix} 2 & 1 & 7 & 6 & 3 \\ 5 & 9 & 0 & 4 & 2 \\ 6 & 8 & 5 & 1 & 4 \end{bmatrix}$$

8. For the following minimization problem,

   (a) state the dual problem;

   (b) solve the problem using the simplex method.

       Minimize $w = 5y_1 + 7y_2$

        subject to: $y_1 \geq 4$

                 $y_1 + y_2 \geq 8$

                 $y_1 + 2y_2 \geq 10$

        with $y_1 \geq 0,\ y_2 \geq 0.$

9. State the dual problem for the following minimization problem.

       Minimize $w = 3y_1 + 4y_2$

        subject to: $y_1 + 2y_2 \geq 8$

               $2y_1 + 2y_2 \geq 10$

               $y_1 + 4y_2 \geq 12$

        with $y_1 \geq 0,\ y_2 \geq 0.$

[4.4]

10. Rewrite the following system of inequalities, adding slack variables or subtracting surplus variables as necessary.

      $x_1 + 3x_2 + 2x_3 \leq 42$

      $3x_1 + x_3 \geq 20$

      $2x_1 + x_2 + 2x_3 \geq 13$

11. Convert the following problem into a maximization problem.

Minimize $w = 8y_1 + 5y_2 + 3y_3$

subject to: $\quad y_1 + 3y_2 + 2y_3 \geq 90$

$3y_1 + y_2 + y_3 \geq 75$

$2y_1 \qquad + 3y_3 \geq 60$

with $\qquad y_1 \geq 0,\ y_2 \geq 0,\ y_3 \geq 0.$

12. Use the simplex method to solve the following problem.

Maximize $z = 6x_1 - 2x_2$

subject to: $\quad x_1 + x_2 \leq 10$

$3x_1 + 2x_2 \geq 24$

with $\qquad x_1 \geq 0,\ x_2 \geq 0.$

Chapter 4 Test Answers

1. (a) 3 slack variables are needed.   **(b)**

$$
\begin{aligned}
x_1 + 2x_2 + s_1 \quad\quad\quad &= 32 \\
3x_1 + 4x_2 \quad\quad + s_2 \quad\quad &= 84 \\
x_2 \quad\quad\quad + s_3 &= 12
\end{aligned}
$$

2. (a)

x_1	x_2	s_1	s_2	z	
6	2	1	0	0	36
2	4	0	1	0	32
−10	−5	0	1	1	0

(b) $s_1 = 36$, $s_2 = 32$, $z = 0$

(c) Column one has the most negative indicator. The smallest nonnegative quotient occurs in row 1, since $\frac{36}{6}$ is smaller than $\frac{32}{2}$. Pivot on the 6.

3. $x_1 = 6$, $x_2 = 14$, $x_3 = 0$, $s_1 = 0$, $s_2 = 12$, $s_3 = 0$, $z = 40$

4. The maximum value is 28,000 when $x_1 = 0$, $x_2 = 80$, and $x_3 = 40$.

5. The maximum vlaue is 85 when $x_1 = \frac{5}{2}$, $x_2 = \frac{20}{3}$, and $x_3 = \frac{5}{3}$.

6. They should buy 1000 single-sided from Large Disks, and they should buy 200 single-sided and 1600 double-sided from Doubles Are Us for a minimum cost of $820,000.

7.
$$
\begin{bmatrix}
2 & 5 & 6 \\
1 & 9 & 8 \\
7 & 0 & 5 \\
6 & 4 & 1 \\
3 & 2 & 4
\end{bmatrix}
$$

8. (a)   Maximize $z = 4x_1 + 8x_2 + 10x_3$

subject to:
$$
\begin{aligned}
x_1 + x_2 + x_3 &\le 5 \\
x_2 + 2x_3 &\le 7
\end{aligned}
$$
with $\quad x_1 \ge 0,\ x_2 \ge 0,\ x_3 \ge 0.$

(b) The minimum value is 44 when $y_1 = 6$ and $y_2 = 2$.

9.   Maximize $z = 8x_1 + 10x_2 + 12x_3$

subject to:
$$
\begin{aligned}
x_1 + 2x_2 + x_3 &\le 3 \\
2x_1 + 2x_2 + 4x_3 &\le 4
\end{aligned}
$$
with $\quad x_1 \ge 0,\ x_2 \ge 0,\ x_3 \ge 0.$

10.
$$
\begin{aligned}
x_1 + 3x_2 + 2x_3 + s_1 \quad\quad\quad &= 42 \\
3x_1 \quad\quad + x_3 \quad\quad - s_2 \quad\quad &= 20 \\
2x_1 + x_2 + 2x_3 \quad\quad\quad - s_3 &= 13
\end{aligned}
$$

11. Maximize $z = -w = -8y_1 - 5y_2 - 3y_3$

     subject to: $\quad y_1 + 3y_2 + 2y_3 \geq 90$
$$3y_1 + y_2 + y_3 \geq 75$$
$$2y_1 + 3y_3 \geq 60$$

     with $\quad\quad\quad y_1 \geq 0, \ y_2 \geq 0, \ y_3 \geq 0.$

12. The maximum value is 60 when $x_1 = 10$ and $x_2 = 0$.

MATHEMATICS OF FINANCE

5.1 Simple and Compound Interest

1. The variable r is the annual interest rate, while i is the interest rate per period. The variable t is the number of years, while n is the number of compounding periods per year.

3. The interest rate and number of compounding periods determine the amount of interest on a fixed principal. In the formula $I = Prt$, if P remains constant then the value of I will be affected by the values of r and t.

7. $3850 at 9% for 8 mo

Use the formula for simple interest.

$$I = Prt = 3850(.09)\left(\frac{8}{12}\right) = \$231.00$$

9. $3724 at 8.4% for 11 mo

$$I = Prt = 3724(.084)\left(\frac{11}{12}\right) \approx \$286.75$$

11. $2930.42 at 11.9% for 123 days

$$I = Prt = 2930.42(.119)\left(\frac{123}{360}\right) \approx \$119.15$$

15. $1000 at 7% compounded annually for 10 yr

Use the formula for compound amount with $P = 1000$, $i = .07$, and $n = 10(1) = 10$.

$$A = P(1+i)^n$$
$$= 1000(1.07)^{10}$$
$$\approx 1967.15$$

The compound amount for this deposit is $1967.15.

17. $15,000 at 6% compounded semiannually for 11 yr

Use the formula for compound amount with $P = 15,000$, $i = \frac{.06}{2} = .03$, and $n = 11(2) = 22$.

$$A = P(1+i)^n$$
$$= 15,000(1.03)^{22}$$
$$\approx 28,741.55$$

The compound amount is $28,741.55.

19. $9100 at 8% compounded quarterly for 4 yr

Here $A = 9100$, $i = \frac{.08}{4} = .02$, and $n = 4(4) = 16$.

$$A = P(1+i)^n$$
$$= 9100(1.02)^{16}$$
$$\approx 12,492.35$$

The compound amount is $12,492.35.

21. $27,159.68 at 12.3% compounded annually for 11 yr

Use the formula for present value for compound interest with $A = 27,159.68$, $i = .123$, and $n = 11(1) = 11$.

$$P = \frac{A}{(1+i)^n} = \frac{27,159.68}{(1.123)^{11}} \approx 7581.36$$

The present value is $7581.36.

23. $2000 at 11% compounded semiannually for 8 yr

Here $A = 2000$, $i = \frac{.11}{2} = .055$, and $n = 8(2) = 16$.

$$P = \frac{A}{(1+i)^n} = \frac{2000}{(1.055)^{16}} \approx 849.16$$

The present value is $849.16.

25. $7500 at 12% compounded quarterly for 9 yr

Here $A = 7500$, $i = \frac{.12}{4} = .03$, and $n = 9(4) = 36$.

$$P = \frac{A}{(1+i)^n} = \frac{7500}{(1.03)^{36}} \approx 2587.74$$

The present value is $2587.74.

27. The effective rate is higher than the stated rate because the effective rate includes interest paid on interest already earned.

29. 8% compounded quarterly

Use the formula for effective rate with $r = .08$ and $m = 4$, the number of times interest is compounded per year.

$$r_e = \left(1 + \frac{r}{m}\right)^m - 1$$
$$= \left(1 + \frac{.08}{4}\right)^4 - 1$$
$$= (1.02)^4 - 1$$
$$\approx .08243$$

The effective rate is 8.243%.

31. 10.08% compounded semiannually

Here $r = .1008$ and $m = 2$.

$$r_e = \left(1 + \frac{r}{m}\right)^m - 1$$
$$= \left(1 + \frac{.1008}{2}\right)^2 - 1$$
$$= (1.0504)^2 - 1$$
$$\approx .10334$$

The effective rate is 10.334%.

33. $I = Prt$

$$= 725,896.15(.127)\left(\frac{34}{365}\right)$$
$$\approx 8587.45$$

The penalty is $8587.45.

$$A = P + I$$
$$= 725,896.15 + 8587.45$$
$$= 734,483.60$$

The total amount (tax and penalty) that was paid is $734,483.60.

35. The future value is $7(5104) = 35,728$. Use the formula for present value with $A = 35,728$, $r = .0642$, and $t = \frac{7}{12}$.

$$P = \frac{A}{1 + rt}$$
$$= \frac{35,728}{1 + .0642\left(\frac{7}{12}\right)}$$
$$\approx 34,438.29.$$

They should deposit about $34,438.29.

37. Use the formula for future value for simple interest with $A = 10,000$, $P = 5988.02$, and $t = 10$.

$$A = P(1 + rt)$$
$$10,000 = 5988.02[1 + r(10)]$$
$$1.67 = 1 + 10r$$
$$.67 = 10r$$
$$.067 = r$$

The simple interest rate is 6.7%.

39. Use the formula for compound amount with $P = 10,000$, $i = \frac{.06}{2} = .03$, and $n = 15(2) = 30$.

$$A = P(1 + i)^n$$
$$= 10,000(1.03)^{30}$$
$$\approx 24,272.62$$

The amount on deposit in 15 yr will be $24,272.62.

41. $P = 50,000$, $i = \frac{.12}{12}$, $n = 12(4) = 48$

First find the compound amount.

$$A = P(1 + i)^n$$
$$= 50,000\left(1 + \frac{.12}{12}\right)^{48}$$
$$= 50,000(1.01)^{48}$$
$$\approx \$80,611.30$$

$$\text{Amount of interest} = A - P$$
$$= \$80,611.30 - \$50,000$$
$$= \$30,611.30$$

The business will pay $30,611.30 in interest.

43. Substitute $A = 2.9$ million, $i = \frac{.08}{12}$, and $n = 12(5) = 60$ in the formula for present value with compound interest.

$$P = \frac{A}{(1 + i)^n}$$
$$= \frac{2.9}{\left(1 + \frac{.08}{12}\right)^{60}}$$
$$\approx \frac{2.9}{(1.00667)^{60}}$$
$$\approx 1.946$$

They should invest about $1.946 million now.

45. Substitute $P = 10,000$, $i = \frac{.06}{2}$, and $n = 2(3) = 6$ in the formula for compound amount.

$$A = P(1 + i)^n$$
$$= 10,000\left(1 + \frac{.06}{2}\right)^6$$
$$= 10,000(1.03)^6$$
$$\approx 11,940.52$$

She should contribute about $11,940.52 in 3 yr.

47. Solve $r_e = \left(1 + \frac{r}{m}\right)^m - 1$ for r.

$$r_e + 1 = \left(1 + \frac{r}{m}\right)^m$$
$$(r_e + 1)^{1/m} = 1 + \frac{r}{m}$$
$$(r_e + 1)^{1/m} - 1 = \frac{r}{m}$$
$$r = m[(r_e + 1)^{1/m} - 1]$$

In each case, $m = 4$.

For 5.00%,

$$r = 4[(.0500 + 1)^{1/4} - 1]$$
$$\approx .0491 \quad \text{or} \quad 4.91\%.$$

For 5.30%,

$$r = 4[(.0530 + 1)^{1/4} - 1]$$
$$\approx .0520 \quad \text{or} \quad 5.20\%.$$

For 5.45%,

$$r = 4[(.0545 + 1)^{1/4} - 1]$$
$$\approx .0534 \quad \text{or} \quad 5.34\%.$$

For 5.68%,

$$r = 4[(.0568 + 1)^{1/4} - 1]$$
$$\approx .0556 \quad \text{or} \quad 5.56\%.$$

For 5.75%,

$$r = 4[(.0575 + 1)^{1/4} - 1]$$
$$\approx .0563 \quad \text{or} \quad 5.63\%.$$

49. To find the number of years it will take prices to double at 4% annual inflation, find n in the equation

$$2 = (1 + .04)^n,$$

which simplifies to

$$2 = (1.04)^n.$$

By trying various values of n, find that $n = 18$ is approximately correct, because

$$1.04^{18} \approx 2.0258 \approx 2.$$

Prices will double in about 18 yr.

51. To find the number of years it will be until the generating capacity will need to be doubled, find n in the equation

$$2 = (1 + .06)^n,$$

which simplifies to

$$2 = (1.06)^n.$$

By trying various values of n, find that $n = 12$ is approximately correct, because

$$1.06^{12} \approx 2.0122 \approx 2.$$

The generating capacity will need to be doubled in about 12 yr.

53. Let $P = 150,000$, $i = -2.4\% = -.024$, and $n = 4$.

$$A = P(1 + i)^n$$
$$= 150,000[1 + (-.024)]^4$$
$$= 150,000(.976)^4$$
$$\approx 136,110.16$$

After 4 yr, the amount on deposit will be \$136,110.16.

55. Use the formula

$$A = P(1 + i)^n$$

with $A = 420,000,000$, $P = 100$, and $n = 160$.

$$420,000,000 = 100(1 + r)^{160}$$
$$4,200,000 = (1 + r)^{160}$$

$$4,200,000^{1/160} = 1 + r$$
$$r = 4,200,000^{1/160} - 1$$
$$r \approx .1000$$

The rate used was 10.00%.

57. First use the formula for simple interest where $P = 5200$, $r = .07$, and $t = \frac{10}{12}$.

$$A = P(1 + rt)$$

$$= 5200 \left[1 + .07\left(\frac{10}{12}\right)\right]$$

$$\approx 5503.33$$

Now use the formula for compound interest with $P = 5503.33$, $i = \frac{.063}{4}$, and $n = 5(4) = 20$.

$$A = P(1 + i)^n$$

$$= 5503.33 \left(1 + \frac{.063}{4}\right)^{20}$$

$$= 5503.33(1.01575)^{20}$$

$$\approx 7522.50$$

He will have \$7522.50 at the end of the 5 yr.

5.2 Future Value of an Annuity

1. $a = 3$; $r = 2$

The first five terms are

$$3, \ 3(2), \ 3(2)^2, \ 3(2)^3, \ 3(2)^4$$

or

$$3, \ 6, \ 12, \ 24, \ 48.$$

The fifth term is 48.

Or, use the formula $a_n = ar^{n-1}$ with $n = 5$.

$$a_5 = ar^{5-1} = 3(2)^4 = 3(16) = 48$$

3. $a = -8$; $r = 3$; $n = 5$

$$a_5 = ar^{5-1} = -8(3)^4 = -8(81) = -648$$

The fifth term is -648.

5. $a = 1$; $r = -3$; $n = 5$

$$a_5 = ar^{5-1} = 1(-3)^4 = 81$$

The fifth term is 81.

7. $a = 1024$; $r = \frac{1}{2}$; $n = 5$

$$a_5 = ar^{5-1} = 1024\left(\frac{1}{2}\right)^4 = 1024\left(\frac{1}{16}\right) = 64$$

The fifth term is 64.

9. $a = 1$; $r = 2$; $n = 4$

To find the sum of the first 4 terms, S_4, use the formula for the sum of the first n terms of a geometric sequence.

$$S_n = \frac{a(r^n - 1)}{r - 1}$$

$$S_4 = \frac{1(2^4 - 1)}{2 - 1} = \frac{16 - 1}{1} = 15$$

11. $a = 5$; $r = \frac{1}{5}$; $n = 4$

$$S_n = \frac{a(r^n - 1)}{r - 1}$$

$$S_4 = \frac{5\left[\left(\frac{1}{5}\right)^4 - 1\right]}{\frac{1}{5} - 1} = \frac{5\left(-\frac{624}{625}\right)}{-\frac{4}{5}}$$

$$= \frac{-\frac{624}{125}}{-\frac{4}{5}} = \left(-\frac{624}{125}\right)\left(-\frac{5}{4}\right) = \frac{156}{25}$$

13. $a = 128$; $r = -\frac{3}{2}$; $n = 4$

$$S_n = \frac{a(r^n - 1)}{r - 1}$$

$$S_4 = \frac{128\left[\left(-\frac{3}{2}\right)^4 - 1\right]}{-\frac{3}{2} - 1} = \frac{128\left(\frac{65}{16}\right)}{-\frac{5}{2}}$$

$$= -208$$

15. $s_{\overline{n}|i} = \frac{(1 + i)^n - 1}{i}$

$$s_{\overline{12}|.05} = \frac{(1 + .05)^{12} - 1}{.05} \approx 15.91713$$

17. $s_{\overline{n}|i} = \frac{(1 + i)^n - 1}{i}$

$$s_{\overline{16}|.043} = \frac{(1 + .043)^{16} - 1}{.043} \approx 22.35633$$

21. $R = 100$; $i = .06$; $n = 4$

Use the formula for the future value of an ordinary annuity.

$$S = R\left[\frac{(1 + i)^n - 1}{i}\right]$$

$$= 100\left[\frac{(1.06)^4 - 1}{.06}\right]$$

$$= 100\left(\frac{1.262477 - 1}{.06}\right)$$

$$\approx 437.46$$

The future value is $437.46.

23. $R = 46,000$, $i = .063$, $n = 32$

$$S = R\left[\frac{(1 + i)^n - 1}{i}\right]$$

$$= 46,000\left[\frac{(1.063)^{32} - 1}{.063}\right]$$

$$\approx 4,427,846.13$$

The future value is $4,427,846.13.

25. $R = 9200$; 10% interest compounded semiannually for 7 yr

Interest of $\frac{10\%}{2} = 5\%$ is earned semiannually, so $i = .05$. In 7 yr, there are $7(2) = 14$ semiannual periods, so $n = 14$.

$$S = R\left[\frac{(1 + i)^n - 1}{i}\right]$$

$$= 9200\left[\frac{(1.05)^{14} - 1}{.05}\right]$$

$$\approx 180,307.41$$

The future value is $180,307.41.

27. $R = 800$; 6.51% interest compounded semiannually for 12 yr

Interest of $\frac{6.51\%}{2}$ is earned semiannually, so $i = \frac{.0651}{2} = .03255$. In 12 yr, there are $12(2) = 24$ semiannual periods, so $n = 24$.

$$S = R\left[\frac{(1 + i)^n - 1}{i}\right]$$

$$= 800\left[\frac{(1 + .03255)^{24} - 1}{.03255}\right]$$

$$\approx 28,438.21$$

The future value is $28,438.21.

29. $R = 15,000$; 12.1% interest compounded quarterly for 6 yr

Interest of $\frac{12.1\%}{4}$ is earned quarterly, so $i = \frac{.121}{4} = .03025$. In 6 yr, there are $6(4) = 24$ quarterly periods, so $n = 24$.

$$S = R\left[\frac{(1+i)^n - 1}{i}\right]$$

$$= 15,000\left[\frac{(1+.03025)^{24} - 1}{.03025}\right]$$

$$\approx 518,017.56$$

The future value is $518,017.56.

31. $R = 600$, $i = .06$, $n = 8$

Use the formula for the future value of an annuity due.

$$S = R\left[\frac{(1+i)^{n+1} - 1}{i}\right] - R$$

$$= 600\left[\frac{(1.06)^9 - 1}{.06}\right] - 600$$

$$\approx 6294.79$$

The future value is $6294.79.

33. $R = 20,000$, $i = .08$, $n = 6$

$$S = R\left[\frac{(1+i)^{n+1} - 1}{i}\right] - R$$

$$= 20,000\left[\frac{(1.08)^7 - 1}{.08}\right] - 20,000$$

$$\approx 158,456.07$$

The future value is $158,456.07.

35. $R = 1000$, $i = \frac{.0815}{2} = .04075$, $n = 9(2) = 18$

$$S = R\left[\frac{(1+i)^{n+1} - 1}{i}\right] - R$$

$$= 1000\left[\frac{(1+.04075)^{19} - 1}{.04075}\right] - 1000$$

$$\approx 26,874.97$$

The future value is $26,874.97.

37. $R = 100$, $i = \frac{.124}{4} = .031$, $n = 9(4) = 36$

$$S = R\left[\frac{(1+i)^{n+1} - 1}{i}\right] - R$$

$$= 100\left[\frac{(1+.031)^{37} - 1}{.031}\right] - 100$$

$$\approx 6655.99$$

The future value is $6655.99.

39. $S = \$1000$; interest is 5% compounded annually; payments are made at the end of each year for 12 yr

This is a sinking fund. Use the formula for an ordinary annuity with $S = 10,000$, $i = .05$, and $n = 12$ to find the value of R, the amount of each payment.

$$10,000 = Rs_{\overline{12}|.05}$$

$$R = \frac{10,000}{s_{\overline{12}|.05}}$$

$$= \frac{10,000}{\frac{(1+.05)^{12} - 1}{.05}}$$

$$\approx 628.25$$

The required periodic payment is $628.25.

43. $2000; money earns 6% compounded annually; 5 annual payments

Let R be the amount of each payment.

$$2000 = Rs_{\overline{5}|.06}$$

$$R = \frac{2000}{s_{\overline{5}|.06}}$$

$$= \frac{2000}{\frac{(1.06)^5 - 1}{.06}}$$

$$\approx 354.79$$

The amount of each payment is $354.79.

45. $25,000; money earns 5.7% compounded quarterly for $3\frac{1}{2}$ yr

Thus, $i = \frac{.057}{4} = .01425$ and $n = \left(3\frac{1}{2}\right)4 = 14$.

$$R = \frac{25,000}{s_{\overline{14}|.01425}}$$

$$= \frac{25,000}{\frac{(1+.01425)^{14} - 1}{.01425}}$$

$$\approx 1626.16$$

The amount of each payment is $1626.16.

47. $9000; money earns 12.23% compounded monthly for $2\frac{1}{2}$ yr

Thus, $i = \frac{.1223}{12} = .010191\overline{6}$ and $n = \left(2\frac{1}{2}\right)12 = 30$.

$$R = \frac{9000}{s_{\overline{30}|.010191\overline{6}}}$$

$$= \frac{9000}{\frac{(1+.010191\overline{6})^{30}-1}{.010191\overline{6}}}$$

$$\approx 257.99$$

The amount of each payment is \$257.99.

49. Use the formula for the future value of an ordinary annuity with $R = 60$, $i = \frac{.08}{12}$, and $n = 12(3) = 36$.

$$S = R\left[\frac{(1+i)^n - 1}{i}\right]$$

$$= 60\left[\frac{\left(1+\frac{.08}{12}\right)^{36} - 1}{\frac{.08}{12}}\right]$$

$$\approx 2432.13$$

The amount in the account after 3 yr is \$2432.13.

51. \$55 is invested each month at 4.8% compounded monthly for 40 yr. Thus, $i = \frac{.048}{12} = .004$ and $n = 40(12) = 480$. Use the formula for the future value of an ordinary annuity.

$$S = R\left[\frac{(1+i)^n - 1}{i}\right]$$

$$= 55\left[\frac{(1+.004)^{480} - 1}{.004}\right]$$

$$\approx 79,679.68$$

The account would be worth \$79,679.68.

53. For the first 15 yr, we have an ordinary annuity with $R = 1000$, $i = \frac{.08}{4} = .02$, and $n = 15(4) = 60$. The amount on deposit after 15 yr is

$$S = R\left[\frac{(1+i)^n - 1}{i}\right]$$

$$= 1000\left[\frac{(1+.02)^{60} - 1}{.02}\right]$$

$$\approx 114,051.54.$$

For the remaining 5 yr, this amount earns compound interest at 8% compounded quarterly. To find the final amount on deposit, use the formula

for the compound amount with $P = 114,051.54$, $i = \frac{.08}{4} = .02$, and $n = 5(4) = 20$.

$$A = P(1+i)^n$$

$$= 114,051.54(1.02)^{20}$$

$$\approx 169,474.59$$

The man will have about \$169,474.59 in the account when he retires.

55. For the first 8 yr, we have an annuity due with $R = 2435$, $i = \frac{.06}{2} = .03$, and $n = 8(2) = 16$. The amount on deposit after 8 yr is

$$S = R\left[\frac{(1+i)^{n+1} - 1}{i}\right] - R$$

$$= 2435\left[\frac{(1+.03)^{17} - 1}{.03}\right]$$

$$\approx 50,554.47.$$

For the remaining 5 yr, this amount, \$50,554.47, earns compound interest at 6% compounded semiannually. To find the final amount on deposit, use the formula for the compound amount with $P = 50,554.47$, $i = \frac{.06}{2} = .03$, and $n = 5(2) = 10$.

$$A = P(1+i)^n$$

$$= 50,554.47(1.03)^{10}$$

$$\approx 67,940.98$$

The final amount on deposit will be about \$67,940.98.

57. (a) This is a sinking fund with $S = 10,000$, $i = \frac{.08}{4} = .02$, and $n = 8(4) = 32$. Let R represent the amount of each payment.

$$S = Rs_{\overline{n}|i}$$

$$10,000 = Rs_{\overline{32}|.02}$$

$$R = \frac{10,000}{s_{\overline{32}|.02}}$$

$$= \frac{10,000(.02)}{(1+.02)^{32} - 1}$$

$$\approx 226.11$$

If the money is deposited at 8% compounded quarterly, Berkowitz's quarterly deposit will need to be about \$226.11.

(b) Here $S = 10,000$, $i = \frac{.06}{4} = .015$, and $n = 8(4) = 32$. Let R represent the amount of each payment.

$$S = Rs_{\overline{n}|i}$$

$$10,000 = Rs_{\overline{32}|.015}$$

$$R = \frac{10,000}{s_{\overline{32}|.015}}$$

$$= \frac{10,000(.015)}{(1 + .015)^{32} - 1}$$

$$\approx 245.77$$

If the money is deposited at 6% compounded quarterly, Berkowitz's quarterly deposit will need to be about $245.77.

59. $S = 18,000$, $i = \frac{.05}{4} = .0125$, $n = 6(4) = 24$

Let R represent the amount of each payment.

$$S = Rs_{\overline{n}|i}$$

$$18,000 = Rs_{\overline{24}|.0125}$$

$$R = \frac{18,000}{s_{\overline{24}|.0125}}$$

$$= \frac{18,000(.0125)}{(1 + .0125)^{24} - 1}$$

$$\approx 647.76$$

She must deposit about $647.76 at the end of each quarter.

61. $R = \frac{2000}{2} = 1000$, $i = \frac{.08}{2} = .04$, $n = 25(2) = 50$

$$S = R\left[\frac{(1 + i)^n - 1}{i}\right]$$

$$= 1000\left[\frac{(1 + .04)^{50} - 1}{.04}\right]$$

$$\approx 152,667.08$$

There will be about $152,667.08 in the IRA.

63. $R = \frac{2000}{2} = 1000$, $i = \frac{.10}{2} = .05$, $n = 25(2) = 50$

$$S = R\left[\frac{(1 + i)^n - 1}{i}\right]$$

$$= 1000\left[\frac{(1 + .05)^{50} - 1}{.05}\right]$$

$$\approx 209,348.00$$

There will be about $209,348 in the IRA.

65. Let $x =$ the annual interest rate.

$$n = 30(12) = 360$$

Graph $y_1 = 330,000$ and

$$y_2 = 250\left[\frac{\left(1 + \frac{x}{12}\right)^{360} - 1}{\frac{x}{12}}\right].$$

The x-coordinate of the point of intersection is .0739706. Thus, she would need to earn an annual interest rate of about 7.397%.

67. This exercise should be solved by graphing calculator or computer methods. The answers, which may vary slightly, are as follows.

(a) The buyer's quarterly interest payment will be $1200.

(b) The buyer's semiannual payments into the sinking fund will be $3511.58 for each of the first 13 payments and $3511.59 for the last payment. A table showing the amount in the sinking fund after each deposit is as follows.

Payment Number	Amount of Deposit	Interest Earned	Total
1	$3511.58	$0	$3511.58
2	$3511.58	$105.35	$7128.51
3	$3511.58	$213.86	$10,853.95
4	$3511.58	$325.62	$14,691.15
5	$3511.58	$440.73	$18,643.46
6	$3511.58	$559.30	$22,714.34
7	$3511.58	$681.43	$26,907.35
8	$3511.58	$807.22	$31,226.15
9	$3511.58	$936.78	$35,674.51
10	$3511.58	$1070.24	$40,256.33
11	$3511.58	$1207.69	$44,975.60
12	$3511.58	$1349.27	$49,836.45
13	$3511.58	$1495.09	$54,843.12
14	$3511.59	$1645.29	$60,000.00

5.3 Present Value of an Annuity; Amortization

1. $\dfrac{1 - (1 + i)^{-n}}{i}$

is represented by $a_{\overline{n}|i}$, and it is choice (c).

3. $a_{\overline{n}|i} = \dfrac{1-(1+i)^{-n}}{i}$

  $a_{\overline{15}|.06} = \dfrac{1-(1+.06)^{-15}}{.06}$

  $= \dfrac{1-(1.06)^{-15}}{.06}$

  ≈ 9.71225

5. $a_{\overline{n}|i} = \dfrac{1-(1+i)^{-n}}{i}$

  $a_{\overline{18}|.045} = \dfrac{1-(1.045)^{-18}}{.045} \approx 12.15999$

9. $R = 1400$, $i = .08$, $n = 8$

Use the formula for the present value of an annuity.

$$P = Ra_{\overline{n}|i}$$

$$P = 1400a_{\overline{8}|.08}$$

$$= 1400\left[\frac{1-(1+.08)^{-8}}{.08}\right]$$

$$\approx 8045.30$$

The present value is \$8045.30.

11. $R = 50{,}000$, $i = \frac{.08}{4} = .02$, $n = 10(4) = 40$

$$P = Ra_{\overline{n}|i}$$

$$P = 50{,}000a_{\overline{40}|.02}$$

$$= 50{,}000\left[\frac{1-(1+.02)^{-40}}{.02}\right]$$

$$\approx 1{,}367{,}774$$

The present value is \$1,367,774.

13. $R = 18{,}579$, $i = \frac{.094}{2} = .047$, $n = 8(2) = 16$

$$P = Ra_{\overline{n}|i}$$

$$P = 18{,}579a_{\overline{16}|.047}$$

$$= 18{,}579\left[\frac{1-(1+.047)^{-16}}{.047}\right]$$

$$\approx 205{,}724.40$$

The present value is \$205,724.40.

15. We want the present value, P, of an annuity with $R = 10{,}000$, $i = .06$, and $n = 15$.

$$P = R\left[\frac{1-(1+i)^{-n}}{i}\right]$$

$$= 10{,}000\left[\frac{1-(1.06)^{-15}}{.06}\right]$$

$$\approx 97{,}122.49$$

The required lump sum is \$97,122.49.

17. \$2500, 8% compounded quarterly, 6 quarterly payments

Use the formula for amortization payments with $P = 2500$, $i = \frac{.08}{4} = .02$, and $n = 6$.

$$R = \frac{Pi}{1-(1+i)^{-n}}$$

$$= \frac{2500(.02)}{1-(1.02)^{-6}} \approx 446.31$$

Payments of \$446.31 each are necessary to amortize this loan.

19. \$90,000, 8% compounded annually, 12 annual payments

Use the formula for amortization payments with $P = 90{,}000$, $i = .08$, and $n = 12$.

$$R = \frac{Pi}{1-(1+i)^{-n}}$$

$$= \frac{90{,}000(.08)}{1-(1.08)^{-12}} \approx 11{,}942.55$$

Payments of \$11,942.55 are necessary to amortize this loan.

21. \$7400, 8.2% compounded semiannually, 18 semiannual payments

Use the formula for amortization payments with $P = 7400$, $i = \frac{.082}{2} = .041$, and $n = 18$.

$$R = \frac{Pi}{1-(1+i)^{-n}}$$

$$= \frac{7400(.041)}{1-(1.041)^{-18}} \approx 589.31$$

Payments of \$589.31 are necessary to amortize this loan.

23. Look at the entry for payment number 4 under the heading "Interest for Period." The amount of interest included in the fourth payment is \$7.61.

25. To find the amount of interest paid in the first 4 mo of the loan, add the entries for payments 1, 2, 3, and 4 under the heading "Interest for Period."

$$\$10.00 + 9.21 + 8.42 + 7.61 = \$35.24$$

In the first 4 mo of the loan, $35.24 of interest is paid.

27. First, find the value of the annuity at the end of 8 yr. Use the formula for future value of an ordinary annuity.

$$S = R\left[\frac{(1+i)^n - 1}{i}\right]$$

$$= 1000\left[\frac{(1+.06)^8 - 1}{.06}\right]$$

$$\approx 9897.47$$

The future value of the annuity is $9897.47.

Now find the present value of $9897.47 at 5% compounded annually for 8 yr. Use the formula for present value for compound interest.

$$P = \frac{A}{(1+i)^n} = \frac{9897.47}{(1.05)^8} \approx 6699.00$$

The required amount is $6699.

29. $149,560 at 7.75% for 25 yr

Use the formula for amortization payments with $P = 149,560$, $i = \frac{.0775}{12}$, and $n = 25(12) = 300$.

$$R = \frac{Pi}{1 - (1+i)^{-n}}$$

$$= \frac{149,560\left(\frac{.0775}{12}\right)}{1 - \left(1 + \frac{.0775}{12}\right)^{-300}}$$

$$\approx 1129.67$$

A monthly payment of $1129.67 will amortize this loan.

31. $153,762 at 8.45% for 30 yr

Here $P = 153,762$, $i = \frac{.0845}{12}$, and $n = 30(12) = 360$.

$$R = \frac{Pi}{1 - (1+i)^{-n}}$$

$$= \frac{153,762\left(\frac{.0845}{12}\right)}{1 - \left(1 + \frac{.0845}{12}\right)^{-360}}$$

$$\approx 1176.85$$

A monthly payment of $1176.85 will amortize this loan.

33. From Example 5, $P = 78,000$ and $i = \frac{9.6\%}{12} = \frac{.096}{12} = .008$. For a 15-yr loan, $n = 12(15) = 180$.

$$R = \frac{Pi}{1 - (1+i)^{-n}}$$

$$= \frac{78,000(.008)}{1 - (1+.008)^{-180}}$$

$$\approx 819.21$$

The monthly payment would be $819.21. The family makes 180 payments of $819.21 each, for a total of $147,457.80. Since the amount of the loan was $78,000, the total interest paid is

$$\$147,457.80 - 78,000 = \$69,457.80.$$

The payments for the 15-yr loan are

$$\$819.21 - 661.56 = \$157.65$$

more than those for the 30-yr loan in Example 5. However, the total interest paid is

$$\$160,161.60 - 69,457.80 = \$90,703.80$$

less than for the 30-yr loan in Example 5.

35. (a) $P = 6000$. $i = \frac{12}{12} = .01$, $n = 4(12) = 48$

$$R = \frac{Pi}{1 - (1+i)^{-n}}$$

$$= \frac{6000(.01)}{1 - (1+.01)^{-48}}$$

$$\approx 158$$

The amount of each payment is $158.

(b) 48 payments of $158 are made, and 48($158) = $7584. The total amount of interest Le will pay is $7584 - $6000 = $1584.

37. For parts (a) and (b), if $1 million is divided into 20 equal payments, each payment is $50,000.

(a) $i = .05$, $n = 20$

$$P = R\left[\frac{1 - (1+i)^{-n}}{i}\right]$$

$$= 50,000\left[\frac{1 - (1+.05)^{-20}}{.05}\right]$$

$$\approx 623,110.52$$

The present value is $623,110.52.

(b) $i = .09$, $n = 20$

$$P = R\left[\frac{1-(1+i)^{-n}}{i}\right]$$

$$= 50,000\left[\frac{1-(1+.09)^{-20}}{.09}\right]$$

$$\approx 456,427.28$$

The present value is $456,427.28$.

For parts (c) and (d), if \$1 million is divided into 25 equal payments, each payment is \$40,000.

(c) $i = .05$, $n = 25$

$$P = R\left[\frac{1-(1+i)^{-n}}{i}\right]$$

$$= 40,000\left[\frac{1-(1+.05)^{-25}}{.05}\right]$$

$$\approx 563,757.78$$

The present value is $563,757.78$.

(d) $i = .09$, $n = 25$

$$P = R\left[\frac{1-(1+i)^{-n}}{i}\right]$$

$$= 40,000\left[\frac{1-(1+.09)^{-25}}{.09}\right]$$

$$\approx 392,903.18$$

The present value is $392,903.18$.

39. $P = 35,000$ at 7.43% compounded monthly for 20 yr. Thus, $i = \frac{.0743}{12} = .006191\overline{6}$ and $n = 20(12) = 240$.

$$R = \frac{Pi}{1-(1+i)^{-n}}$$

$$= \frac{35,000(.006191\overline{6})}{1-(1+.006191\overline{6})^{-240}}$$

$$\approx 280.46$$

The monthly payment is 280.46. The total interest is given by

$$240(280.46) - 35,000 = 32,310.40.$$

The total interest is $32,310.40$.

41. $P = 72,000, i = \frac{10}{2} = .05, n = 9$

$$R = \frac{72,000}{a_{\overline{9}|.05}} \approx \$10,129.69$$

is the amount of each payment.

Of the first payment, the company owes interest of

$$I = Prt = 72,000(.05)(1) = \$3600.$$

Therefore, from the first payment, $3600 goes to interest and the balance,

$$\$10,129.69 - 3600 = \$6529.69,$$

goes to principal. The principal at the end of this period is

$$\$72,000,000 - 6529.69 = \$65,470.31.$$

The interest for the second payment is

$$I = Prt = 65,470.31(.05)(1) = \$3273.52.$$

Of the second payment, $3273.52 goes to interest and

$$\$10,129.69 - 3273.52 = \$6856.17$$

goes to principal. Continue in this way to complete the amortization schedule for the first four payments.

Payment Number	Amount of Payment	Interest for Period	Portion to Principal	Principal at End of Period
0	——	——	——	$72,000.00
1	$10,129.69	$3600.00	$6529.69	$65,470.31
2	$10,129.69	$3273.52	$6856.17	$58,614.14
3	$10,129.69	$2930.71	$7198.98	$51,415.16
4	$10,129.69	$2570.76	$7558.93	$43,856.23

43. The total amount of the loan is

$$14,000 + 7200 - 1200 = 20,000.$$

We have \$20,000 at 12% compounded semiannually for 5 yr.

(a) $i = \frac{.12}{2} = .06$, $n = 5(2) = 10$

$$R = \frac{Pi}{1-(1+i)^{-n}}$$

$$= \frac{20,000(.06)}{1-(1+.06)^{-10}} \approx 2717.36$$

The amount of each payment is 2717.36.

(b) Graph

$$y_1 = 2717.36\left[\frac{1-1.06^{-(10-x)}}{.06}\right] \text{ and } y_2 = 5000.$$

The x-coordinate of the point of intersection is 7.9923292 or approximately 8. Therefore, 2 payments are left.

45. This is an amortization problem with $P = 25,000$. R represents the amount of each annual withdrawal.

(a) $i = .06$, $n = 8$

$$R = \frac{Pi}{1 - (1+i)^{-n}}$$

$$= \frac{25,000(.06)}{1 - (1 + .06)^{-8}}$$

$$\approx 4025.90$$

She will be able to withdraw about $4025.90/mo for the 8 yr.

(b) $i = .06$, $n = 12$

$$R = \frac{Pi}{1 - (1+i)^{-n}}$$

$$= \frac{25,000(.06)}{1 - (1 + .06)^{-12}}$$

$$\approx 2981.93$$

She will be able to withdraw about $2981.93/mo for the 12 yr.

47. This exercise should be solved by graphing calculator or computer methods. The amortization schedule, which may vary slightly, is as follows.

Payment Number	Amount of Payment	Interest for Period	Portion to Principal	Principal at End of Period
1	$5783.49	$3225.54	$2557.95	$35,389.55
2	$5783.49	$3008.11	$2775.38	$32,614.17
3	$5783.49	$2772.20	$3011.29	$29,602.88
4	$5783.49	$2516.24	$3267.25	$26,335.63
5	$5783.49	$2238.53	$3544.96	$22,790.67
6	$5783.49	$1937.21	$3846.28	$18,944.39
7	$5783.49	$1610.27	$4173.22	$14,771.17
8	$5783.49	$1255.55	$4527.94	$10,243.22
9	$5783.49	$870.67	$4912.82	$5330.41
10	$5783.49	$453.08	$5330.41	$0.00

Chapter 5 Review Exercises

1. $I = Prt$

$$= 15,903(.08)\left(\frac{8}{12}\right)$$

$$= 848.16$$

The simple interest is $848.16.

3. $I = Prt$

$$= 42,368(.0522)\left(\frac{5}{12}\right)$$

$$= 921.50$$

The simple interest is $921.50.

5. For a given amount of money at a given interest rate for a given time period greater than 1, compound interest produces more interest than simple interest.

7. $19,456.11 at 12% compounded semiannually for 7 yr

Use the formula for compound amount with $P = 19,456.11$, $i = \frac{.12}{2} = .06$, and $n = 7(2) = 14$.

$$A = P(1+i)^n$$
$$= 19,456.11(1.06)^{14}$$
$$\approx 43,988.40$$

The compound amount is $43,988.40.

9. $57,809.34 at 12% compounded quarterly for 5 yr

Use the formula for compound amount with $P = 57,809.34$, $i = \frac{.12}{4} = .03$, and $n = 5(4) = 20$.

$$A = P(1+i)^n$$
$$= 57,809.34(1.03)^{20}$$
$$\approx 104,410.10$$

The compound amount is $104,410.10.

11. $12,699.36 at 10% compounded semiannually for 7 yr

Here $P = 12,699.36$, $i = \frac{.10}{2} = .05$, and $n = 7(2) = 14$. First find the compound amount.

$$A = P(1+i)^n$$
$$= 12,699.36(1.05)^{14}$$
$$\approx 25,143.86$$

The compound amount is $25,143.86.

To find the amount of interest earned, subtract the initial deposit from the compound amount. The interest earned is

$$\$25,143.86 - 12,699.36 = \$12,444.50.$$

13. $34,677.23 at 9.72% compounded monthly for 32 mo

Here $P = 34,677.23$, $i = \frac{.0972}{12} = .0081$, and $n = 32$.

$$A = P(1+i)^n$$
$$= 34,677.23(1.0081)^{32}$$
$$\approx 44,891.08$$

The compound amount is $44,891.08.

The interest earned is

$$\$44,891.08 - 34,677.23 = \$10,213.85.$$

15. $42,000 in 7 yr, 12% compounded monthly

Use the formula for present value for compound interest with $A = 42,000$, $i = \frac{12}{12} = .01$, and $n = 7(12) = 84$.

$$P = \frac{A}{(1+i)^n} = \frac{42.000}{(1.01)^{84}} \approx 18,207.65$$

The present value is $18,207.65.

17. $1347.89 in 3.5 yr, 6.77% compounded semiannually

Use the formula for present value for compound interest with $A = 1347.89$, $i = \frac{.0677}{2} = .03385$, and $n = 3.5(2) = 7$.

$$P = \frac{A}{(1+i)^n} = \frac{1347.89}{(1.03385)^7} \approx 1067.71$$

The present value is $1067.71.

19. $a = 2$; $r = 3$

The first five terms are

$$2,\ 2(3),\ 2(3)^2,\ 2(3)^3,\ \text{and}\ 2(3)^4,$$

or

$$2, 6, 18, 54,\ \text{and}\ 162.$$

21. $a = -3$; $r = 2$

To find the sixth term, use the formula $a_n = ar^{n-1}$ with $a = -3$, $r = 2$, and $n = 6$.

$$a_6 = ar^{6-1} = -3(2)^5 = -3(32) = -96$$

23. $a = -3$; $r = 3$

To find the sum of the first 4 terms of this geometric sequence, use the formula $S_n = \frac{a(r^n-1)}{r-1}$ with $n = 4$.

$$S_4 = \frac{-3(3^4-1)}{3-1} = \frac{-3(80)}{2} = \frac{-240}{2} = -120$$

25. $s_{\overline{n}|i} = \dfrac{(1+i)^n - 1}{i}$

$$s_{\overline{30}|.01} = \frac{(1.01)^{30} - 1}{.01} \approx 34.78489$$

29. $1288 deposited at the end of each year for 14 yr; money earns 8% compounded annually

Since deposits are made at the end of each time period, this is an ordinary annuity. Use the formula for future value of an ordinary annuity with $R = 1288$, $i = .08$, and $n = 14$.

$$S = R\left[\frac{(1+i)^n - 1}{i}\right]$$
$$= 1288\left[\frac{(1.08)^{14} - 1}{.08}\right]$$
$$\approx 31,188.82$$

The future value is $31,188.82.

31. $233 deposited at the end of each month for 4 yr; money earns 12% compounded monthly

Since deposits are made at the end of each period, this is an ordinary annuity. Here, $R = 233$, $i = \frac{12}{12} = .01$, and $n = 4(12) = 48$.

$$S = R\left[\frac{(1+i)^n - 1}{i}\right]$$
$$= 233\left[\frac{(1.01)^{48} - 1}{.01}\right]$$
$$\approx 14,264.87$$

The future value is $14,264.87.

33. $11,900 deposited at the beginning of each month for 13 mo; money earns 12% compounded monthly

Since deposits are made at the beginning of each time period, this is an annuity due. Use the formula for future value of an annuity due with $R = 11,900$, $i = \frac{12}{12} = .01$, and $n = 13$.

$$S = R\left[\frac{(1+i)^{n+1} - 1}{i}\right] - R$$
$$= 11,900\left[\frac{(1.01)^{14} - 1}{.01}\right] - 11,900$$
$$= 165.974.31$$

The future value is $165.974.31.

35. $6500; money earns 8% compounded annually; 6 annual payments

$S = 6500$, $i = .08$, $n = 6$
Let R be the amount of each payment.

$$S = Rs_{\overline{n}|i}$$
$$R = \frac{6500}{s_{\overline{6}|.08}}$$
$$= \frac{6500(.08)}{(1 + .08)^6 - 1}$$
$$\approx 886.05$$

The amount of each payment is $886.05.

37. $233,188; money earns 9.7% compounded quarterly for $7\frac{3}{4}$ yr

Here $S = 233,188$, $i = \frac{.097}{4} = .02425$, and $n = 7.75(4) = 31$.

$$S = Rs_{\overline{n}|i}$$
$$R = \frac{233,188}{s_{\overline{31}|.02425}}$$
$$= \frac{233,188(.02425)}{(1 + .02425)^{31} - 1}$$
$$\approx 5132.48$$

The amount of each payment is $5132.48.

39. Deposits of $850 annually for 4 yr at 8% compounded annually

Use the formula for present value of an annuity with $R = 850$, $i = .08$, and $n = 4$.

$$P = R\left[\frac{1 - (1 + i)^{-n}}{i}\right]$$
$$= 850\left[\frac{1 - (1.08)^{-4}}{.08}\right]$$
$$\approx 2815.31$$

The present value is $2815.31.

41. Payments of $4210 semiannually for 8 yr at 8.6% compounded semiannually

Use the formula for present value of an annuity with $R = 4210$, $i = \frac{.086}{2} = .043$, and $n = 8(2) = 16$.

$$P = R\left[\frac{1 - (1 + i)^{-n}}{i}\right]$$
$$= 4210\left[\frac{1 - (1.043)^{-16}}{.043}\right]$$
$$\approx 47,988.11$$

The present value is $47,988.11.

43. Two types of loans that are commonly amortized are home loans and auto loans.

45. $3200 loan; 8% compounded quarterly; 10 quarterly payments

Use the formula for amortization payments with $P = 3200$, $i = \frac{.08}{4} = .02$, and $n = 10$.

$$R = \frac{Pi}{1 - (1 + i)^{-n}}$$
$$= \frac{3200(.02)}{1 - (1.02)^{-10}}$$
$$\approx 356.24$$

The payment necessary to amortize this loan is $356.24.

47. $51,607 loan; 13.6% compounded monthly; 32 monthly payments

Use the formula for amortization payments with $P = 51,607$, $i = \frac{.136}{12}$, and $n = 32$.

$$R = \frac{Pi}{1 - (1 + i)^{-n}}$$
$$= \frac{51,607\left(\frac{.136}{12}\right)}{1 - \left(1 + \frac{.136}{12}\right)^{-32}}$$
$$\approx 1931.82$$

The payment necessary to amortize this loan is $1931.82.

49. $77,110 at 11.45% for 30 yr

Use the formula for amortization payments with $P = 77,110$, $i = \frac{.1145}{12}$, and $n = 30(12) = 360$.

$$R = \frac{Pi}{1 - (1 + i)^{-n}}$$
$$= \frac{77,110\left(\frac{.1145}{12}\right)}{1 - \left(1 + \frac{.1145}{12}\right)^{-360}}$$
$$\approx 760.67$$

The monthly house payment for this mortgage is $760.67.

51. The answer can be found in the table under payment number 12 in the column labeled "Portion to Principal." The amount of principal repayment included in the fifth payment is $132.99.

53. The last entry in the column "Principal at End of Period," $125,464.43, shows the debt remaining at the end of the first year (after 12 payments). Since the original debt (loan principal) was $127,000, the amount by which the debt has been reduced at the end of the first year is

$$\$127,000 - 125,464.43 = \$1535.57.$$

55. Here $P = 9820$, $r = 12.1\% = .121$, and $t = \frac{7}{12}$.

$$I = Prt$$

$$= 9820(.121)\left(\frac{7}{12}\right)$$

$$\approx 693.13$$

The interest he will pay is $693.13. The total amount he will owe in 7 mo is

$$\$9820 + 693.13 = \$10,513.13.$$

57. $P = 84,720$, $t = \frac{7}{12}$, $I = 4055.46$

Substitute these values into the formula for simple interest to find the value of r.

$$I = Prt$$

$$4055.46 = 84,720r\left(\frac{7}{12}\right)$$

$$4055.46 = 49,420r$$
$$.0821 \approx r$$

The interest rate is 8.21%.

59. Use the formula for present value for compound interest with $A = 2800$, $i = \frac{.06}{12} = .005$, and $n = 17$.

$$P = \frac{A}{(1+i)^n} = \frac{2800}{(1.005)^{17}} \approx 2572.38$$

A lump sum of $2572.38 must be invested.

61. $R = 5000$, $i = \frac{.10}{2} = .05$, $n = 7\frac{1}{2}(2) = 15$

This is an ordinary annuity.

$$S = R\left[\frac{(1+i)^n - 1}{i}\right]$$

$$S = 5000\left[\frac{(1+.05)^{15} - 1}{.05}\right]$$

$$\approx 107,892.82$$

The future value is $107,892.82. The amount of interest earned is

$$\$107,892.82 - 15(5000) = \$32,892.82.$$

63. Use the formula for amortization payments with $P = 48,000$, $i = .10$ and $n = 7$.

$$R = \frac{Pi}{1 - (1+i)^{-n}}$$

$$= \frac{48,000(.10)}{1 - (1+.10)^{-7}}$$

$$\approx 9859.46$$

The owner should deposit about $9859.46 at the end of each year.

65. Use the formula for amortization payments with $P = 115,700$, $i = \frac{.105}{12} = .00875$, and $n = 300$.

$$R = \frac{Pi}{1 - (1+i)^{-n}}$$

$$= \frac{115,700(.00875)}{1 - (1+.00875)^{-300}}$$

$$\approx 1092.42$$

Each monthly payment will be about $1092.42. The total amount of interest will be

$$300(\$1092.42) - 115,700 = \$212,026.$$

67. The death benefit grows to

$$10,000(1.05)^7 \approx 14,071.$$

This 14,071 is the present value of an annuity due with $P = 14,071$, $i = \frac{.03}{12} = .0025$, and $n = 120$. Let X represent the amount of each monthly payment.

$$P = R \cdot a_{\overline{n+1}|i} - R$$

$$14,071 = X \cdot a_{\overline{121}|.0025} - X$$

$$14,071 = (a_{\overline{121}|.0025} - 1)X$$

$$14,071 \approx (104.301 - 1)X$$
$$14,071 = 103.301X$$
$$135 \approx X$$

Each payment is about $135, which corresponds to choice (d).

Extended Application: Time, Money, and Polynomials

1.

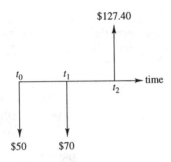

The polynomial equation is

$$50(1+i)^2 + 70(1+i) - 127.40 = 0.$$

Let $x = 1 + i$. The equation becomes

$$50x^2 + 70x - 127.40 = 0.$$

Use the quadratic formula to solve the equation for x.

$$x = \frac{-b \pm \sqrt{b^2 - 4ac}}{2a}$$

$$x = \frac{-70 \pm \sqrt{70^2 - 4(50)(-127.40)}}{2(50)}$$

Reject $x = \frac{-70 - \sqrt{30,380}}{100}$ because it is negative. Thus,

$$x = \frac{-70 + \sqrt{30,380}}{100} \approx 1.04298.$$

Since $x = 1 + i$,

$$1 + i = 1.04298$$
$$i \approx .043.$$

Thus, the YTM is 4.3%.

2. (a)

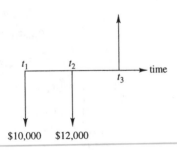

(b) $A = 1.05(10,000) + .045(1.05)(10,000)$
$\qquad + 1.045(12,000)$
$\qquad = 23,512.5$

At the end of the second year, $23,512.50 was in the account.

(c) The polynomial equation is

$$10,000(1+i)^2 + 12,000(1+i) = 23,512.50 = 0.$$

Let $x = 1 + i$. The equation becomes

$$10,000x^2 + 12,000x - 23,512.50 = 0.$$

Use the quadratic formula to solve for x.

$$x = \frac{-12,000 \pm \sqrt{12,000^2 - 4(10,000)(-23,512.50)}}{2(10,000)}$$

Reject

$$x = \frac{-12,000 - \sqrt{12,000^2 - 4(10,000)(-23,512.50)}}{20,000}$$

because it is negative. Thus,

$$x = \frac{-12,000 + \sqrt{12,000^2 - 4(10,000)(-23,512.50)}}{20,000}$$

$$\approx 1.04658.$$

Since $x = 1 + i$,

$$1 + i = 1.04658$$
$$i = .04658.$$

Thus, the YTM is 4.7%. As might be expected, the YTM is between 4.5% and 5%.

3. (a)

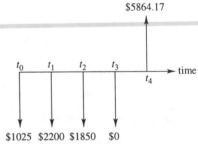

(b) The polynomial equation is

$$1025(1+i)^4 + 2200(1+i)^3$$
$$+ 1850(1+i)^2 - 5864.17 = 0.$$

Let $x = 1 + i$. The equation becomes

$$1025x^4 + 2200x^3 + 1850x^2 - 5864.17 = 0.$$

Let $f(x) = 1025x^4 + 2200x^3 + 1850x^2 - 5864.17$. Since $0 < i < 1$, then $1 < x < 2$ and

$$f(1) = -789.17;$$
$$f(1.1) = 803.2325;$$
$$f(1.05) = -31.8761;$$
$$f(1.052) = .00172.$$

The YTM is approximately 5.2%.

4. (a)

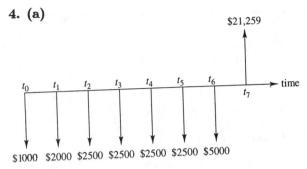

$21,259

$1000 $2000 $2500 $2500 $2500 $2500 $5000

(b) The polynomial equation is

$$1000(1+i)^7 + 2000(1+i)^6 + 2500(1+i)^5$$
$$+2500(1+i)^4 + 2500(1+i)^3 + 2500(1+i)^2$$
$$+5000(1+i) - 21,259 = 0.$$

(c) Let $x = 1 + i$ and

$$f(x) = 1000(x^7 + 2x^6 + 2.5x^5 + 2.5x^4 + 2.5x^3$$
$$+ 2.5x^2 + 5x - 21.259).$$

Then

$$f(1.0507) = 7.1216;$$
$$f(1.0505) = -6.9040.$$

Since $f(1.0507)$ is positive and $f(1.0505)$ is negative, the value of x that makes $f(x)$ zero is between 1.0507 and 1.0505.

$$1.0505 < 1 + i < 1.0507$$
$$.0505 < i < .0507$$
$$5.05\% < i < 5.07\%$$

(d) Graph $y_1 = f(x)$. The graph intersects the x-axis at $x = 1.0505985$. Therefore, $i = .0505985$ or 5.06%.

5. (a)

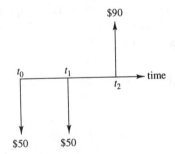

$90

$50 $50

The polynomial equation is

$$50(1+i)^2 + 50(1+i) - 90 = 0.$$

Let $x = 1 + i$. The equation becomes

$$50x^2 + 50x - 90 = 0$$
$$5x^2 + 5x - 9 = 0.$$

Solve for x using the quadratic formula.

$$x = \frac{-5 \pm \sqrt{5^2 - 4(5)(-9)}}{2(5)}$$

$$x = \frac{-5 \pm \sqrt{205}}{10}$$

$$x = \frac{-5 + \sqrt{205}}{10} \quad \text{or} \quad x = \frac{-5 - \sqrt{205}}{10}$$

$$\approx .93178 \qquad\qquad \approx -1.93178$$

Then

$$\begin{aligned}
1 + i &= .93178 & \text{or} \quad 1 + i &= -1.93178 \\
i &= -.06822 & i &= -2.93178 \\
i &\approx -6.8\% \quad \text{or} & i &\approx -293.2\%.
\end{aligned}$$

(b) The reasonable answer is -6.8%.

Chapter 5 Test

[5.1]

1. Find the simple interest on $1252 at 5% for 11 months.

2. Using a 360 day year, find the simple interest on $12,000 at 6.25% for 170 days.

3. Find the compound amount if $7000 is deposited for 8 years in an account paying 6% per year compounded quarterly.

4. Find the amount that should be invested now to accumulate $8000 at 8.1% compounded monthly for 2 years.

5. Find the effective rate (to the nearest hundredth of a percent) corresponding to a nominal rate of 6.5% compounded monthly.

6. To the nearest tenth of a year, how long will it take for an investment to double if it is invested at 5.86% compounded monthly?

[5.2]

7. Find the fifth term of the geometric sequence with $a = 4.7$, $r = 2$.

8. Find the sum of the first eight terms of the geometric sequence with $a = 14$ and $r = \frac{1}{2}$.

9. Find the future value of an ordinary annuity in which $750 is deposited at the end of each quarter for 6 years, with 12% interest compounded quarterly.

10. Find the future value of an annuity due in which $1375 is deposited at the beginning of each six months for 12 years, with 8% interest compounded semiannually.

11. Ralph deposits $100 at the end of each month for 3 years in an account paying 6% compounded monthly. He then uses the money from this account to buy a certificate of deposit which pays 7.5% compounded annually. If he redeems the certificate 4 years later, how much money will he receive?

12. What amount must be deposited into a sinking fund at the end of each quarter at 8% compounded quarterly to have $10,000 in 10 years?

[5.3]

13. Find the present value of an ordinary annuity with payments of $800 per month at 6% compounded monthly for 4 years.

14. Ms Morroco borrows $12,000 at 8% compounded quarterly, to be paid off with equal quarterly payments over 2 years.

    (a) What quarterly payment is needed to amortize this loan?
    (b) Prepare an amortization table for the first 4 payments.

Chapter 5 Test Answers

1. $57.38

2. $354.17

3. $11,272.27

4. $6807.23

5. 6.70%

6. 11.9 years

7. 75.2

8. $\frac{1785}{64}$

9. $25,819.85

10. $55,888.12

11. $5253.21

12. $165.56

13. $34,064.25

14. **(a)** $1638

   (b)

Payment Number	Amount of Payment	Interest for Period	Portion to Principal	Principal at End of Period
0	—	—	—	$12,000.00
1	$1638.12	$240.00	$1398.12	$10,601.88
2	$1638.12	$212.04	$1426.08	$9175.80
3	$1638.12	$183.52	$1454.60	$7721.19
4	$1638.12	$154.42	$1483.70	$6237.49

Chapter 6

SETS AND PROBABILITY

6.1   Sets

1. $3 \in \{2, 5, 7, 9, 10\}$

The number 3 is not an element of the set, so the statement is false.

3. $9 \notin \{2, 1, 5, 8\}$

Since 9 is not an element of the set, the statement is true.

5. $\{2, 5, 8, 9\} = \{2, 5, 9, 8\}$

The sets contain exactly the same elements, so they are equal. The statement is true.

7. {All whole numbers greater than 7 and less than 10} = $\{8, 9\}$

Since 8 and 9 are the only such numbers, the statement is true.

9. $0 \in \emptyset$

The empty set has no elements. The statement is false.

In Exercises 11-21,

$$A = \{2, 4, 6, 8, 10, 12\},$$
$$B = \{2, 4, 8, 10\},$$
$$C = \{4, 8, 12\},$$
$$D = \{2, 10\},$$
$$E = \{6\},$$
$$\text{and} \quad U = \{2, 4, 6, 8, 10, 12, 14\}.$$

11. Since every element of A is also an element of U, A is a subset of U, written $A \subseteq U$.

13. A contains elements that do not belong to E, namely 2, 4, 8, 10, and 12, so A is not a subset of E, written $A \nsubseteq E$.

15. The empty set is a subset of every set, so $\emptyset \subseteq A$.

17. Every element of D is also an element of B, so D is a subset of B, $D \subseteq B$.

19. A set with n distinct elements has 2^n subsets. A has $n = 6$ elements, so there are exactly $2^6 = 64$ subsets of A.

21. A set with n distinct elements has 2^n subsets, and C has $n = 3$ elements. Therefore, there are exactly $2^3 = 8$ subsets of C.

25. $\{8, 11, 15\} \cap \{8, 11, 19, 20\} = \{8, 11\}$

$\{8, 11\}$ is the set of all elements belonging to both of the first two sets, so it is the intersection of those sets.

27. $\{6, 12, 14, 16\} \cap \{6, 14, 19\} = \{6, 14\}$

$\{6, 14\}$ is the set of all elements belonging to both of the first two sets, so it is the intersection of those sets.

29. $\{3, 5, 9, 10\} \cup \emptyset = \{3, 5, 9, 10\}$

The empty set contains no elements, so the union of any set with the empty set will result in an answer set that is identical to the original set. (On the other hand, $\{3, 5, 9, 10\} \cap \emptyset = \emptyset$.)

31. $\{1, 2, 4\} \cup \{1, 2\} = \{1, 2, 4\}$

The answer set $\{1, 2, 4\}$ consists of all elements belonging to the first set, to the second set, or to both sets, and therefore it is the union of the first two sets. (On the other hand, $\{1, 2, 4\} \cap \{1, 2\} = \{1, 2\}$.)

In Exercises 33-41,

$$U = \{2, 3, 4, 5, 7, 9\},$$
$$X = \{2, 3, 4, 5\},$$
$$Y = \{3, 5, 7, 9\},$$
$$\text{and} \quad Z = \{2, 4, 5, 7, 9\}.$$

33. $X \cap Y$, the intersection of X and Y, is the set of elements belonging to both X and Y. Thus,

$$X \cap Y = \{2, 3, 4, 5\} \cap \{3, 5, 7, 9\}$$
$$= \{3, 5\}.$$

35. X', the complement of X, consists of those elements of U that are not in X. Thus,

$$X' = \{7, 9\}.$$

37. From Exercise 35, $X' = \{7,9\}$; from Exercise 36, $Y' = \{2,4\}$. There are no elements common to both X' and Y' so

$$X' \cap Y' = \emptyset.$$

39. $X = \{2,3,4,5\}$ and $Y \cap Z = \{5,7,9\}$.

Hence, the union of these two sets is

$$X \cup (Y \cap Z) = \{2,3,4,5,7,9\}$$
$$= U.$$

41. From Exercise 35, $X' = \{7,9\}$; from Exercise 36, $Y' = \{2,4\}$.

$X' \cap (Y' \cup Z)$
$= \{7,9\} \cap (\{2,4\} \cup \{2,4,5,7,9\})$
$= \{7,9\} \cap \{2,4,5,7,9\}$
$= \{7,9\}$

43. (a) $(A \cap B) \cup (A \cap B')$
$= (\{3,6,9\} \cap \{2,4,6,8\}) \cup (\{3,6,9\}$
$\quad \cap \{0,1,3,5,7,9,10\})$
$= \{6\} \cup \{3,9\}$
$= \{3,6,9\}$
$= A$

45. $M \cup N$ is the set of all students in this school taking this course or taking accounting.

47. $N' \cap P'$ is the set of all students in this school not taking accounting and not taking zoology.

49. Disjoint sets have no elements in common. Since each pair of sets has at least one element in common, none of the pairs are disjoint.

51. $A \cap B$ is the set of all stocks with a high price greater than \$100 and a last price between \$75 and \$100. Since no stocks fall into both categories, $A \cap B = \emptyset$.

53. $(A \cup C)'$ is the set of all stocks on the list that do not have a high price greater than \$100 or a positive price change.

$$(A \cup C)' = \{\text{ATT, GnMill, PepsiCo}\}$$

55. $U = \{s,d,c,g,i,m,h\}$ and $N = \{s,d,c,g\}$, so

$$N' = \{i,m,h\}.$$

57. $N \cup O = \{s,d,c,g\} \cup \{i,m,h,g\}$
$= \{s,d,c,g,i,m,h\}$
$= U$

59. The number of subsets of a set with 51 elements (50 states plus the District of Columbia) is

$$2^{51} \approx 2.522 \times 10^{15}.$$

61. The number of people in $F \cup B$ is given by the number of people in F plus the number of people in B minus the number of people in $F \cap B$ (since these cannot be counted twice). This is

$$51.1 + 26.8 - 3.7 = 74.2.$$

Therefore, the number of people in $F \cup B$ is 74.2 million.

63. $E' \cap (B \cup G) = (F \cup G) \cap (B \cup G)$ since $E' = F \cup G$.

The entries in $F \cup G$ are 40.8, 3.7, 5.0, 1.6, 28.6, 1.5, 2.7, and .6.

The entries in $B \cup G$ are 21.6, 3.7, 28.6, 1.5, 2.7, and .6.

$(F \cup G) \cap (B \cup G)$ is the sum of the common values or

$$3.7 + 28.6 + 1.5 + 2.7 + .6 = 37.1.$$

Therefore, the number of people in $E' \cap (B \cup G)$ is 37.1 million.

65. $F' \cap (A' \cap C') = (E \cup G) \cap (B \cup D)$ since $F' = E \cup G$ and $A' \cap C' = B \cup D$.

The entries in $E \cup G$ are 124.5, 21.6, 23.9, 7.0, 28.6, 1.5, 2.7, and .6.

The entries in $B \cup D$ are 21.6, 3.7, 1.5, 7.0, 1.6, and .6.

$(E \cup G) \cap (B \cup D)$ is the sum of the common values or

$$21.6 + 1.5 + 7.0 + .6 = 30.7.$$

Therefore, the number of people in $F' \cap (A' \cap C')$ is 30.7 million.

67. The number of subsets of a set containing 9 elements is $2^9 = 512$.

69. The only network in the table that shows cartoons is the Disney Channel. Therefore,

$$G = \{\text{The Disney Channel}\}.$$

71. The networks that have more than 12,000 viewers and show cartoons is the set

$$F \cap G = \{\text{HBO, The Disney Channel}\}$$
$$\cap \{\text{The Disney Channel}\}$$
$$= \{\text{The Disney Channel}\}.$$

73. F' is the set of networks with less than 12,000 viewers. Since $F = \{\text{HBO, The Disney Channel}\}$ and $U = \{\text{HBO, The Disney Channel, Showtime/The Movie Channel, Spice, Cinemax}\}$,

$$F' = \{\text{Showtime/The Movie Channel,}$$
$$\text{Spice, Cinemax}\}.$$

6.2 Applications of Venn Diagrams

1. $B \cap A'$ is the set of all elements in B and not in A. See the Venn diagram in the back of the textbook.

3. $A' \cup B$ is the set of all elements that do not belong to A or that do belong to B, or both. See the Venn diagram in the back of the textbook.

5. $B' \cup (A' \cap B')$

First find $A' \cap B'$, the set of elements not in A and not in B.

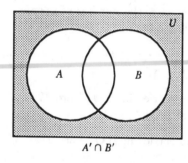

$A' \cap B'$

For the union, we want those elements in B' or $(A' \cap B')$, or both.

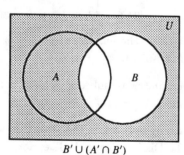

$B' \cup (A' \cap B')$

7. U' is the empty set $\emptyset$. See the Venn diagram in the back of the textbook.

9. Three sets divide the universal set into 8 regions. (Examples of this situation will be seen in Exercises 11-17.)

11. $(A \cap C') \cup B$

First find $A \cap C'$, the region in A and not in C.

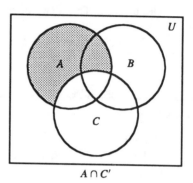

$A \cap C'$

For the union, we want the region in $(A \cap C')$ or in B, or both.

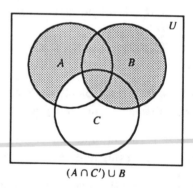

$(A \cap C') \cup B$

13. $A' \cap (B \cap C)$

First find A', the region not in A.

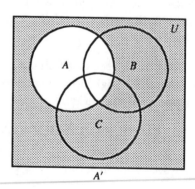

A'

Then find $B \cap C$, the region where B and C overlap.

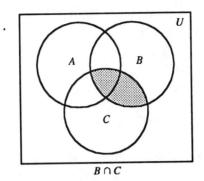

$B \cap C$

Now intersect these regions.

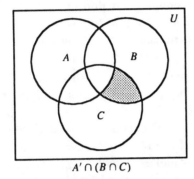

$A' \cap (B \cap C)$

15. $(A \cap B') \cup C$

First find $A \cap B'$, the region in A *and* not in B.

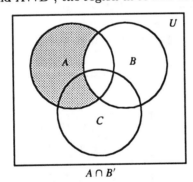

$A \cap B'$

For the union, we want the region in $(A \cap B')$ *or* in C, or both.

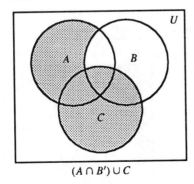

$(A \cap B') \cup C$

17. $A' \cap (B' \cup C)$

First find A'.

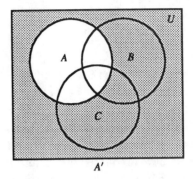

A'

Then find $B' \cup C$, the region not in B *or* in C, or both.

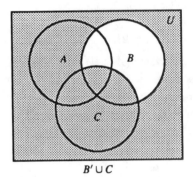

$B' \cup C$

Now intersect these regions.

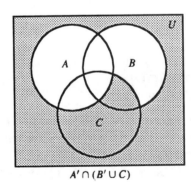

$A' \cap (B' \cup C)$

19. $n(A \cup B) = n(A) + n(B) - n(A \cap B)$
$$= 5 + 8 - 4$$
$$= 9$$

21. $n(A \cup B) = n(A) + n(B) - n(A \cap B)$
$$20 = n(A) + 7 - 3$$
$$20 = n(A) + 4$$
$$16 = n(A)$$

For Exercises 23-29, see the Venn diagrams in the back of the textbook.

23. $n(U) = 38$
$n(A) = 16$
$n(A \cap B) = 12$
$n(B') = 20$

First put 12 in $A \cap B$. Since $n(A) = 16$, and 12 are in $A \cap B$, there must be 4 elements in A that are not in $A \cap B$. $n(B') = 20$, so there are 20 not in B. We already have 4 not in B (but in A), so there must be another 16 outside B *and* outside A. So far we have accounted for 32, and $n(U) = 38$, so 6 must be in B but not in any region yet identified. Thus $n(A' \cap B) = 6$.

25. $n(A \cup B) = 17$
$n(A \cap B) = 3$
$n(A) = 8$
$n(A' \cup B') = 21$

Start with $n(A \cap B) = 3$. Since $n(A) = 8$, there must be 5 more in A not in B. $n(A \cup B) = 17$; we already have 8, so 9 more must be in B not yet counted. $A' \cup B'$ consists of all the region not in $A \cap B$, where we have 3. So far $5 + 9 = 14$ are in this region, so another $21 - 14 = 7$ must be outside both A and B.

27. $n(A) = 28$
$n(B) = 34$
$n(C) = 25$
$n(A \cap B) = 14$
$n(B \cap C) = 15$
$n(A \cap C) = 11$
$n(A \cap B \cap C) = 9$
$n(U) = 59$

We start with $n(A \cap B \cap C) = 9$. If $n(A \cap B) = 14$, an additional 5 are in $A \cap B$ but not in $A \cap B \cap C$. Similarly, $n(B \cap C) = 15$, so $15 - 9 = 6$ are in $B \cap C$ but not in $A \cap B \cap C$. Also, $n(A \cap C) = 11$, so $11 - 9 = 2$ are in $A \cap C$ but not in $A \cap B \cap C$.

Now we turn our attention to $n(A) = 28$. So far we have $2 + 9 + 5 = 16$ in A; there must be another $28 - 16 = 12$ in A not yet counted. Similarly, $n(B) = 34$; we have $5 + 9 + 6 = 20$ so far, and $34 - 20 = 14$ more must be put in B. For C, $n(C) = 25$; we have $2 + 9 + 6 = 17$ counted so far. Then there must be 8 more in C not yet counted. The count now stands at 56, and $n(U) = 59$, so 3 must be outside the three sets.

29. $n(A \cap B) = 6$
$n(A \cap B \cap C) = 4$
$n(A \cap C) = 7$
$n(B \cap C) = 4$
$n(A \cap C') = 11$
$n(B \cap C') = 8$
$n(C) = 15$
$n(A' \cap B' \cap C') = 5$

Start with $n(A \cap B) = 6$ and $n(A \cap B \cap C) = 4$ to get $6 - 4 = 2$ in that portion of $A \cap B$ outside of C. From $n(B \cap C) = 4$, there are $4 - 4 = 0$ elements in that portion of $B \cap C$ outside of A. Use $n(A \cap C) = 7$ to get $7 - 4 = 3$ elements in that portion of $A \cap C$ outside of B.

Since $n(A \cap C') = 11$, there are $11 - 2 = 9$ elements in that part of A outside of B and C. Use $n(B \cap C') = 8$ to get $8 - 2 = 6$ elements in that part of B outside of A and C. Since $n(C) = 15$, there are $15 - 3 - 4 - 0 = 8$ elements in C outside of A and B. Finally, 5 must be outside all three sets, since $n(A' \cap B' \cap C') = 5$.

31. $(A \cup B)' = A' \cap B'$

For $(A \cup B)'$, first find $A \cup B$.

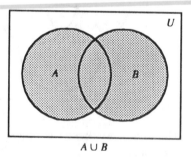

$A \cup B$

Now find $(A \cup B)'$, the region outside $A \cup B$.

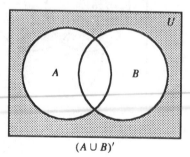

$(A \cup B)'$

For $A' \cap B'$, first find A' and B' individually.

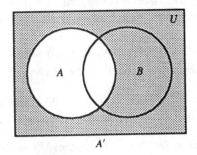

A'

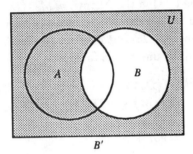

B'

Then $A' \cap B'$ is the region where A' and B' overlap, which is the entire region outside $A \cup B$ (the same result as in the second diagram). Therefore,

$$(A \cup B)' = A' \cap B'.$$

33. $A \cap (B \cup C) = (A \cap B) \cup (A \cap C)$

First find A and $B \cup C$ individually.

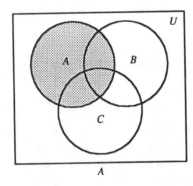

A

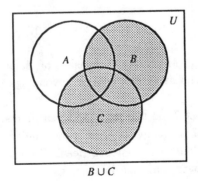

$B \cup C$

Then $A \cap (B \cup C)$ is the region where the above two diagrams overlap.

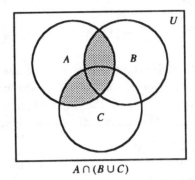

$A \cap (B \cup C)$

Next find $A \cap B$ and $A \cap C$ individually.

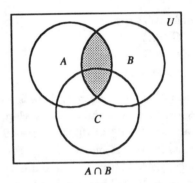

$A \cap B$

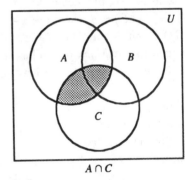

$A \cap C$

Then $(A \cap B) \cup (A \cap C)$ is the union of the above two diagrams.

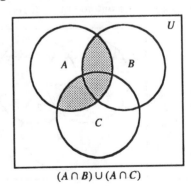

$(A \cap B) \cup (A \cap C)$

The Venn diagram for $A \cap (B \cup C)$ is identical to the Venn diagram for $(A \cap B) \cup (A \cap C)$, so conclude that

$$A \cap (B \cup C) = (A \cap B) \cup (A \cap C).$$

35. Let M be the set of those who use a microwave oven, E be the set of those who use an electric range, and G be the set of those who use a gas range. We are given the following information.

$n(U) = 140$
$n(M) = 58$
$n(E) = 63$
$n(G) = 58$
$n(M \cap E) = 19$
$n(M \cap G) = 17$
$n(G \cap E) = 4$
$n(M \cap G \cap E) = 1$
$n(M' \cap G' \cap E') = 2$

Since $n(M \cap G \cap E) = 1$, there is 1 element in the region where the three sets overlap.
Since $n(M \cap E) = 19$, there are $19 - 1 = 18$ elements in $M \cap E$ but not in $M \cap G \cap E$.
Since $n(M \cap G) = 17$, there are $17 - 1 = 16$ elements in $M \cap G$ but not in $M \cap G \cap E$.
Since $n(G \cap E) = 4$, there are $4 - 1 = 3$ elements in $G \cap E$ but not in $M \cap G \cap E$.
Now consider $n(M) = 58$. So far we have $16 + 1 + 18 = 35$ in M; there must be another $58 - 35 = 23$ in M not yet counted.
Similarly, $n(E) = 63$; we have $18 + 1 + 3 = 22$ counted so far. There must be $63 - 22 = 41$ more in E not yet counted.
Also, $n(G) = 58$; we have $16 + 1 + 3 = 20$ counted so far. There must be $58 - 20 = 38$ more in G not yet counted.
Lastly, $n(M' \cap G' \cap E') = 2$ indicates that there are 2 elements outside of all three sets.

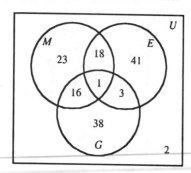

Note that the numbers in the Venn diagram add up to 142 even though $n(U) = 140$. Jeff has made some error, and he should definitely be reassigned.

37. **(a)** $n(Y \cap R) = 40$ since 40 is the number in the table where the Y row and the R column meet.

(b) $n(M \cap D) = 30$ since 30 is the number in the table where the M row and the D column meet.

(c) $n(D \cap Y) = 15$ and $n(M) = 80$ since that is the total in the M row. $n(M \cap (D \cap Y)) = 0$ since no person can simultaneously have an age in the range 21-25 *and* have an age in the range 26-35. By the union rule for sets,

$n(M \cup (D \cap Y))$
  $= n(M) + n(D \cap Y) - n(M \cap (D \cap Y))$
  $= 80 + 15 - 0$
  $= 95.$

(d) $Y' \cap (D \cup N)$ consists of all people in the D column or in the N column who are at the same time not in the Y row. Therefore,

$n(Y' \cap (D \cup N)) = 30 + 50 + 20 + 10$
   $= 110.$

(e) $n(N) = 45$
$n(O) = 70$
$n(O') = 220 - 70 = 150$
$n(O' \cap N) = 15 + 20 = 35$

By the union rule,

$n(O' \cup N) = n(O') + n(N) - n(O' \cap N)$
    $= 150 + 45 - 35$
    $= 160.$

(f) $M' \cap (R' \cap N')$ consists of all people who are not in the R column and not in the N column and who are at the same time not in the M row. Therefore,

$$n(M' \cap (R' \cap N')) = 15 + 50 = 65.$$

(g) $M \cup (D \cap Y)$ consists of all people age 21-25 who drink diet cola *or* anyone age 26-35.

39. Let T be the set of tall pea plants, G be the set of plants with green peas, and S be the set of plants with smooth peas. We are given the following information.

$n(T) = 22$
$n(G) = 25$
$n(S) = 39$
$n(T \cap G) = 9$
$n(T \cap S) = 17$
$n(G \cap S) = 20$
$n(T \cap G \cap S) = 6$
$n(T' \cap G' \cap S') = 4$

Start with the last two restricted regions, $T \cap G \cap S$ and $T' \cap G' \cap S'$. With $n(G \cap S) = 20$, there are $20 - 6 = 14$ yet to be labeled; $n(T \cap S) = 17$ puts $17 - 6 = 11$ in $T \cap S \cap G'$; also, $n(T \cap G) = 9$ puts $9 - 6 = 3$ in $T \cap G \cap S'$. Now fill in $T \cap (G' \cap S') = 22 - 20 = 2$, $G \cap (T' \cap S') = 25 - 23 = 2$, and $S \cap (T' \cap G') = 39 - 31 = 8$.

$$n(U) = 2 + 3 + 2 + 11 + 6 + 14 + 8 + 4$$
$$= 50$$

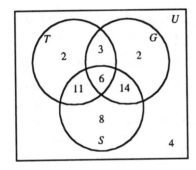

(a) $n(U) = 50$

(b) $n(T \cap S' \cap G') = 2$

(c) $n(T' \cap S \cap G) = 14$

41. First fill in the Venn diagram, starting with the region common to all three sets.

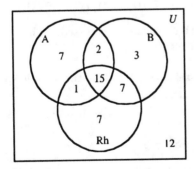

(a) The total of these numbers in the diagram is 54.

(b) $7 + 3 + 7 = 17$ had only one antigen.

(c) $1 + 2 + 7 = 10$ had exactly two antigens.

(d) A person with O-positive blood has only the Rh antigen, so this number is 7.

(e) A person with AB-positive blood has all three antigens, so this number is 15.

(f) A person with B-negative blood has only the B antigen, so this number is 3.

(g) A person with O-negative blood has none of the antigens. There are 12 such people.

(h) A person with A-positive blood has the A and Rh antigens, but not the B-antigen. The number is 1.

43. Let W be the set of women, C be the set of those who speak Cantonese, and F be the set of those who set off firecrackers. We are given the following information.

$$n(W) = 120$$
$$n(C) = 150$$
$$n(F) = 170$$
$$n(W' \cap C) = 108$$
$$n(W' \cap F') = 100$$
$$n(W \cap C' \cap F) = 18$$
$$n(W' \cap C' \cap F') = 78$$
$$n(W \cap C \cap F) = 30$$

Note that

$$n(W' \cap C \cap F')$$
$$= n(W' \cap F') - n(W' \cap C' \cap F')$$
$$= 100 - 78 = 22.$$

Furthermore,

$$n(W' \cap C \cap F)$$
$$= n(W' \cap C) - n(W' \cap C \cap F')$$
$$= 108 - 22 = 86.$$

We now have

$$n(W \cap C \cap F')$$
$$= n(C) - n(W' \cap C \cap F)$$
$$- n(W \cap C \cap F) - n(W' \cap C \cap F')$$
$$= 150 - 86 - 30 - 22 = 12.$$

With all of the overlaps of W, C, and F determined, we can now compute $n(W \cap C' \cap F') = 60$ and $n(W' \cap C' \cap F) = 36$.

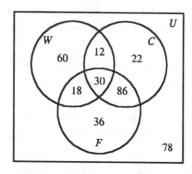

(a) Adding up the disjoint components, we find the total attendance to be

$$60 + 12 + 18 + 30 + 22 + 86 + 36 + 78 = 342.$$

(b) $n(C') = 342 - n(C)$
$$= 342 - 150 = 192$$

(c) $n(W \cap F') = 60 + 12 = 72$

(d) $n(W' \cap C \cap F) = 86$

6.3 Introduction To Probability

3. The sample space is the set of the twelve months, {January, February, March, ..., December}.

5. The possible number of points earned could be any whole number from 0 to 80. The sample space is the set

$$\{0, 1, 2, 3, \ldots, 80\}.$$

7. The possible decisions are to go ahead with a new oil shale plant or to cancel it. The sample space is the set {go ahead, cancel}.

9. Let h = heads and t = tails for the coin; the die can display 6 different numbers. There are 12 possible outcomes in the sample space, which is the set

$$\{(h, 1), (t, 1), (h, 2), (t, 2), (h, 3), (t, 3),$$
$$(h, 4), (t, 4), (h, 5), (t, 5), (h, 6), (t, 6)\}.$$

13. Use the first letter of each name. The sample space is the set

{AB, AC, AD, AE, BC, BD, BE, CD, CE, DE}.

(a) One of the committee members must be Chinn. This event is {AC, BC, CD, CE}.

(b) Alam, Bartolini, and Chinn may be on any committee; Dickson and Ellsberg may not be on the same committee. This event is

{AB, AC, AD, AE, BC, BD, BE, CD, CE}.

(c) Both Alam and Chinn are on the committee. This event is {AC}.

15. Each outcome consists of two of the numbers 1, 2, 3, 4, and 5, without regard for order. For example, let $(2, 5)$ represent the outcome that the slips of paper marked with 2 and 5 are drawn. There are 10 pairs in this sample space, which is

$$\{(1, 2), (1, 3), (1, 4), (1, 5), (2, 3),$$
$$(2, 4), (2, 5), (3, 4), (3, 5), (4, 5)\}.$$

(a) Both numbers in the outcome pair are even. This event is $\{(2, 4)\}$, which is called a simple event since it consists of only one outcome.

(b) One number in the pair is even and the other number is odd. This event is

$$\{(1, 2), (1, 4), (2, 3), (2, 5), (3, 4), (4, 5)\}.$$

(c) Each slip of paper has a different number written on it, so it is not possible to draw two slips marked with the same number. This event is $\emptyset$, which is called an impossible event since it contains no outcomes.

17. $S = \{$HH, THH, HTH, TTHH, THTH, HTTH, TTTH, TTHT, THTT, HTTT, TTTT$\}$

(a) The coin is tossed four times. This event is written {TTHH, THTH, HTTH, TTTH, TTHT, THTT, HTTT, TTTT}.

(b) Exactly two heads are tossed. This event is written {HH, THH, HTH, TTHH, THTH, HTTH}.

(c) No heads are tossed. This event is written {TTTT}.

For Exercises 19-23, use the sample space

$$S = \{1, 2, 3, 4, 5, 6\}$$

with $n(S) = 6$.

19. "Getting a 2" is the event $E = \{2\}$, so $n(E) = 1$.

If all the outcomes in a sample space S are equally likely, then the probability of an event E is

$$P(E) = \frac{n(E)}{n(S)}.$$

7. $\{3, 6, 9, 10\} \subseteq \{3, 9, 11, 13\}$

10 is an element of $\{3, 6, 9, 10\}$, but 10 is not an element of $\{3, 9, 11, 13\}$. Therefore, $\{3, 6, 9, 10\}$ is not a subset of $\{3, 9, 11, 13\}$. The statement is false.

9. $\{2, 8\} \not\subseteq \{2, 4, 6, 8\}$

Since both 2 and 8 are elements of $\{2, 4, 6, 8\}$, $\{2, 8\}$ is a subset of $\{2, 4, 6, 8\}$. This statement is false.

In Exercises 11-19,

$$U = \{a, b, c, d, e, f, g\},$$
$$K = \{c, d, f, g\},$$
$$\text{and} \quad R = \{a, c, d, e, g\}.$$

11. K has 4 elements, so it has $2^4 = 16$ subsets.

13. K' (the complement of K) is the set of all elements of U that do *not* belong to K.

$$K' = \{a, b, e\}$$

15. $K \cap R$ (the intersection of K and R) is the set of all elements belonging to both set K and set R.

$$K \cap R = \{c, d, g\}$$

17. $(K \cap R)' = \{a, b, e, f\}$ since these elements are in U but not in $K \cap R$. (See Exercise 15.)

19. $\emptyset = U$

21. $A \cap C$ is the set of all female employees in the K.O. Brown Company who are in the accounting department.

23. $A \cup D$ is the set of all employees in the K.O. Brown Company who are in the accounting department *or* have MBA degrees.

25. $B' \cap C'$ is the set of all male employees who are not in the sales department.

27. $A \cup B'$ is the set of all elements which belong to A or do not belong to B, or both. See the Venn diagram in the back of the textbook.

29. $(A \cap B) \cup C$

First find $A \cap B$.

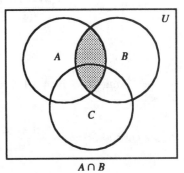

$A \cap B$

Now find the union of this region with C.

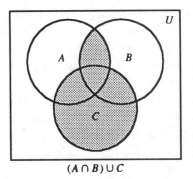

$(A \cap B) \cup C$

31. The sample space for rolling a die is

$$S = \{1, 2, 3, 4, 5, 6\}.$$

33. The sample space of the possible weights is

$$S = \{0, .5, 1, 1.5, 2, \ldots, 299.5, 300\}.$$

35. The sample space consists of all ordered pairs (a, b) where a can be 3, 5, 7, 9, or 11, and b is either R (red) or G(green). Thus,

$$S = \{(3,R), (3,G), (5,R), (5,G), (7,R),$$
$$(7,G), (9,R), (9,G), (11,R), (11,G)\}.$$

37. The event F that the second ball is green is

$$F = \{(3,G), (5,G), (7,G), (9,G), (11,G)\}.$$

39. There are 13 hearts out of 52 cards in a deck. Thus,

$$P(\text{heart}) = \frac{13}{52} = \frac{1}{4}.$$

41. There are 3 face cards in each suit (jack, queen, and king) and there are 4 suits, so there are $3 \cdot 4 = 12$ face cards out of the 52 cards. Thus,

$$P(\text{face card}) = \frac{12}{52} = \frac{3}{13}.$$

43. There are 4 queens of which 2 are red, so

$$P(\text{red}|\text{queen}) = \frac{n(\text{red and queen})}{n(\text{queen})}$$

$$= \frac{2}{4} = \frac{1}{2}.$$

45. There are 4 kings of which all 4 are face cards. Thus,

$$P(\text{face card}|\text{king}) = \frac{n(\text{face card and king})}{n(\text{king})}$$

$$= \frac{4}{4} = 1.$$

51. Marilyn vos Savant's answer is that the contestant should switch doors. To understand why, recall that the puzzle begins with the contestant choosing door 1 and then the host opening door 3 to reveal a goat. When the host opens door 3 and shows the goat, that does not affect the probability of the car being behind door 1; the contestant had a $\frac{1}{3}$ probability of being correct to begin with, and he still has a $\frac{1}{3}$ probability after the host opens door 3.

The contestant knew that the host would open another door regardless of what was behind door 1, so opening either other door gives no new information about door 1. The probability of the car being behind door 1 is still $\frac{1}{3}$; with the goat behind door 3, the only other place the car could be is behind door 2, so the probability that the car is behind door 2 is now $\frac{2}{3}$. By switching to door 2, the contestant can double his chances of winning the car.

53. Let E represent the event "draw a black jack." $P(E) = \frac{2}{52} = \frac{1}{26}$ and then $P(E') = \frac{25}{26}$. The odds in favor of drawing a black jack are

$$\frac{P(E)}{P(E')} = \frac{\frac{1}{26}}{\frac{25}{26}} = \frac{1}{25},$$

or 1 to 25.

55. The sum is 8 for each of the 5 outcomes 2-6, 3-5, 4-4, 5-3, and 6-2. There are 36 outcomes in all in the sample space.

$$P(\text{sum is 8}) = \frac{5}{36} \approx .139$$

57. $P(\text{sum is at least 10})$
$$= P(\text{sum is 10}) + P(\text{sum is 11})$$
$$\quad + P(\text{sum is 12})$$
$$= \frac{3}{36} + \frac{2}{36} + \frac{1}{36}$$
$$= \frac{6}{36} = \frac{1}{6} \approx .167$$

59. The sum can be 9 or 11. $P(\text{sum is 9}) = \frac{4}{36}$ and $P(\text{sum is 11}) = \frac{2}{36}$.

$P(\text{sum is odd number greater than 8})$
$$= \frac{4}{36} + \frac{2}{36}$$
$$= \frac{6}{36} = \frac{1}{6} \approx .167$$

61. Consider the reduced sample space of the 11 outcomes in which at least one die is a four. Of these, 2 have a sum of 7, 3-4 and 4-3. Therefore,

$P(\text{sum is 7}|\text{at least one die is a 4})$
$$= \frac{2}{11} \approx .182.$$

63. $P(E) = .51$, $P(F) = .37$, $P(E \cap F) = .22$

(a) $P(E \cup F) = P(E) + P(F) - P(E \cap F)$
$$= .51 + .37 - .22$$
$$= .66$$

(b) Draw a Venn diagram.

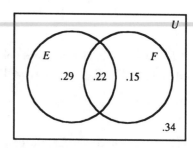

$E \cap F'$ is the portion of the diagram that is inside E and outside F.

$$P(E \cap F') = .29$$

(c) $E' \cup F$ is outside E or inside F, or both.

$$P(E' \cup F) = .22 + .15 + .34 = .71.$$

(d) $E' \cap F'$ is outside E and outside F.

$$P(E' \cap F') = .34$$

65. The probability that the ball came from box B, given that it is red, is

$P(B|\text{red})$

$$= \frac{P(B) \cdot P(\text{red}|B)}{P(B) \cdot P(\text{red}|B) + P(A) \cdot P(\text{red}|A)}$$

$$= \frac{\frac{5}{8}\left(\frac{2}{5}\right)}{\frac{5}{8}\left(\frac{2}{5}\right) + \frac{3}{8}\left(\frac{5}{6}\right)}$$

$$= \frac{4}{9} \approx .444.$$

67. Let C represent "competent shop" and R represent "able to repair appliance." Draw a tree diagram and label the given information.

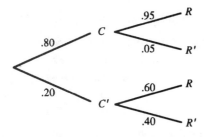

The probability that an appliance that was repaired correctly was repaired by an incompetent shop is

$$P(C'|R) = \frac{P(C' \cap R)}{P(R)}$$

$$= \frac{P(C') \cdot P(R|C')}{P(C) \cdot P(R|C) + P(C') \cdot P(R|C')}$$

$$= \frac{.20(.60)}{.80(.95) + .20(.60)}$$

$$= \frac{.12}{.76 + .12} = \frac{.12}{.88} = \frac{12}{88}$$

$$= \frac{3}{22} \approx .136.$$

69. See the tree diagram in Exercise 67. The probability that an appliance that was repaired incorrectly was repaired by an incompetent shop is

$$P(C'|R') = \frac{P(C' \cap R')}{P(R')}$$

$$= \frac{.20(.40)}{.20(.40) + .80(.05)}$$

$$= \frac{.08}{.12} = \frac{8}{12} = \frac{2}{3}.$$

71. (a) $P(\text{no more than 3 defects})$
$$= P(0) + P(1) + P(2) + P(3)$$
$$= .31 + .25 + .18 + .12$$
$$= .86$$

(b) $P(\text{at least 3 defects})$
$$= P(3) + P(4) + P(5)$$
$$= .12 + .08 + .06$$
$$= .26$$

73. (a)

Car Type	Satisfied	Not Satisfied	Totals
New	300	100	400
Used	450	150	600
Totals	750	250	1000

(b) 1000 buyers were surveyed.

(c) 300 bought a new car and were satisfied.

(d) 250 were not satisfied.

(e) 600 bought used cars.

(f) 150 who were not satisfied had bought a used car.

(g) The event is "those who purchased a used car given that the buyer is not satisfied."

(h) $P(\text{used car}|\text{not satisfied})$

$$= \frac{n(\text{used car and not satisfied})}{n(\text{not satisfied})}$$

$$= \frac{150}{250} = \frac{3}{5}$$

(i) $P(\text{used car and not satisfied})$

$$= \frac{n(\text{used car and not satisfied})}{n(\text{buyers})}$$

$$= \frac{150}{600} = \frac{1}{4}$$

75. Let M be the event "voter was male" and D be the event "voter voted Democratic." Draw a tree diagram.

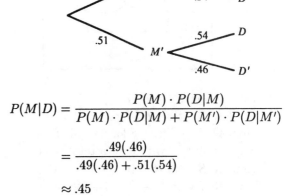

$$P(M|D) = \frac{P(M) \cdot P(D|M)}{P(M) \cdot P(D|M) + P(M') \cdot P(D|M')}$$

$$= \frac{.49(.46)}{.49(.46) + .51(.54)}$$

$$\approx .45$$

77. (a) $P(\text{answer yes})$
$$= P(\text{answer } B) \cdot P(\text{answer yes}|\text{answer } B)$$
$$+ P(\text{answer } A) \cdot P(\text{answer yes}|\text{answer } A)$$

Divide by $P(\text{answer } B)$.

$$\frac{P(\text{answer yes})}{P(\text{answer } B)} = P(\text{answer yes}|\text{answer } B)$$
$$+ \frac{P(\text{answer } A) \cdot P(\text{answer yes}|\text{answer } A)}{P(\text{answer } B)}$$

Solve for $P(\text{answer yes}|\text{answer } B)$.

$P(\text{answer yes}|\text{answer } B)$

$$= \frac{P(\text{answer yes}) - P(\text{answer } A) \cdot P(\text{answer yes}|\text{ answer } A)}{P(\text{answer } B)}$$

(b) Using the formula from part (a),

$$\frac{.6 - \frac{1}{2}\left(\frac{1}{2}\right)}{\frac{1}{2}} = \frac{7}{10}.$$

79. Let $W =$ the set of western states;
$S =$ the set of small states;
$E =$ the set of early states.

We are given the following information.

$n(W) = 22$
$n(S) = 22$
$n(E) = 26$
$n(W \cap E) = 11$
$n(W \cap S) = 13$
$n(E \cap S) = 10$
$n(E \cap S \cap W) = 5$

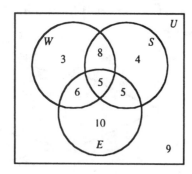

First put 5 in $E \cap S \cap W$.
Complete $E \cap S$ with 5 for a total of 10.
Complete $W \cap S$ with 8 for a total of 13.
Complete $W \cap E$ with 6 for a total of 11.
Complete E with $26 - (6 + 5 + 5) = 10$.
Complete S with $22 - (8 + 5 + 5) = 4$.

Complete W with $22 - (8 + 5 + 6) = 3$.
Now

$$50 - (3 + 8 + 5 + 6 + 4 + 5 + 10) = 9.$$

This is the number in $(W \cup S \cup E)'$.

(a) $n(W \cap S' \cap E') = 3$ states

(b) $n(W' \cap S' \cap E') = 9$ states

81. (a) $P(\text{double miss}) = .05(.05) = .0025$

(b) $P(\text{specific silo destroyed})$
$$= 1 - P(\text{double miss})$$
$$= 1 - .0025$$
$$= .9975$$

(c) $P(\text{all ten destroyed}) = (.9975)^{10} \approx .9753$

(d) $P(\text{at least one survived})$
$$= 1 - P(\text{none survived})$$
$$= 1 - P(\text{all ten destroyed})$$
$$= 1 - .9753$$
$$= .0247 \quad \text{or} \quad 2.47\%$$

This does not agree with the quote of a 5% chance that at least one would survive.

(e) The events that each of the two bombs hit their targets are assumed to be independent. The events that each silo is destroyed are assumed to be independent.

Extended Application: Medical Diagnosis

1. Using Bayes' theorem,

$P(H_2|C_1)$

$$= \frac{P(C_1|H_2) \cdot P(H_2)}{P(C_1|H_1)P(H_1) + P(C_1|H_2)P(H_2) + P(C_1|H_3)P(H_3)}$$

$$= \frac{.4(.15)}{.9(.8) + .4(.15) + .1(.05)}$$

$$= \frac{.06}{.785} \approx .076.$$

2. Using Bayes' theorem,

$P(H_1|C_2)$

$$= \frac{P(C_2|H_1) \cdot P(H_1)}{P(C_2|H_1)P(H_1) + P(C_2|H_2)P(H_2) + P(C_2|H_3)P(H_3)}$$

$$= \frac{.2(.8)}{.2(.8) + .8(.15) + .3(.05)}$$

$$= \frac{.16}{.295} \approx .542.$$

3. Using Bayes' theorem,

$P(H_3|C_2)$

$$= \frac{P(C_2|H_3) \cdot P(H_3)}{P(C_2|H_1)P(H_1) + P(C_2|H_2)P(H_2) + P(C_2|H_3)P(H_3)}$$

$$= \frac{.3(.05)}{.2(.8) + .8(.15) + .3(.05)}$$

$$= \frac{.015}{.295} \approx .051.$$

Chapter 6 Test

[6.1]

1. Write *true* or *false* for each statement.

(a) $3 \in \{1, 5, 7, 9\}$ **(b)** $\{1, 3\} \not\subset \{0, 1, 2, 3, 4\}$ **(c)** $\emptyset \subset \{2\}$

(d) A set of 6 distinct elements has exactly 64 subsets.

2. Let $U = \{1, 2, 3, 4, 5, 6, 7, 8, 9\}$, $A = \{1, 3, 4, 5\}$, $B = \{2, 4, 5\}$, and $C = \{1, 3, 5, 7\}$.
List the members of each of the following sets, using set braces.

(a) $A \cap B'$ **(b)** $A \cap (B \cup C')$

[6.2]

3. Draw a Venn diagram and shade the region that represents $A \cap (B \cup C')$.

4. Draw a Venn diagram and fill in the number of elements in each region given
that $n(U) = 25, n(A) = 11, n(B \cap A') = 9$, and $n(A \cap B) = 6$.

5. A survey of 70 children obtained the following results:

> 32 play soccer;
> 29 play basketball;
> 13 play tennis only;
> 6 play all three sports;
> 18 play soccer and basketball;
> 15 play soccer and tennis;
> 10 play basketball and tennis;
> 14 play none of the three sports.

Use a Venn diagram to answer the following questions.

(a) How many children play basketball only?

(b) How many children play tennis and basketball, but not soccer?

(c) How many children play soccer and tennis, but not basketball?

(d) How many children play exactly one of the three sports?

[6.3]

6. A single fair die is rolled. Find the probabilities of the following events.

(a) Getting a 4

(b) Getting an even number

(c) Getting a number less than 5

(d) Getting a number greater than 1

(e) Getting any number except 4 or 5

(f) Getting a number less than 8

[6.4]

7. Suppose that for events A and B, $P(A) = .4$, $P(B) = .3$ and $P(A \cup B) = .68$. Find each of the following probabilities.

   (a) $P(A \cap B)$ (b) $P(A')$

   (c) $P(A \cap B')$ (d) $P(A' \cap B')$

8. An urn contains 4 red, 3 blue, and 2 yellow marbles. A single marble is drawn.

   (a) Find the odds in favor of drawing a red marble.

   (b) Find the probability that a red or a blue marble is drawn.

[6.5]

9. For events E and F, $P(E) = .4$, $P(F) = .5$, and $P(E \cup F) = .8$. Find

   (a) $P(E|F)$ (b) $P(F|E')$.

10. Three cards are drawn without replacement from a standard deck of 52.

   (a) What is the probability that all three are spades?

   (b) What is the probability that all three are spades, given that the first card drawn is a spade?

11. The probability of passing the University of Waterloo's physical fitness test is .3. If you fail the first time, your chances of passing on the second try drop to .1. Draw a tree diagram and compute the probability that a person will pass on the first or second try.

[6.6]

12. The Magnum Opus Publishing Company uses three printers to put out its lengthy tomes. Printer A produces 40% of their books with a 20% failure rate. Printer B produces 25% with a 10% failure rate. Printer C produces the remainder with a 40% failure rate. Given that a book is badly printed, what is the probability that it was printed by Printer C?

Chapter 6 Test Answers

1. (a) False   **(b)** False   **(c)** True   **(d)** True

2. (a) $\{1,3\}$ **(b)** $\{4,5\}$

3.

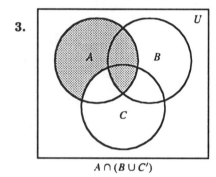

$A \cap (B \cup C')$

4.

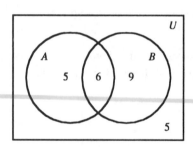

5. (a) 7   **(b)** 4   **(c)** 9   **(d)** 25

6. (a) $\frac{1}{6}$ **(b)** $\frac{1}{2}$ **(c)** $\frac{2}{3}$ **(d)** $\frac{5}{6}$ **(e)** $\frac{2}{3}$ **(f)** 1

7. (a) .02   **(b)** .6   **(c)** .38   **(d)** .32

8. (a) 4 to 5   **(b)** $\frac{7}{9}$

9. (a) $\frac{1}{5}$ **(b)** $\frac{2}{3}$

10. (a) $\left(\frac{13}{52}\right)\left(\frac{12}{51}\right)\left(\frac{11}{50}\right) \approx .013$ **(b)** $\left(\frac{12}{51}\right)\left(\frac{11}{50}\right) \approx .052$

11.

$$P(\text{pass}) = .3 + .07 = .37$$

12. $\dfrac{(.35)(.4)}{(.35)(.4)+(.4)(.2)+(.25)(.1)} \approx .57$

Chapter 7

COUNTING PRINCIPLES; FURTHER PROBABILITY TOPICS

7.1 The Multiplication Principle; Permutations

1. $6! = 6 \cdot 5 \cdot 4 \cdot 3 \cdot 2 \cdot 1 = 720$

3. $15! = 15 \cdot 14 \cdot 13 \cdot 12 \cdot 11 \cdot 10 \cdot 9 \cdot 8 \cdot 7$
$\cdot 6 \cdot 5 \cdot 4 \cdot 3 \cdot 2 \cdot 1$
$\approx 1.308 \cdot 10^{12}$

5. $P(13, 2) = \dfrac{13!}{(13 - 2)!} = \dfrac{13!}{11!}$
$= \dfrac{13 \cdot 12 \cdot 11!}{11!}$
$= 156$

7. $P(38, 17) = \dfrac{38!}{(38 - 17)!} = \dfrac{38!}{21!}$
$\approx 1.024 \cdot 10^{25}$

9. $P(n, 0) = \dfrac{n!}{(n - 0)!} = \dfrac{n!}{n!} = 1$

11. $P(n, 1) = \dfrac{n!}{(n - 1)!} = \dfrac{n(n - 1)!}{(n - 1)!} = n$

13. By the multiplication principle, there will be $5 \cdot 3 \cdot 2 = 30$ different home types available.

15. There are 3 choices for the first name and 5 choices for the middle name, so, by the multiplication principle, there are $3 \cdot 5 = 15$ possible arrangements.

19. In Example 7, there are only 3 unordered 2-letter subsets of letters A, B, and C. They are AB, AC, and BC.

21. Use the formula for distinguishable permutations. The number of different "words" is
$$\frac{n!}{n_1! n_2! n_3! n_4!} = \frac{13!}{5! 4! 2! 2!} = 540, 540.$$

23. (a) Since there are 14 distinguishable objects to be arranged, use permutations. The number of arrangements is
$$P(14, 14) = 14! = 87, 178, 291, 000$$
$$\text{or } 8.7178291 \cdot 10^{10}.$$

(b) There are $3!$ ways to arrange the pyramids among themselves, $4!$ ways to arrange the cubes, and $7!$ ways to arrange the spheres. We must also consider the number of ways to arrange the order of the three groups of shapes. This can be done in $3!$ ways. Using the multiplication principle, the number of arrangements is
$$3! 4! 7! 3! = 6 \cdot 24 \cdot 5040 \cdot 6$$
$$= 4, 354, 560.$$

(c) In this case, all of the objects that are the same shape are indistinguishable. Use the formula for distinguishable permutations. The number of distinguishable arrangements is
$$\frac{n!}{n_1! n_2! n_3!} = \frac{14!}{3! 4! 7!} = 120, 120.$$

(d) There are 3 choices for the pyramid, 4 for the cube, and 7 for the sphere. The total number of ways is
$$3 \cdot 4 \cdot 7 = 84.$$

25. $10! = 10 \cdot 9!$

To find the value of $10!$, multiply the value of $9!$ by 10.

27. 5 of the 11 drugs can be administered in
$$P(11, 5) = \frac{11!}{(11 - 5)!} = \frac{11!}{6!}$$
$$= \frac{11 \cdot 10 \cdot 9 \cdot 8 \cdot 7 \cdot 6!}{6!}$$
$$= 55, 440$$

different sequences.

29. 7 of 10 monkeys can be arranged in
$$P(10, 7) = \frac{10!}{(10 - 7)!} = \frac{10!}{3!}$$
$$= \frac{10 \cdot 9 \cdot 8 \cdot 7 \cdot 6 \cdot 5 \cdot 4 \cdot 3!}{3!}$$
$$= 604, 800$$

different ways.

31. A ballot would consist of a list of the 3 candidates for office 1 and a list of the 5 candidates for office 2.

The number of ways to list candidates for office 1 is $P(3,3) = 3! = 6$. The number of ways to list candidates for office 2 is $P(5,5) = 5! = 120$. There are two ways to choose which office goes first. By the multiplication principle, the number of different ballots is

$$6 \cdot 120 \cdot 2 = 1440.$$

33. Pick any 4 of the 380 nonmathematical courses. The number of possible schedules is

$$P(380, 4) = \frac{380!}{(380 - 4)!} = \frac{380!}{376!}$$
$$= \frac{380 \cdot 379 \cdot 378 \cdot 377 \cdot 376!}{376!}$$
$$= 20,523,714,120$$
$$\approx 2.05237 \cdot 10^{10}.$$

35. The number of ways to select the 4 officers is

$$P(35, 4) = \frac{35!}{(35 - 4)!} = \frac{35!}{31!}$$
$$= \frac{35 \cdot 34 \cdot 33 \cdot 32 \cdot 31!}{31!}$$
$$= 1,256,640.$$

37. (a) Pick one of the 5 traditional numbers followed by an arrangement of the remaining total of 7. The program can be arranged in

$$P(5, 1) \cdot P(7, 7) = 5 \cdot 7! = 25,200$$

different ways.

(b) Pick one of the 3 original Cajun compositions to play last, preceded by an arrangement of the remaining total of 7. This program can be arranged in

$$P(7, 7) \cdot P(3, 1) = 7! \cdot 3 = 15,120$$

different ways.

39. (a) There are 4 tasks to be performed in selecting 4 letters for the call letters. The first task may be done in 2 ways, the second in 25, the third in 24, and the fourth in 23. By the multiplication principle, there will be

$$2 \cdot 25 \cdot 24 \cdot 23 = 27,600$$

different call letter names possible.

(b) With repeats possible, there will be

$$2 \cdot 26 \cdot 26 \cdot 26 = 2 \cdot 26^3 \quad \text{or} \quad 35,152$$

call letter names possible.

(c) To start with W or K, make no repeats, and end in R, there will be

$$2 \cdot 24 \cdot 23 \cdot 1 = 1104$$

possible call letter names.

41. (a) Our number system has ten digits, which are 1 through 9 and 0.

There are 3 tasks to be performed in selecting 3 digits for the area code. The first task may be done in 8 ways, the second in 2, and the third in 10. By the multiplication principle, there will be

$$8 \cdot 2 \cdot 10 = 160$$

different area codes possible.

There are 7 tasks to be performed in selecting 7 digits for the telephone number. The first task may be done in 8 ways, and the other 6 tasks may each be done in 10 ways. By the multiplication principle, there will be

$$8 \cdot 10^6 = 8,000,000$$

different telephone numbers possible within each area code.

(b) Some numbers, such as 911, 800, and 900, are reserved for special purposes and are therefore unavailable for use as area codes.

43. (a) There were

$$26^3 \cdot 10^3 = 17,576,000$$

license plates possible that had 3 letters followed by 3 digits.

(b) There were

$$10^3 \cdot 26^3 = 17,576,000$$

new license plates possible when plates were also issued having 3 digits followed by 3 numbers.

(c) There were

$$26 \cdot 10^3 \cdot 26^3 = 456,976,000$$

new license plates possible when plates were also issued having 1 letter followed by 3 digits and then 3 letters.

45. If there are no restrictions on the digits used, there would be

$$10^5 = 100,000$$

different 5-digit zip codes possible.

If the first digit is not allowed to be 0, there would be

$$9 \cdot 10^4 = 90,000$$

zip codes possible.

47. There are 3 possible numbers of shapes on each card, 3 possible colors, 3 possible styles, and 3 possible shapes. Therefore, the total number of cards is

$$3 \cdot 3 \cdot 3 \cdot 3 = 81.$$

49. There are 3 possible answers for the first question and 2 possible answers for each of the 19 other questions. The number of possible objects is

$$3 \cdot 2^{19} = 1,572,864.$$

7.2   Combinations

1. To evaluate $\binom{8}{3}$, use the formula

$$\binom{n}{r} = \frac{n!}{(n-r)!r!}$$

with $n = 8$ and $r = 3$.

$$\binom{8}{3} = \frac{8!}{(8-3)!3!}$$
$$= \frac{8!}{5!3!}$$
$$= \frac{8 \cdot 7 \cdot 6 \cdot 5!}{5! \cdot 3 \cdot 2 \cdot 1} = 56$$

3. To evaluate $\binom{44}{20}$, use the formula

$$\binom{n}{r} = \frac{n!}{(n-r)!r!}$$

with $n = 44$ and $r = 20$.

$$\binom{44}{20} = \frac{44!}{(44-20)!20!}$$
$$= \frac{44!}{24!20!}$$
$$= 1.761 \cdot 10^{12}$$

5. $\binom{n}{0} = \dfrac{n!}{(n-0)!0!}$

$$= \frac{n!}{n! \cdot 1}$$
$$= 1$$

7. $\binom{n}{1} = \dfrac{n!}{(n-1)!1!}$

$$= \frac{n(n-1)!}{(n-1)! \cdot 1}$$
$$= n$$

9. There are 13 clubs, from which 6 are to be chosen. The number of ways in which a hand of 6 clubs can be chosen is

$$\binom{13}{6} = \frac{13!}{7!6!} = 1716.$$

11. (a) The number of ways to select a committee of 4 from a club with 30 members is

$$\binom{30}{4} = 27,405.$$

(b) If the committee must have at least 1 member and at most 3 members, it must have 1, 2, or 3 members. The number of committees is

$$\binom{30}{1} + \binom{30}{2} + \binom{30}{3} = 30 + 435 + 4060$$
$$= 4525.$$

13. (a) With repetition permitted, the tree diagram shows 16 different pairs.

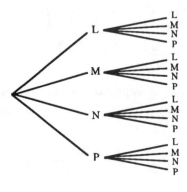

(b) If repetition is not permitted, one branch is missing from each of the clusters of second branches, for a total of 12 different pairs.

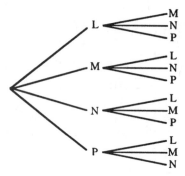

(c) Find the number of combinations of 4 elements taken 2 at a time.

$$\binom{4}{2} = 6$$

No repetitions are allowed, so the answer cannot equal that for part (a). However, since order does not matter, our answer is only half of the answer for part (b). For example, LM and ML are distinct in (b) but not in (c). Thus, the answer differs from both (a) and (b).

17. Since order is not important, the answers are combinations.

(a) If there are at least 4 women, there will be either 4 women and 1 man or 5 women and no men. The number of such committees is

$$\binom{11}{4}\binom{8}{1} + \binom{11}{5}\binom{8}{0} = 2640 + 462$$
$$= 3102.$$

(b) If there are no more than 2 men, there will be either no men and 5 women, 1 man and 4 women, or 2 men and 3 women. The number of such committees is

$$\binom{8}{0}\binom{11}{5} + \binom{8}{1}\binom{11}{4} + \binom{8}{2}\binom{11}{3}$$
$$= 462 + 2640 + 4620$$
$$= 7722.$$

19. Order is important, so use permutations. The number of ways in which the children can find seats is

$$P(12,11) = \frac{12!}{(12-11)!} = \frac{12!}{1!}$$
$$= 12!$$
$$= 479,001,600.$$

21. Since order does not matter, use combinations.

(a) There are

$$\binom{25}{3} = 2300$$

possible samples of 3 apples.

(b) There are

$$\binom{5}{3} = 10$$

possible samples of 3 rotten apples.

(c) There are

$$\binom{5}{1}\binom{20}{2} = 950$$

possible samples with exactly 1 rotten apple.

23. Since order is important, use a permutation. The plants can be arranged in

$$P(9,5) = 9 \cdot 8 \cdot 7 \cdot 6 \cdot 5 = 15,120$$

different ways.

25. Use combinations since order does not matter.

(a) First consider how many pairs of circles there are. This number is

$$\binom{6}{2} = \frac{6!}{2!4!} = 15.$$

Each pair intersects in two points. The total number of intersection points is $2 \cdot 15 = 30$.

(b) The number of pairs of circles is

$$\binom{n}{2} = \frac{n!}{(n-2)!2!}$$
$$= \frac{n(n-1)(n-2)!}{(n-2)! \cdot 2}$$
$$= \frac{1}{2}n(n-1).$$

Each pair intersects in two points. The total number of points is

$$2 \cdot \frac{1}{2}n(n-1) = n(n-1).$$

27. Order is important in arranging a schedule, so use permutations.

(a) $P(6,6) = \dfrac{6!}{0!} = 6! = 720$

She can arrange her schedule in 720 ways if she calls on all 6 prospects.

(b) $P(6,4) = \dfrac{6!}{2!} = 360$

She can arrange her schedule in 360 ways if she calls on only 4 of the 6 prospects.

29. There are 2 types of meat and 6 types of extras. Order does not matter here, so use combinations.

(a) There are $\binom{2}{1}$ ways to choose one type of meat and $\binom{6}{3}$ ways to choose exactly three extras. By the multiplication principle, there are

$$\binom{2}{1}\binom{6}{3} = 2 \cdot 20 = 40$$

different ways to order a hamburger with exactly three extras.

(b) There are

$$\binom{6}{3} = 20$$

different ways to choose exactly three extras.

(c) "At least five extras" means "5 extras or 6 extras." There are $\binom{6}{5}$ different ways to choose exactly 5 extras and $\binom{6}{6}$ ways to choose exactly 6 extras, so there are

$$\binom{6}{5} + \binom{6}{6} = 6 + 1 = 7$$

different ways to choose at least five extras.

31. Select 8 of the 16 smokers and 8 of the 20 non-smokers; order does not matter in the group, so use combinations. There are

$$\binom{16}{8}\binom{20}{8} = 1,621,233,900$$

different ways to select the study group. In scientific notation, this answer would be written as $1.6212 \cdot 10^9$.

33. Order does not matter in choosing a delegation, so use combinations. This committee has $5 + 4 = 9$ members.

(a) There are

$$\binom{9}{3} = \frac{9!}{6!3!}$$
$$= \frac{9 \cdot 8 \cdot 7 \cdot 6!}{6! \cdot 3 \cdot 2 \cdot 1}$$
$$= 84 \text{ possible delegations.}$$

(b) To have all Democrats, the number of possible delegations is

$$\binom{5}{3} = 10.$$

(c) To have 2 Democrats and 1 Republican, the number of possible delegations is

$$\binom{5}{2}\binom{4}{1} = 10 \cdot 4 = 40.$$

(d) We have previously calculated that there are 84 possible delegations, of which 10 consist of all Democrats. Those 10 delegations are the only ones with no Republicans, so the remaining $84 - 10 = 74$ delegations include at least one Republican.

35. Since order does not matter, use combinations.

$$\binom{52}{13} = \frac{52!}{(52-13)!13!}$$
$$= \frac{52!}{39!13!}$$
$$= 635,013,559,600$$
$$\approx 6.3501 \cdot 10^{11}$$

37. Since order does not matter, use combinations.

2 good hitters: $\binom{5}{2}\binom{4}{1} = 10 \cdot 4 = 40$

3 good hitters: $\binom{5}{3}\binom{4}{0} = 10 \cdot 1 = 10$

The total number of ways is $40 + 10 = 50$.

39. Since order does not matter, use combinations.

(a) There are

$$\binom{20}{5} = 15,504$$

different ways to select 5 of the orchids.

(b) If 2 special orchids must be included in the show, that leaves 18 orchids from which the other 3 orchids for the show must be chosen. This can be done in

$$\binom{18}{3} = 816$$

different ways.

41. In the lottery, 6 different numbers are to be chosen from the 99 numbers.

(a) There are

$$\binom{99}{6} = \frac{99!}{93!6!} = 1,120,529,256$$

different ways to choose 6 numbers if order is not important. In scientific notation, this answer would be written as $1.1205 \cdot 10^9$.

(b) There are

$$P(99,6) = \frac{99!}{93!} = 806,781,064,320$$

different ways to choose 6 numbers if order matters. In scientific notation, this answer would be written as $8.0678 \cdot 10^{11}$.

43. In each column, there are 15 numbers that can be used in 5 positions, except for the middle column with 4 positions. The number of possibilities for each column is given below.

Column 1: $15 \cdot 14 \cdot 13 \cdot 12 \cdot 11$

Column 2: $15 \cdot 14 \cdot 13 \cdot 12 \cdot 11$

Column 3 (middle column): $15 \cdot 14 \cdot 13 \cdot 12$

Column 4: $15 \cdot 14 \cdot 13 \cdot 12 \cdot 11$

Column 5: $15 \cdot 14 \cdot 13 \cdot 12 \cdot 11$

The total number of different Bingo cards is the product of these numbers, that is,

$$(15 \cdot 14 \cdot 13 \cdot 12 \cdot 11)^4 (15 \cdot 14 \cdot 13 \cdot 12)$$
$$\approx 5.524 \cdot 10^{26}.$$

7.3   Probability Applications of Counting Principles

1. There are $\binom{10}{3}$ samples of 3 apples.

$$\binom{10}{3} = \frac{10 \cdot 9 \cdot 8}{3 \cdot 2 \cdot 1} = 120$$

There are $\binom{6}{3}$ samples of 3 red apples.

$$\binom{6}{3} = \frac{6 \cdot 5 \cdot 4}{3 \cdot 2 \cdot 1} = 20$$

Thus,

$$P(\text{all red apples}) = \frac{20}{120} = \frac{1}{6}.$$

3. There are $\binom{4}{2}$ samples of 2 yellow apples.

$$\binom{4}{2} = \frac{4 \cdot 3}{2 \cdot 1} = 6$$

There are $\binom{6}{1} = 6$ samples of 1 red apple. Thus, there are $6 \cdot 6 = 36$ samples of 3 in which 2 are yellow and 1 red. Thus,

$$P(\text{2 yellow and 1 red apple}) = \frac{36}{120} = \frac{3}{10}.$$

5. The number of 2-card hands is

$$\binom{52}{2} = \frac{52 \cdot 51}{2 \cdot 1} = 1326.$$

7. There are $\binom{52}{2} = 1326$ different 2-card hands. The number of 2-card hands with exactly one ace is

$$\binom{4}{1}\binom{48}{2} = 4 \cdot 48 = 192.$$

The number of 2-card hands with two aces is

$$\binom{4}{2} = 6.$$

Thus there are 198 hands with at least one ace. Therefore,

$P(\text{the 2-card hand contains an ace})$

$$= \frac{198}{1326} = \frac{33}{221} \approx .149.$$

9. There are $\binom{52}{2} = 1326$ different 2-card hands. There are $\binom{13}{2} = 78$ ways to get a 2-card hand where both cards are of a single named suit, but there are 4 suits to choose from. Thus,

$P(\text{two cards of same suit})$

$$= \frac{4 \cdot \binom{13}{2}}{\binom{52}{2}} = \frac{312}{1326} = \frac{52}{221} \approx .235.$$

11. There are $\binom{52}{2} = 1326$ different 2-card hands. There are 12 face cards in a deck, so there are 40 cards that are not face cards. Thus,

$P(\text{no face cards})$

$$= \frac{\binom{40}{2}}{\binom{52}{2}} = \frac{780}{1326} = \frac{130}{221} \approx .588.$$

13. There are 26 choices for each slip pulled out, and there are 5 slips pulled out, so there are

$$26^5 = 11,881,376$$

different "words" that can be formed from the letters. If the "word" must be "chuck," there is only one choice for each of the 5 letters (the first slip must contain a "c," the second an "h," and so on). Thus,

$P(\text{word is "chuck"})$

$$= \frac{1^5}{26^5} = \left(\frac{1}{26}\right)^5 \approx 8.42 \cdot 10^{-8}.$$

15. There are $26^5 = 11,881,376$ different "words" that can be formed. If the "word" is to have no repetition of letters, then there are 26 choices for the first letter, but only 25 choices for the second (since the letters must all be different), 24 choices for the third, and so on. Thus,

$P(\text{all different letters})$

$$= \frac{26 \cdot 25 \cdot 24 \cdot 23 \cdot 22}{26^5}$$

$$= \frac{1 \cdot 25 \cdot 24 \cdot 23 \cdot 22}{26^4}$$

$$= \frac{303,600}{456,976}$$

$$= \frac{18,975}{28,561} \approx .664.$$

19. $P(\text{at least 2 presidents have the same birthday})$
$$= 1 - P(\text{no 2 presidents have the same birthday})$$

The number of ways that 41 people can have the same birthday is $(365)^{41}$. The number of ways that 41 people can have all different birthdays is the number of permutations of 365 things taken 41 at a time or $P(365, 41)$. Thus,

$P(\text{at least 2 presidents have the same birthday})$

$$= 1 - \frac{P(365, 41)}{365^{41}}.$$

(Be careful to realize that the symbol P is sometimes used to indicate permutations and sometimes used to indicate probability; in this solution, the symbol is used both ways.)

21. Since there are 435 members of the House of Representatives, and there are only 365 days in a year, it is a certain event that at least 2 people will have the same birthday. Thus,

$P(\text{at least 2 members have the same birthday}) = 1.$

23. Each of the 4 people can choose to get off at any one of the 7 floors, so there are 7^4 ways the four people can leave the elevator. The number of ways the people can leave at different floors is the number of permutations of 7 things (floors) taken 4 at a time or

$$P(7, 4) = 7 \cdot 6 \cdot 5 \cdot 4 = 840.$$

Thus, the probability that no 2 passengers leave at the same floor is

$$\frac{P(7, 4)}{7^4} = \frac{840}{2401} \approx .3499.$$

(Note the similarity of this problem and the "birthday problem.")

25. $P(\text{at least one } \$100\text{-bill})$

$$= P(1 \ \$100\text{-bill}) + P(2 \ \$100\text{-bills})$$

$$= \frac{\binom{2}{1}\binom{4}{1}}{\binom{6}{2}} + \frac{\binom{2}{2}\binom{4}{0}}{\binom{6}{2}}$$

$$= \frac{8}{15} + \frac{1}{15}$$

$$= \frac{9}{15} = \frac{3}{5}$$

$$P(\text{no } \$100\text{-bill}) = \frac{\binom{2}{0}\binom{4}{2}}{\binom{6}{2}} = \frac{6}{15} = \frac{2}{5}$$

It is more likely to get at least one $100-bill.

27. The number of orders of the three types of birds is $P(3, 3)$. The number of arrangements of the crows is $P(3, 3)$, of the bluejays is $P(4, 4)$, and of the starlings is $P(5, 5)$. The total number of arrangements of all the birds is $P(12, 12)$.

$P(\text{all birds of same type are sitting together})$

$$= \frac{P(3, 3) \cdot P(3, 3) \cdot P(4, 4) \cdot P(5, 5)}{P(12, 12)}$$

$$\approx 2.165 \cdot 10^{-4}$$

29. There are $\binom{7}{2}$ possible ways to choose 2 nondefective typewriters out of the $\binom{9}{2}$ possible ways of choosing any 2. Thus,

$$P(\text{no defective}) = \frac{\binom{7}{2}}{\binom{9}{2}} = \frac{21}{36} = \frac{7}{12}.$$

31. There are $\binom{7}{4}$ possible ways to choose 4 nondefective typewriters out of the $\binom{9}{4}$ possible ways of choosing any 4. Thus,

$$P(\text{no defective}) = \frac{\binom{7}{4}}{\binom{9}{4}} = \frac{35}{126} = \frac{5}{18}.$$

33. There are $\binom{12}{5} = 792$ ways to pick a sample of 5. It will be shipped if all 5 are good. There are $\binom{10}{5} = 252$ ways to pick 5 good ones, so

$$P(\text{all good}) = \frac{252}{792} = \frac{7}{22} \approx .318.$$

35. There are 20 people in all, so the number of possible 5-person committees is $\binom{20}{5} = 15,504$. Thus, in parts (a)-(g), $n(S) = 15,504$.

(a) There are $\binom{10}{3}$ ways to choose the 3 men and $\binom{10}{2}$ ways to choose the 2 women. Thus,

$P(3 \text{ men and } 2 \text{ women})$

$$= \frac{\binom{10}{3}\binom{10}{2}}{\binom{20}{5}} = \frac{120 \cdot 45}{15,504} = \frac{225}{646} \approx .348.$$

(b) There are $\binom{6}{3}$ ways to choose the 3 Miwoks and $\binom{9}{2}$ ways to choose the 2 Pomos. Thus,

$P(\text{exactly } 3 \text{ Miwoks and } 2 \text{ Pomos})$

$$= \frac{\binom{6}{3}\binom{9}{2}}{\binom{20}{5}} = \frac{20 \cdot 36}{15,504} = \frac{15}{323} \approx .046.$$

(c) Choose 2 of the 6 Miwoks, 2 of the 5 Hoopas, and 1 of the 9 Pomos. Thus,

$P(2 \text{ Miwoks}, 2 \text{ Hoopas, and a Pomo})$

$$= \frac{\binom{6}{2}\binom{5}{2}\binom{9}{1}}{\binom{20}{5}} = \frac{15 \cdot 10 \cdot 9}{15,504} = \frac{225}{2584} \approx .087.$$

(d) There cannot be 2 Miwoks, 2 Hoopas, and 2 Pomos, since only 5 people are to be selected. Thus,

$P(2 \text{ Miwoks}, 2 \text{ Hoopas, and } 2 \text{ Pomos}) = 0.$

(e) Since there are more women then men, there must be 3, 4, or 5 women.

$P(\text{more women than men})$

$$= \frac{\binom{10}{3}\binom{10}{2} + \binom{10}{4}\binom{10}{1} + \binom{10}{5}\binom{10}{0}}{\binom{20}{5}}$$

$$= \frac{7752}{15,504} = \frac{1}{2}$$

(f) Choose 3 of 5 Hoopas and any 2 of the 15 non-Hoopas.

$P(\text{exactly } 3 \text{ Hoopas})$

$$= \frac{\binom{5}{3}\binom{15}{2}}{\binom{20}{5}} = \frac{175}{2584} \approx .068$$

(g) There can be 2 to 5 Pomos, the rest chosen from the 11 nonPomos.

$P(\text{at least } 2 \text{ Pomos})$

$$= \frac{\binom{9}{2}\binom{11}{3} + \binom{9}{3}\binom{11}{2} + \binom{9}{4}\binom{11}{1} + \binom{9}{5}\binom{11}{0}}{\binom{20}{5}}$$

$$= \frac{503}{646} \approx .779$$

37. A flush could start with an ace, 2, 3, 4, ..., 7, 8, or 9. This gives 9 choices in each of 4 suits, so there are 36 choices in all. Thus,

$$P(\text{straight flush}) = \frac{35}{\binom{52}{5}} = \frac{36}{2,598,960}$$

$$\approx .00001385$$
$$= 1.385 \cdot 10^{-5}.$$

39. A straight could start with an ace, 2, 3, 4, 5, 6, 7, 8, 9, or 10 as the low card, giving 40 choices. For each succeeding card, only the suit may be chosen. Thus, the number of straights is

$$40 \cdot 4^4 = 10,240.$$

But this also counts the straight flushes, of which there are 36 (see Exercise 37), and the 4 royal flushes. There are thus 10,200 straights that are not also flushes, so

$$P(\text{straight}) = \frac{10,200}{2,598,960} \approx .00392.$$

41. There are 13 different values of cards and 4 cards of each value. Choose 2 values out of the 13 for the values of the pairs. The number of ways to select the 2 values is $\binom{13}{2}$. The number of ways to select a pair for each value is $\binom{4}{2}$. There are $52 - 8 = 44$ cards that are neither of these 2 values, so the number of ways to select the fifth card is $\binom{44}{1}$. Thus,

$$P(\text{two pairs}) = \frac{\binom{13}{2}\binom{4}{2}\binom{4}{2}\binom{44}{1}}{\binom{52}{5}}$$

$$= \frac{123,552}{2,598,960} \approx .04754.$$

43. There are $\binom{52}{13}$ different 13-card bridge hands. Since there are only 13 hearts, there is exactly one way to get a bridge hand containing only hearts. Thus,

$$P(\text{only hearts}) = \frac{1}{\binom{52}{13}} \approx 1.575 \cdot 10^{-12}.$$

45. There are $\binom{4}{3}$ ways to obtain 3 aces, $\binom{4}{3}$ ways to obtain 3 kings, and $\binom{44}{7}$ ways to obtain the 7 remaining cards. Thus,

$$P(\text{exactly 3 aces and exactly 3 kings})$$

$$= \frac{\binom{4}{3}\binom{4}{3}\binom{44}{7}}{\binom{52}{13}} \approx 9.655 \cdot 10^{-4}.$$

47. There are 21 books, so the number of selections of any 6 books is

$$\binom{21}{6} = 54,264.$$

(a) The probability that the selection consisted of 3 Hughes and 3 Morrison books is

$$\frac{\binom{9}{3}\binom{7}{3}}{\binom{21}{6}} = \frac{85 \cdot 35}{54,264} = \frac{2940}{54,264} \approx .054.$$

(b) A selection containing exactly 4 Baldwin books will contain 2 of the 16 books by the other authors, so the probability is

$$\frac{\binom{5}{4}\binom{16}{2}}{\binom{21}{6}} = \frac{5 \cdot 120}{54,264} = \frac{600}{54,264} \approx .011.$$

(c) The probability of a selection consisting of 2 Hughes, 3 Baldwin, and 1 Morrison book is

$$\frac{\binom{9}{2}\binom{5}{3}\binom{7}{1}}{\binom{21}{6}} = \frac{36 \cdot 10 \cdot 7}{54,264} = \frac{2520}{54,264} \approx .046.$$

(d) A selection consisting of at least 4 Hughes books may contain 4, 5, or 6 Hughes books, with any remaining books by the other authors. Therefore, the probability is

$$\frac{\binom{9}{4}\binom{12}{2} + \binom{9}{5}\binom{12}{1} + \binom{9}{6}\binom{12}{0}}{\binom{21}{6}}$$

$$= \frac{126 \cdot 66 + 126 \cdot 12 + 84}{54,264}$$

$$= \frac{8316 + 1512 + 84}{54,264}$$

$$= \frac{9912}{54,264} \approx .183.$$

(e) Since there are 9 Hughes books and 5 Baldwin books, there are 14 books written by males. The probability of a selection with exactly 4 books written by males is

$$\frac{\binom{14}{4}\binom{7}{2}}{\binom{21}{6}} = \frac{1001 \cdot 21}{54,264} = \frac{21,021}{54,264} \approx .387.$$

(f) A selection with no more than 2 books written by Baldwin may contain 0, 1, or 2 books by Baldwin, with the remaining books by the other authors. Therefore, the probability is

$$\frac{\binom{5}{0}\binom{16}{6} + \binom{5}{1}\binom{16}{5} + \binom{5}{2}\binom{16}{4}}{\binom{21}{6}}$$

$$= \frac{8008 + 5 \cdot 4368 + 10 \cdot 1820}{54,264}$$

$$= \frac{8008 + 21,840 + 18,200}{54,264}$$

$$= \frac{48,048}{54,264} \approx .885.$$

49. To find the probability of picking 5 of the 6 lottery numbers correctly, we must recall that the total number of ways to pick the 6 lottery numbers is $\binom{99}{6} = 1,120,529,256$. To pick 5 of the 6 winning numbers, we must also pick 1 of the 93 losing numbers. Therefore, the number of ways of picking 5 of the 6 winning numbers is

$$\binom{6}{5}\binom{93}{1} = 558.$$

Thus, the probability of picking 5 of the 6 numbers correctly is

$$\frac{\binom{6}{5}\binom{93}{1}}{\binom{99}{6}} \approx 5.0 \cdot 10^{-7}.$$

51. $P(\text{saying "Math class is tough."})$

$$= \frac{\binom{1}{1}\binom{269}{3}}{\binom{270}{4}} \approx .0148$$

The correct figure is 1.48%.

7.4 Binomial Probability

1. This is a Bernoulli trial problem with $P(\text{success}) = P(\text{girl}) = \frac{1}{2}$. The probability of exactly x successes in n trials is

$$\binom{n}{x} p^x (1-p)^{n-x},$$

where p is the probability of success in a single trial. We have $n = 5$, $x = 2$, and $p = \frac{1}{2}$. Note that

$$1 - p = 1 - \frac{1}{2} = \frac{1}{2}.$$

$$P(\text{exactly 2 girls and 3 boys}) = \binom{5}{2}\left(\frac{1}{2}\right)^2\left(\frac{1}{2}\right)^3$$

$$= \frac{10}{32} = \frac{5}{16} \approx .313$$

3. We have $n = 5$, $x = 0$, $p = \frac{1}{2}$, and $1 - p = \frac{1}{2}$.

$$P(\text{no girls}) = \binom{5}{0}\left(\frac{1}{2}\right)^0\left(\frac{1}{2}\right)^5 = \frac{1}{32} \approx .031$$

5. "At least 4 girls" means either 4 or 5 girls.

$P(\text{at least 4 girls})$

$$= \binom{5}{4}\left(\frac{1}{2}\right)^4\left(\frac{1}{2}\right)^1 + \binom{5}{5}\left(\frac{1}{2}\right)^5\left(\frac{1}{2}\right)^0$$

$$= \frac{5}{32} + \frac{1}{32} = \frac{6}{32} = \frac{3}{16} \approx .188$$

7. $P(\text{no more than 3 boys})$
$= 1 - P(\text{at least 4 boys})$
$= 1 - P(4 \text{ boys or 5 boys})$
$= 1 - [P(4 \text{ boys}) + P(5 \text{ boys})]$

$$= 1 - \left(\frac{5}{32} + \frac{1}{32}\right)$$

$$= 1 - \frac{6}{32}$$

$$= 1 - \frac{3}{16} = \frac{13}{16} \approx .813$$

9. On one roll, $P(1) = \frac{1}{6}$. We have $n = 12$, $x = 12$, and $p = \frac{1}{6}$. Note that $1 - p = \frac{5}{6}$. Thus,

$$P(\text{exactly 12 ones}) = \binom{12}{12}\left(\frac{1}{6}\right)^{12}\left(\frac{5}{6}\right)^0$$

$$\approx 4.6 \cdot 10^{-10}.$$

11. $P(\text{exactly 1 one}) = \binom{12}{1}\left(\frac{1}{6}\right)^1\left(\frac{5}{6}\right)^{11} \approx .269$

13. "No more than 3 ones" means 0, 1, 2, or 3 ones. Thus,

$P(\text{no more than 3 ones})$
$= P(0 \text{ ones}) + P(1 \text{ one}) + P(2 \text{ ones})$
$\quad + P(3 \text{ ones})$

$$= \binom{12}{0}\left(\frac{1}{6}\right)^0\left(\frac{5}{6}\right)^{12} + \binom{12}{1}\left(\frac{1}{6}\right)^1\left(\frac{5}{6}\right)^{11}$$

$$+ \binom{12}{2}\left(\frac{1}{6}\right)^2\left(\frac{5}{6}\right)^{10} + \binom{12}{3}\left(\frac{1}{6}\right)^3\left(\frac{5}{6}\right)^9$$

$$\approx .875.$$

15. Each time the coin is tossed, $P(\text{head}) = \frac{1}{2}$. We have $n = 6$, $x = 6$, $p = \frac{1}{2}$, and $1 - p = \frac{1}{2}$. Thus,

$$P(\text{all heads}) = \binom{6}{6}\left(\frac{1}{2}\right)^6\left(\frac{1}{2}\right)^0$$

$$= \frac{1}{64} \approx .016.$$

17. $P(\text{no more than 3 heads})$
$= P(0 \text{ heads}) + P(1 \text{ head}) + P(2 \text{ heads})$
$\quad + P(3 \text{ heads})$

$$= \binom{6}{0}\left(\frac{1}{2}\right)^0\left(\frac{1}{2}\right)^6 + \binom{6}{1}\left(\frac{1}{2}\right)^1\left(\frac{1}{2}\right)^5$$

$$+ \binom{6}{2}\left(\frac{1}{2}\right)^2\left(\frac{1}{2}\right)^4 + \binom{6}{3}\left(\frac{1}{2}\right)^3\left(\frac{1}{2}\right)^3$$

$$= \frac{42}{64} = \frac{21}{32} \approx .656$$

21. $\binom{n}{r} + \binom{n}{r+1}$

$$= \frac{n!}{r!(n-r)!} + \frac{n!}{(r+1)![n-(r+1)]!}$$

$$= \frac{n!(r+1)}{r!(r+1)(n-r)!} + \frac{n!(n-r)}{(r+1)![n-(r+1)]!(n-r)}$$

$$= \frac{rn! + n!}{(r+1)!(n-r)!} + \frac{n(n!) - rn!}{(r+1)!(n-r)!}$$

$$= \frac{rn! + n! + n(n!) - rn!}{(r+1)!(n-r)!}$$

$$= \frac{n!(n+1)}{(r+1)!(n-r)!}$$

$$= \frac{(n+1)!}{(r+1)![(n+1)-(r+1)]!}$$

$$= \binom{n+1}{r+1}$$

23. We define a success to be the event that a customer overpays. In this situation, $n = 15$, $x = 0$, $p = \frac{1}{10}$, and $1 - p = \frac{9}{10}$.

$P(\text{customer does not overpay for any item})$

$$= \binom{15}{0}\left(\frac{1}{10}\right)^0\left(\frac{9}{10}\right)^{15}$$

$$\approx .2059$$

25. In Exercise 23, we defined a success to be the event that a customer overpays. In this situation, $n = 15$; $x = 2, 3, 4, \ldots, 15$; $p = \frac{1}{10}$; and $1 - p = \frac{9}{10}$.

P(a customer overpays on at least 2 items)

$= 1 - P$(a customer overpays on 0 or 1 item)

$$= 1 - \binom{15}{0}\left(\frac{1}{10}\right)^0 \left(\frac{9}{10}\right)^{15} - \binom{15}{1}\left(\frac{1}{10}\right)^1 \left(\frac{9}{10}\right)^{14}$$

$$\approx .4510$$

27. We have $n = 10$, $x = 10$, $p = .9$, and $1 - p = .1$, so

$$P(x = 10) = \binom{10}{10}(.9)^{10}(.1)^0 \approx .349.$$

29. We have

$$P(x \geq 9) = P(x = 9) + P(x = 10)$$

$$= \binom{10}{9}(.9)^9(.1)^1 + \binom{10}{10}(.9)^{10}(.1)^0$$

$$\approx .387 + .349 = .736.$$

31. We have $n = 6$, $x = 2$, $p = \frac{1}{5}$, and $1 - p = \frac{4}{5}$. Thus,

$$P(\text{exactly 2 correct}) = \binom{6}{2}\left(\frac{1}{5}\right)^2 \left(\frac{4}{5}\right)^4 \approx .246.$$

33. We have

P(at least 4 correct)

$= P(4 \text{ correct}) + P(5 \text{ correct}) + P(6 \text{ correct})$

$$= \binom{6}{4}\left(\frac{1}{5}\right)^4 \left(\frac{4}{5}\right)^2 + \binom{6}{5}\left(\frac{1}{5}\right)^5 \left(\frac{4}{5}\right)^1$$

$$+ \binom{6}{6}\left(\frac{1}{5}\right)^6 \left(\frac{4}{5}\right)^0$$

$$\approx .017.$$

35. We have $n = 3$, $x = 1$, $p = .1$, and $1 - p = .9$. Thus,

$$P(\text{exactly 1 loses everything}) = \binom{3}{1}(.1)^1(.9)^2$$

$$\approx .243.$$

37. We have $n = 20$, $p = .05$, and $1 - p = .95$. Thus,

P(at most 2 defective transistors)

$$= P(x \leq 2)$$
$$= P(x = 0) + P(x = 1) + P(x = 2)$$

$$= \binom{20}{0}(.05)^0(.95)^{20} + \binom{20}{1}(.05)^1(.95)^{19}$$

$$+ \binom{20}{2}(.05)^2(.95)^{18}$$

$$\approx .925.$$

39. $n = 58$, $p = .7$

(a) The probability that all 58 like the product is

$$P(x = 58) = \binom{58}{58}(.7)^{58}(.3)^0 \approx 0.$$

(b) The probability that from 28 to 30 people (inclusive) like the product is

$$P(x = 28) + P(x = 29) + P(x = 30)$$

$$= \binom{59}{28}(.7)^{28}(.3)^{30} + \binom{58}{29}(.7)^{29}(.3)^{29}$$

$$+ \binom{58}{30}(.7)^{30}(.3)^{28}$$

$$\approx .002438.$$

41. We have $n = 20$, $x = 17$, $p = .7$, and $1 - p = .3$. Thus,

$$P(\text{exactly 17 cured}) = \binom{20}{17}(.7)^{17}(.3)^3$$

$$= 1140(.7)^{17}(.3)^3$$

$$\approx .072.$$

43. P(at least 18 cured)

$= P(\text{exactly 18 cured}) + P(\text{exactly 19 cured})$
$\quad + P(20 \text{ cured})$

$$= \binom{20}{18}(.7)^{18}(.3)^2 + \binom{20}{19}(.7)^{19}(.3)^1 + (.7)^{20}$$

$$+ \binom{20}{20}(.7)^{20}(.3)^0$$

$$= 190(.7)^{18}(.3)^2 + 20(.7)^{19}(.3) + (.7)^{20}$$

$$\approx .035$$

45. We have $n = 100$, $p = .012$, and $1 - p = .988$. Thus,

$P(x \leq 2)$

$= P(x = 0) + P(x = 1) + P(x = 2)$

$$= \binom{100}{0}(.012)^0(.988)^{100}$$

$$+ \binom{100}{1}(.012)^1(.988)^{99}$$

$$+ \binom{100}{2}(.012)^2(.988)^{98}$$

$$= (.012)^0(.988)^{100} + 100(.012)^1(.988)^{99}$$
$$+ 4950(.012)^2(.988)^{98}$$

$$\approx .881.$$

47. We have $n = 6$, $x = 3$, $p = .70$, and $1 - p = .30$. Thus,

$$P(\text{exactly 3 recover}) = \binom{6}{3}(.7)^3(.3)^3$$

$$\approx .185.$$

49. $P(\text{no more than 3 recover})$
$= P(0 \text{ recover}) + P(1 \text{ recovers})$
$\quad + P(2 \text{ recover}) + P(3 \text{ recover})$

$$= \binom{6}{0}(.7)^0(.3)^6 + \binom{6}{1}(.7)^1(.3)^5$$

$$+ \binom{6}{2}(.7)^2(.3)^4 + \binom{6}{3}(.7)^3(.3)^3$$

$$\approx .256$$

51. We have $n = 10,000$, $p = 2.5 \cdot 10^{-7} = .00000025$, and $1 - p = .99999975$. Thus,

$P(\text{at least 1 mutation occurs})$
$\quad = 1 - P(\text{none occurs})$

$$= 1 - \binom{10,000}{0}p^0(1-p)^{10,000}$$

$$= 1 - (.99999975)^{10,000} \approx .0025.$$

53. We define a success to be the event that an inoculated person gets the flu. In this situation, $n = 83$, $p = .2$, and $1 - p = .8$.

(a) $P(10 \text{ successes}) = \binom{83}{10}(.2)^{10}(.8)^{73}$

$$\approx .0210$$

(b) $P(\text{no more than 4 successes})$

$= P(0 \text{ successes}) + P(1 \text{ success})$
$\quad + P(2 \text{ successes}) + P(3 \text{ successes})$
$\quad + P(4 \text{ successes})$

$$= \binom{83}{0}(.2)^0(.8)^{83} + \binom{83}{1}(.2)^1(.8)^{82}$$

$$+ \binom{83}{2}(.2)^2(.8)^{81} + \binom{83}{3}(.2)^3(.8)^{80}$$

$$+ \binom{83}{4}(.2)^4(.83)^{79}$$

$$\approx 8.004 \cdot 10^{-5}$$

(c) $P(\text{none of the inoculated people get the flu})$
$= P(\text{no successes})$

$$= \binom{83}{0}(.2)^0(.8)^{83}$$

$$\approx 9.046 \cdot 10^{-9}$$

55. We define a success to be the event that a woman would prefer to work part-time rather than full-time. In this situation, $n = 10$; $x = 3, 4, 5, \ldots,$ 10; $p = .33$; and $1 - p = .67$.

$P(x = 3, 4, 5, \ldots, \text{ or } 10)$

$\quad = 1 - P(x = 0, 1, \text{ or } 2)$

$\quad = 1 - [P(x = 0) + P(x = 1) + P(x = 2)]$
$\quad = 1 - P(x = 0) - P(x = 1) - P(x = 2)$

$$= 1 - \binom{10}{0}(.33)^0(.67)^{10} - \binom{10}{1}(.33)^1(.67)^9$$

$$- \binom{10}{2}(.33)^2(.67)^8$$

$$\approx .6930$$

57. We have $n = 10$, $x = 7$, $p = .2$, and $1 - p = .8$. Thus,

$$P(x = 7) = \binom{10}{7}(.2)^7(.8)^3 \approx .00079.$$

59. We have

$P(x < 8)$

$\quad = 1 - P(x \geq 8)$
$\quad = 1 - [P(x = 8) + P(x = 9) + P(x = 10)]$
$\quad = 1 - P(x = 8) - P(x = 9) - P(x = 10)$

$$= 1 - \binom{10}{8}(.2)^8(.8)^2 - \binom{10}{9}(.2)^9(.8)^1$$

$$- \binom{10}{10}(.2)^{10}(.8)^0$$

$$= 1 - .000074 - .000004 - .0000001$$
$$= .999922.$$

7.5   Probability Distributions; Expected Value

1. Let x denote the number of heads observed. Then x can take on 0, 1, 2, 3, or 4 as values. The probabilities are as follows.

$$P(x = 0) = \binom{4}{0}\left(\frac{1}{2}\right)^0\left(\frac{1}{2}\right)^4 = \frac{1}{16}$$

$$P(x = 1) = \binom{4}{1}\left(\frac{1}{2}\right)^1\left(\frac{1}{2}\right)^3 = \frac{4}{16} = \frac{1}{4}$$

$$P(x = 2) = \binom{4}{2}\left(\frac{1}{2}\right)^2\left(\frac{1}{2}\right)^2 = \frac{6}{16} = \frac{3}{8}$$

$$P(x = 3) = \binom{4}{3}\left(\frac{1}{2}\right)^3\left(\frac{1}{2}\right)^1 = \frac{4}{16} = \frac{1}{4}$$

$$P(x = 4) = \binom{4}{4}\left(\frac{1}{2}\right)^4\left(\frac{1}{2}\right)^0 = \frac{1}{16}$$

Therefore, the probability distribution is as follows.

Number of Heads	0	1	2	3	4
Probability	$\frac{1}{16}$	$\frac{1}{4}$	$\frac{3}{8}$	$\frac{1}{4}$	$\frac{1}{16}$

3. Let x denote the number of aces drawn. Then x can take on values 0, 1, 2, or 3. The probabilities are as follows.

$$P(x = 0) = \binom{3}{0}\left(\frac{48}{52}\right)\left(\frac{47}{51}\right)\left(\frac{46}{50}\right) \approx .783$$

$$P(x = 1) = \binom{3}{1}\left(\frac{4}{52}\right)\left(\frac{48}{51}\right)\left(\frac{47}{50}\right) \approx .204$$

$$P(x = 2) = \binom{3}{2}\left(\frac{4}{52}\right)\left(\frac{3}{51}\right)\left(\frac{48}{50}\right) \approx .013$$

$$P(x = 3) = \binom{3}{3}\left(\frac{4}{52}\right)\left(\frac{3}{51}\right)\left(\frac{2}{50}\right) \approx .0002$$

Therefore, the probability distribution is as follows.

Number of Aces	0	1	2	3
Probability	.783	.204	.013	.0002

For Exercises 5 and 7, see the histograms in the answer section of the textbook.

5. Use the probabilities that were calculated in Exercise 1. Draw a histogram with 5 rectangles, corresponding to $x = 0$, $x = 1$, $x = 2$, $x = 3$, and $x = 4$. $P(x \leq 2)$ corresponds to

$$P(x = 0) + P(x = 1) + P(x = 2),$$

so shade the first 3 rectangles in the histogram.

7. Use the probabilities that were calculated in Exercise 3. Draw a histogram with 4 rectangles, corresponding to $x = 0$, $x = 1$, $x = 2$, and $x = 3$. $P(\text{at least one ace}) = P(x \geq 1)$ corresponds to

$$P(x = 1) + P(x = 2) + P(x = 3),$$

so shade the last 3 rectangles.

9. $E(x) = 2(.1) + 3(.4) + 4(.3) + 5(.2)$
   $= 3.6$

11. $E(z) = 9(.14) + 12(.22) + 15(.36) + 18(.18)$
    $\qquad + 21(.10)$
    $= 14.64$

13. It is possible (but not necessary) to begin by writing the histogram's data as a probability distribution, which would look as follows.

x	1	2	3	4
$P(x)$	.2	.3	.1	.4

The expected value of x is

$$E(x) = 1(.2) + 2(.3) + 3(.1) + 4(.4)$$
$$= 2.7.$$

15. The expected value of x is

$$E(x) = 6(.1) + 12(.2) + 18(.4) + 24(.2)$$
$$\qquad + 30(.1)$$
$$= 18.$$

17. Using the data from Example 4, the expected winnings for Mary are

$$E(x) = -.8\left(\frac{1}{4}\right) + .8\left(\frac{1}{4}\right) + .8\left(\frac{1}{4}\right) + (-.8)\left(\frac{1}{4}\right)$$

$$= 0.$$

Yes, it is still a fair game if Mary tosses and Donna calls.

19. (a)

Number of Yellow Marbles	Probability
0	$\dfrac{\binom{3}{0}\binom{4}{3}}{\binom{7}{3}} = \dfrac{4}{35}$
1	$\dfrac{\binom{3}{1}\binom{4}{2}}{\binom{7}{3}} = \dfrac{18}{35}$
2	$\dfrac{\binom{3}{2}\binom{4}{1}}{\binom{7}{3}} = \dfrac{12}{35}$
3	$\dfrac{\binom{3}{3}\binom{4}{0}}{\binom{7}{3}} = \dfrac{1}{35}$

Draw a histogram with four rectangles corresponding to $x = 0, 1, 2,$ and 3. See the histogram in the answer section of the textbook.

(b) Expected number of yellow marbles

$$= 0\left(\frac{4}{35}\right) + 1\left(\frac{18}{35}\right) + 2\left(\frac{12}{35}\right) + 3\left(\frac{1}{35}\right)$$

$$= \frac{45}{35} = \frac{9}{7} \approx 1.286$$

21. The probability that the delegation contains no liberals and 3 conservatives is

$$\frac{\binom{5}{0}\binom{4}{3}}{\binom{9}{3}} = \frac{1 \cdot 4}{84} = \frac{4}{84}.$$

Similarly, use combinations to calculate the remaining probabilities for the probability distribution.

(a) Let x represent the number of liberals on the delegation. The probability distribution of x is as follows.

x	0	1	2	3
$P(x)$	$\frac{4}{84}$	$\frac{30}{84}$	$\frac{40}{84}$	$\frac{10}{84}$

The expected value is

$$E(x) = 0\left(\frac{4}{84}\right) + 1\left(\frac{30}{84}\right) + 2\left(\frac{40}{84}\right) + 3\left(\frac{10}{84}\right)$$
$$= \frac{140}{84} = \frac{5}{3} \approx 1.667 \text{ liberals.}$$

(b) Let y represent the number of conservatives on the committee. The probability distribution of y is as follows.

y	0	1	2	3
$P(y)$	$\frac{10}{84}$	$\frac{40}{84}$	$\frac{30}{84}$	$\frac{4}{84}$

The expected value is

$$E(y) = 0\left(\frac{10}{84}\right) + 1\left(\frac{40}{84}\right) + 2\left(\frac{30}{84}\right) + 3\left(\frac{4}{84}\right)$$
$$= \frac{112}{84} = \frac{4}{3} \approx 1.333 \text{ conservatives.}$$

23. Let x represent the number of junior members on the committee. Use combinations to find the probabilities of 0, 1, 2, and 3 junior members. The probability distribution of x is as follows.

x	0	1	2	3
$P(x)$	$\frac{57}{203}$	$\frac{95}{203}$	$\frac{45}{203}$	$\frac{6}{203}$

The expected value is

$$E(x) = 0\left(\frac{57}{203}\right) + 1\left(\frac{95}{203}\right) + 2\left(\frac{45}{203}\right) + 3\left(\frac{6}{203}\right)$$
$$= 1 \text{ junior member.}$$

25. The probability of drawing 2 diamonds is

$$\frac{\binom{13}{2}}{\binom{52}{2}} = \frac{78}{1326},$$

and the probability of not drawing 2 diamonds is

$$1 - \frac{78}{1326} = \frac{1248}{1326}.$$

Let x represent your net winnings. Then the expected value of the game is

$$E(x) = 4.5\left(\frac{78}{1326}\right) + (-.5)\left(\frac{1248}{1326}\right)$$
$$= -\frac{273}{1326} \approx -\$.21 \quad \text{or} \quad -21\cancel{c}.$$

The game is not fair since your expected winnings are not zero.

29. We first compute the amount of money the company can expect to pay out for each kind of policy. The sum of these amounts will be the total amount the company can expect to pay out. For a single $10,000 policy, we have the following probability distribution.

	Pay	Don't Pay
Outcome	$10,000	$10,000
Probability	.001	.999

$$E(\text{payoff}) = 10,000(.001) + 0(.999)$$
$$= \$10$$

For all 100 such policies, the company can expect to pay out

$$100(10) = \$1000.$$

For a single $5000 policy,

$$E(\text{payoff}) = 5000(.001) + 0(.999)$$
$$= \$5.$$

For all 500 such policies, the company can expect to pay out

$$500(5) = \$2500.$$

Similarly, for all 1000 policies of $1000, the company can expect to pay out

$$1000(1) = \$1000.$$

Thus, the total amount the company can expect to pay out is

$$\$1000 + \$2500 + \$1000 = \$4500.$$

31. Let x represent the number of offspring. We have the following probability distribution.

x	0	1	2	3	4
$P(x)$	.31	.21	.19	.17	.12

$$E(x) = 0(.31) + 1(.21) + 2(.19) + 3(.17) + 4(.12)$$
$$= 1.58$$

33. (a) If the two players are equally skilled, the old pro's expected winnings are

$$\frac{1}{2}(80,000) + \frac{1}{2}(20,000) = \$50,000.$$

(b) If the pro's chance of winning is $\frac{3}{4}$, then his expected winnings are

$$\frac{3}{4}(80,000) + \frac{1}{4}(20,000) = \$65,000.$$

35. (a) We define a success to be the event that a letter was delivered the next day. In this situation, $n = 10$; $x = 0, 1, 2, 3, \ldots, 10$; $p = .83$, and $1 - p = .17$.

Number of Letters Delivered the Next Day	Probability
0	$\binom{10}{0}(.83)^0(.17)^{10} \approx .0000$
1	$\binom{10}{1}(.83)^1(.17)^9 \approx .0000$
2	$\binom{10}{2}(.83)^2(.17)^8 \approx .0000$
3	$\binom{10}{3}(.83)^3(.17)^7 \approx .0003$
4	$\binom{10}{4}(.83)^4(.17)^6 \approx .0024$
5	$\binom{10}{5}(.83)^5(.17)^5 \approx .0141$
6	$\binom{10}{6}(.83)^6(.17)^4 \approx .0573$
7	$\binom{10}{7}(.83)^7(.17)^3 \approx .1600$
8	$\binom{10}{8}(.83)^8(.17)^2 \approx .2929$
9	$\binom{10}{9}(.83)^9(.17)^1 \approx .3178$
10	$\binom{10}{10}(.83)^{10}(.17)^0 \approx .1552$

(b) P(4 or fewer letters would be delivered)
$$= P(x = 0) + P(x = 1) + P(x = 2)$$
$$+ P(x = 3) + P(x = 4)$$
$$\approx .0027$$

(d) Expected number of letters delivered next day
$$\approx 0(.0000) + 1(.0000) + 2(.0000)$$
$$+ 3(.0003) + 4(.0024) + 5(.0141)$$
$$+ 6(.0573) + 7(.1600) + 8(.2929)$$
$$+ 9(.3178) + 10(.1552)$$
$$\approx 8.3$$

37. Reduce each price by the 50¢ cost of the raffle ticket, and multiply by the corresponding probability.

$$E(x) = 999.50 \left(\frac{1}{10,000}\right) + 299.50 \left(\frac{2}{10,000}\right)$$
$$+ 9.50 \left(\frac{20}{10,000}\right) + (-.50) \left(\frac{9977}{10,000}\right)$$
$$= \frac{-3200}{10,000} = -\$.32 \quad \text{or} \quad -32¢$$

This is not a fair game. In a fair game the expected value is 0.

39. The probability of getting exactly 3 of the 4 selections correct and winning this game is

$$\binom{4}{3} \left(\frac{1}{13}\right)^3 \left(\frac{12}{13}\right)^1 \approx .001681.$$

The probability of losing is .998319. If you win, your winnings are $199. Otherwise, you lose $1 (win −$1). If x represents your winnings, then the expected value is

$$E(x) = 199(.001681) + (-1)(.998319)$$
$$= .334519 - .998319$$
$$= -.6638 \approx -\$.66 \quad \text{or} \quad -66¢.$$

41. In this form of roulette,

$$P(\text{even}) = \frac{18}{37} \text{ and } P(\text{noneven}) = \frac{19}{37}.$$

If an even number comes up, you win $1. Otherwise, you lose $1 (win −$1). If x represents your winnings, then the expected value is

$$E(x) = 1 \left(\frac{18}{37}\right) + (-1) \left(\frac{19}{37}\right)$$
$$= -\frac{1}{37} \approx -\$.027 \text{ or } -2.7¢.$$

43. In this form of the game Keno,

$$P(\text{your number comes up}) = \frac{20}{80} = \frac{1}{4}$$

and

$$P(\text{your number doesn't come up}) = \frac{60}{80} = \frac{3}{4}.$$

If your number comes up, you win $2.20. Otherwise, you lose $1 (win −$1). If x represents your winnings, then the expected value is

$$E(x) = 2.20 \left(\frac{1}{4}\right) - 1 \left(\frac{3}{4}\right)$$
$$= .55 - .75 = -\$.20 \quad \text{or} \quad -20¢.$$

45. At any one restaurant, your expected winnings are

$$E(x) = 100,000 \left(\frac{1}{176,402,500}\right) + 25,000 \left(\frac{1}{39,200,556}\right)$$
$$+ 5000 \left(\frac{1}{17,640,250}\right) + 1000 \left(\frac{1}{1,568,022}\right)$$
$$+ 100 \left(\frac{1}{288,244}\right) + 5 \left(\frac{1}{7056}\right) + 1 \left(\frac{1}{588}\right)$$

$$= .00488.$$

Going to 25 restaurants gives you expected earnings of $25(.00488) = .122$. Since you spent \$1, you lose 87.8¢ on the average, so your expected value is -87.8¢.

Chapter 7 Review Exercises

1. 6 shuttle vans can line up at the airport in

$$P(6,6) = 6! = 720$$

different ways.

3. 3 oranges can be taken from a bag of 12 in

$$\binom{12}{3} = \frac{12!}{9!3!} = \frac{12 \cdot 11 \cdot 10}{3 \cdot 2 \cdot 1} = 220$$

different ways.

5. 2 pictures from a group of 5 different pictures can be arranged in

$$P(5,2) = 5 \cdot 4 = 20$$

different ways.

7. (a) There are 2! ways to arrange the landscapes, 3! ways to arrange the puppies, and 2 choices whether landscapes or puppies come first. Thus, the pictures can be arranged in

$$2!3! \cdot 2 = 24$$

different ways.

(b) The pictures must be arranged puppy, landscape, puppy, landscape, puppy. Arrange the puppies in 3! or 6 ways. Arrange the landscapes in 2! or 2 ways. In this scheme, the pictures can be arranged in $6 \cdot 2 = 12$ different ways.

9. (a) There are $7 \cdot 5 \cdot 4 = 140$ different groups of 3 representatives possible.

(b) $7 \cdot 5 \cdot 4 = 140$ is the number of groups with 3 representatives. For 2 representatives, the number of groups is

$$7 \cdot 5 + 7 \cdot 4 + 5 \cdot 4 = 83.$$

For 1 representative, the number of groups is

$$7 + 5 + 4 = 16.$$

The total number of these groups is

$$140 + 83 + 16 = 239$$

groups.

13. It is impossible to draw 3 blue balls, since there are only 2 blue balls in the basket; hence,

$$P(\text{all blue balls}) = 0.$$

15. $P(\text{exactly 2 black balls})$

$$= \frac{\binom{4}{2}\binom{7}{1}}{\binom{11}{3}} = \frac{42}{165} = \frac{14}{55} \approx .255$$

17. $P(\text{2 green balls and 1 blue ball})$

$$= \frac{\binom{5}{2}\binom{2}{1}}{\binom{11}{3}} = \frac{20}{165} = \frac{4}{33} \approx .121$$

19. Let x represent the number of girls. We have $n = 6$, $x = 6$, $p = \frac{1}{2}$, and $1 - p = \frac{1}{2}$, so

$$P(\text{all girls}) = \binom{6}{6}\left(\frac{1}{2}\right)^6\left(\frac{1}{2}\right)^0 = \frac{1}{64} \approx .016.$$

21. Let x represent the number of boys, and then $p = \frac{1}{2}$ and $1 - p = \frac{1}{2}$. We have

$$P(\text{no more than 2 boys})$$
$$= P(x \leq 2)$$
$$= P(x = 0) + P(x = 1) + P(x = 2)$$
$$= \binom{6}{0}\left(\frac{1}{2}\right)^0\left(\frac{1}{2}\right)^6 + \binom{6}{1}\left(\frac{1}{2}\right)^1\left(\frac{1}{2}\right)^5$$
$$+ \binom{6}{2}\left(\frac{1}{2}\right)^2\left(\frac{1}{2}\right)^4$$
$$= \frac{11}{32} \approx .344.$$

23. $P(\text{2 spades}) = \dfrac{\binom{13}{2}}{\binom{52}{2}} = \dfrac{78}{1326} = \dfrac{1}{17} \approx .059$

25. $P(\text{exactly 1 face card})$

$$= \frac{\binom{12}{1}\binom{40}{1}}{\binom{52}{2}} = \frac{480}{1326} = \frac{80}{221} \approx .3620$$

27. $P(\text{at most 1 queen})$
$$= P(\text{0 queens}) + P(\text{1 queen})$$

$$= \frac{\binom{48}{2}}{\binom{52}{2}} + \frac{\binom{4}{1}\binom{48}{1}}{\binom{52}{2}}$$

$$= \frac{1128}{1326} + \frac{192}{1326}$$

$$= \frac{1320}{1326} = \frac{220}{221} \approx .9955$$

29. (a) There are $n = 36$ possible outcomes. Let x represent the sum of the dice, and note that the possible values of x are the whole numbers from 2 to 12. The probability distribution is as follows.

x	2	3	4	5	6
$P(x)$	$\frac{1}{36}$	$\frac{2}{36} = \frac{1}{18}$	$\frac{3}{36} = \frac{1}{12}$	$\frac{4}{36} = \frac{1}{9}$	$\frac{5}{36}$

x	7	8	9	10	11	12
$P(x)$	$\frac{6}{36} = \frac{1}{6}$	$\frac{5}{36}$	$\frac{4}{36} = \frac{1}{9}$	$\frac{3}{36} = \frac{1}{12}$	$\frac{2}{36} = \frac{1}{18}$	$\frac{1}{36}$

(b) The histogram consists of 11 rectangles. See the histogram in the answer section of the textbook.

(c) The expected value is

$$E(x) = 2\left(\frac{1}{36}\right) + 3\left(\frac{2}{36}\right) + 4\left(\frac{3}{36}\right) + 5\left(\frac{4}{36}\right)$$
$$+ 6\left(\frac{5}{36}\right) + 7\left(\frac{6}{36}\right) + 8\left(\frac{5}{36}\right) + 9\left(\frac{4}{36}\right)$$
$$+ 10\left(\frac{3}{36}\right) + 11\left(\frac{2}{36}\right) + 12\left(\frac{1}{36}\right)$$
$$= \frac{252}{36} = 7.$$

31. The probability that corresponds to the shaded region of the histogram is the total of the shaded areas, that is,

$$1(.1) + 1(.3) + 1(.2) = .6.$$

33. Let x represent the number of girls. The probability distribution is as follows.

x	0	1	2	3	4	5
$P(x)$	$\frac{1}{32}$	$\frac{5}{32}$	$\frac{10}{32}$	$\frac{10}{32}$	$\frac{5}{32}$	$\frac{1}{32}$

The expected value is

$$E(x) = 0\left(\frac{1}{32}\right) + 1\left(\frac{5}{32}\right) + 2\left(\frac{10}{32}\right) + 3\left(\frac{10}{32}\right)$$
$$+ 4\left(\frac{5}{32}\right) + 5\left(\frac{1}{32}\right)$$
$$= \frac{80}{32} = 2.5 \text{ girls.}$$

35. $P(3 \text{ clubs}) = \dfrac{\binom{13}{3}}{\binom{52}{3}} = \dfrac{286}{22,100} \approx .0129$

Thus,

$P(\text{win}) = .0129$ and
$P(\text{lose}) = 1 - .0129 = .9871.$

Let x represent the amount you should pay. Your net winnings are $100 - x$ if you win and $-x$ if you lose. If it is a fair game, your expected winnings will be 0. Thus, $E(x) = 0$ becomes

$$.0129(100 - x) + .9871(-x) = 0$$
$$1.29 - .0129x - .9871x = 0$$
$$1.29 - x = 0$$
$$x = 1.29.$$

You should pay $1.29.

37. (a) Given a set with n elements, the number of subsets of size

$$0 \text{ is } \binom{n}{0} = 1,$$
$$1 \text{ is } \binom{n}{1} = n,$$
$$2 \text{ is } \binom{n}{2} = \frac{n(n-1)}{2}, \text{ and}$$
$$n \text{ is } \binom{n}{n} = 1.$$

(b) The total number of subsets is

$$\binom{n}{0} + \binom{n}{1} + \binom{n}{2} + \cdots + \binom{n}{n}.$$

(d) Let $n = 4$.

$$\binom{4}{0} + \binom{4}{1} + \binom{4}{2} + \binom{4}{3} + \binom{4}{4}$$
$$= 1 + 4 + 6 + 4 + 1$$
$$= 16$$
$$= 2^4 = 2^n$$

Let $n = 5$.

$$\binom{5}{0} + \binom{5}{1} + \binom{5}{2} + \binom{5}{3} + \binom{5}{4} + \binom{5}{5}$$
$$= 1 + 5 + 10 + 10 + 5 + 1$$
$$= 32$$
$$= 2^5 = 2^n$$

(e) The sum of the elements in row n of Pascal's triangle is 2^n.

39. This is a Bernoulli trial problem with $n = 20$, $p = .01$, and $x = 4$.

$$P(\text{exactly 4 defective}) = \binom{20}{4}(.01)^4(.99)^{16}$$
$$\approx .00004$$

41. $P(\text{no more than 4 defective})$
$= P(0, 1, 2, 3, \text{ or } 4 \text{ defective})$

$= \binom{20}{0}(.01)^0(.99)^{20} + \binom{20}{1}(.01)^1(.99)^{19}$

$\quad + \binom{20}{2}(.01)^2(.99)^{18} + \binom{20}{3}(.01)^3(.99)^{17}$

$\quad + \binom{20}{4}(.01)^4(.99)^{16}$

$\approx .81791 + .16523 + .01586 + .00096 + .00004$
$= 1.0000$

43. $E(x) = 16{,}000(.7) + (-9000)(.3)$
$\qquad = 8500$

The expected profit is $8500.

45. If a box is good (probability .9) and the merchant samples an excellent piece of fruit from that box (probability .80), then he will accept the box and earn a $200 profit on it.

If a box is bad (probability .1) and he samples an excellent piece of fruit from the box (probability .30), then he will accept the box and earn a −$1000 profit on it.

If the merchant ever samples a nonexcellent piece of fruit, he will not accept the box. In this case he pays nothing and earns nothing, so the profit will be $0.

Let x represent the merchant's earnings. Note that
$$.9(.80) = .72,$$
$$.1(.30) = .03,$$
$$\text{and } 1 - (.72 + .03) = .25.$$

The probability distribution is as follows.

x	200	−1000	0
$P(x)$	.72	.03	.25

The expected value when the merchant samples the fruit is

$E(x) = 200(.72) + (-1000)(.03) + 0(.25)$
$\qquad = 144 - 30 + 0$
$\qquad = \$114.$

We must also consider the case in which the merchant does not sample the fruit. Let x again represent the merchant's earnings. The probability distribution is as follows.

x	200	−1000
$P(x)$	.9	.1

The expected value when the merchant does not sample the fruit is

$E(x) = 200(.9) + (-1000)(.1)$
$\qquad = 180 - 100$
$\qquad = \$80.$

Combining these two results, the expected value of the right to sample is $114 − $80 = $34, which corresponds to choice (c).

47. **(a)** Define a success to be the event that a red M&M is selected. In this situation, $n = 4$; $x = 0, 1, 2, 3,$ or 4; $p = .2$; and $1 - p = .8$.

Number of Red M&M's	Probability
0	$\binom{4}{0}(.2)^0(.8)^4 = .4096$
1	$\binom{4}{1}(.2)^1(.8)^3 = .4096$
2	$\binom{4}{2}(.2)^2(.8)^2 = .1536$
3	$\binom{4}{3}(.2)^3(.8)^1 = .0256$
4	$\binom{4}{4}(.2)^4(.8)^0 = .0016$

(b) Draw a histogram with 5 rectangles. See the histogram in the answer section of the textbook.

(c) Expected number of red M&M's
$\qquad = np = 4(.2) = .8$

49. **(a)**

Number of African-Americans	Probability
0	$\dfrac{\binom{2}{0}\binom{6}{3}}{\binom{8}{3}} = \dfrac{20}{56} = \dfrac{10}{28}$
1	$\dfrac{\binom{2}{1}\binom{6}{2}}{\binom{8}{3}} = \dfrac{30}{56} = \dfrac{15}{28}$
2	$\dfrac{\binom{2}{2}\binom{6}{1}}{\binom{8}{3}} = \dfrac{6}{56} = \dfrac{3}{28}$

(b) Draw a histogram with 3 rectangles. See the histogram in the answer section of the textbook.

(c) Expected number of African-Americans

$= 0\left(\dfrac{10}{28}\right) + 1\left(\dfrac{15}{28}\right) + 2\left(\dfrac{3}{28}\right)$

$= \dfrac{21}{28} = \dfrac{3}{4}$

51. Let x represent the net winnings for a person who buys one ticket.

x	\$4999	\$999	\$99	$-$\$1
$P(x)$	$\frac{1}{10,000}$	$\frac{2}{10,000}$	$\frac{2}{10,000}$	$\frac{9995}{10,000}$

$$E(x) = \frac{4999 + 2(999) + 2(99) - 9995}{10,000}$$

$$= \frac{-2800}{10,000} = -\$.28 \quad \text{or} \quad -28\cancel{c}$$

53. (a) First, we assume 4 numbers are picked. You win if 2, 3, or 4 numbers match.

$$P(\text{win}) = \frac{\binom{20}{2}\binom{60}{2} + \binom{20}{3}\binom{60}{1} + \binom{20}{4}\binom{60}{0}}{\binom{20}{0}\binom{60}{4} + \binom{20}{1}\binom{60}{3} + \binom{20}{2}\binom{60}{2} + \binom{20}{3}\binom{60}{1} + \binom{20}{4}\binom{60}{0}}$$

$$= \frac{409,545}{1,581,580}$$

$$\approx \frac{1}{3.86}$$

Next, we assume 5 numbers are picked. You win if 3, 4, or 5 numbers match.

$$P(\text{win}) = \frac{\binom{20}{3}\binom{60}{2} + \binom{20}{4}\binom{60}{1} + \binom{20}{5}\binom{60}{0}}{\binom{20}{0}\binom{60}{5} + \binom{20}{1}\binom{60}{4} + \binom{20}{2}\binom{60}{3} + \binom{20}{3}\binom{60}{2} + \binom{20}{4}\binom{60}{1} + \binom{20}{5}\binom{60}{0}}$$

$$= \frac{2,324,004}{24,040,016}$$

$$\approx \frac{1}{10.34}$$

(b) Expected value when you pick 4

$$= 1 \cdot \frac{\binom{20}{2}\binom{60}{2}}{1,581,580} + 5 \cdot \frac{\binom{20}{3}\binom{60}{1}}{1,581,580} + 55 \cdot \frac{\binom{20}{4}\binom{60}{0}}{1,581,580} - 1(1)$$

$$\approx -\$.4026.$$

Expected value when you pick 5

$$= 2 \cdot \frac{\binom{20}{3}\binom{60}{2}}{24,040,016} + 20 \cdot \frac{\binom{20}{4}\binom{60}{1}}{24,040,016} + 300 \cdot \frac{\binom{20}{5}\binom{60}{0}}{24,040,016} - 1(1)$$

$$\approx -\$.3968.$$

Extended Application: Optimal Inventory For a Service Truck

1. (a) $C(M_0)$
$$= NL[1 - (1 - p_1)(1 - p_2)(1 - p_3)]$$
$$= 3(54)[1 - (.91)(.76)(.83)]$$
$$= \$69.01$$

(b) $C(M_2)$
$$= H_2 + NL[1 - (1 - p_1)(1 - p_3)]$$
$$= 40 + 3(54)[1 - (.91)(.83)]$$
$$= \$79.64$$

(c) $C(M_3)$
$$= H_3 + NL[1 - (1 - p_1)(1 - p_2)]$$
$$= 9 + 3(54)[1 - (.91)(.76)]$$
$$= \$58.96$$

(d) $C(M_{12})$
$$= H_1 + H_2 + NL[1 - (1 - p_3)]$$
$$= 15 + 40 + 3(54)[1 - .83]$$
$$= \$82.54$$

(e) $C(M_{13})$
$$= H_1 + H_3 + NL[1 - (1 - p_2)]$$
$$= 15 + 9 + 3(54)[1 - .76]$$
$$= \$62.88$$

(f) $C(M_{123})$
$$= H_1 + H_2 + H_3 + NL[1 - 1]$$
$$= 15 + 40 + 9$$
$$= \$64.00$$

2. Policy M_3, stocking only part 3 on the truck, leads to the lowest expected cost.

3. It is not necessary for the probabilities to add up to 1 because it is possible that no parts will be needed. That is, the events of needing parts 1, 2, and 3 are not the only events in the sample space.

4. For 3 different parts we have 8 different policies: 1 with no parts, 3 with 1 part, 3 with 2 parts, and 1 with 3 parts. The number of different policies, $8 = 2^3$, is the number of subsets of a set containing 3 distinct elements. If there are n different parts, the number of policies is the number of subsets of a set containing n distinct elements which we showed in Chapter 6 to be 2^n.

Chapter 7 Test

[7.1]

1. The 24 members of the 3rd grade Mitey Mites hockey team must select a head basher, a second basher, and a designated tripper for their team. How many ways can this be done?

2. In Ohio, most license plates consist of three letters followed by three digits. How many different plates are possible in the following situations?

    (a) Repeats of letters and digits are allowed.

    (b) Repeats of letters, but not digits are allowed.

    (c) The first two letters must be AR, and repeats are allowed.

[7.1–7.2]

3. Evaluate each of the following.

    (a) $8!$ (b) $P(6,4)$ (c) $\binom{10}{4}$ (d) $\binom{17}{17}$

[7.2]

4. A basketball team consists of 7 good shooters and 5 poor shooters.

    (a) In how many ways can a 5-person team be selected?

    (b) In how many ways can a team consisting of only good shooters be selected?

    (c) In how many ways can a team consisting of 3 good and 2 poor shooters be selected?

5. A bag contains 4 red, 5 blue, and 8 white marbles. A sample of 6 marbles is drawn from the bag. In how many ways is it possible to get the following results?

    (a) 3 blue and 3 white marbles

    (b) 5 blue marbles and 1 white marble

    (c) 2 red, 2 blue, and 2 white marbles

[7.3]

6. A three-card hand is drawn from a standard deck of 52 cards. Set up the probability of each of the following events. (Do not calculate the answers.)

    (a) The hand contains exactly 2 hearts.

    (b) The hand contains fewer than 2 hearts.

    (c) The hand contains exactly 2 queens.

7. A mathematics class has 15 female students and 8 male students. Five students are randomly selected. Find the probability that there will be at least one male in the group. (Round your answer to four decimal places.)

[7.4]

8. A shipment of bolts has 20% of the bolts defective.

   (a) What is the probability of finding 2 or fewer defective bolts in a sample of 15?

   (b) What is the probability that if 9 bolts are sampled, 2 defectives will be found?

   (c) How many bolts must be sampled to ensure a probability of at least .6 that a defective bolt is found?

[7.5]

9. A nickel, a dime, and a quarter are tossed simultaneously. Let the random variable x denote the number of heads observed in the experiment, and prepare a probability distribution for this experiment.

10. Two dice are rolled, and the total number of points is recorded. Find the expected value.

11. There is a game called Double or Nothing, and it costs $1 to play. If you draw the ace of spades, you are paid $2, but you are paid nothing if you draw any other card. Is this a fair game?

12. At a large university, 62% of the students enrolled are female. A sample of 3 students are selected and the number of female students is noted.

    (a) Give the probability distribution for the number of females.

    (b) Sketch its histogram.

    (c) Find the expected value.

Chapter 7 Test Answers

1. 12,144

2. **(a)** 17,576,000   **(b)** 12,654,720   **(c)** 26,000

3. **(a)** 40,320   **(b)** 360   **(c)** 210   **(d)** 1

4. **(a)** 792   **(b)** 21   **(c)** 350

5. **(a)** 560   **(b)** 8   **(c)** 1680

6. **(a)** $\dfrac{\binom{13}{2}\binom{39}{1}}{\binom{52}{3}}$ **(b)** $\dfrac{\binom{13}{0}\binom{39}{3} + \binom{13}{1}\binom{39}{2}}{\binom{52}{3}}$ **(c)** $\dfrac{\binom{4}{2}\binom{48}{1}}{\binom{52}{3}}$

7. .9108

8. **(a)** $\binom{15}{0}(.2)^0(.8)^{15} + \binom{15}{1}(.2)^1(.8)^{14} + \binom{15}{2}(.2)^2(.8)^{13} \approx .3980$ **(b)** $\binom{9}{2}(.2)^2(.8)^7 \approx .3020$ **(c)** 5

9.

x	0	1	2	3
$P(x)$	$\frac{1}{8}$	$\frac{3}{8}$	$\frac{3}{8}$	$\frac{1}{8}$

10. 7

11. No, it is not a fair game.

12. **(a)**

Number of Females	0	1	2	3
Probability	.0549	.2686	.4382	.2383

   (b)

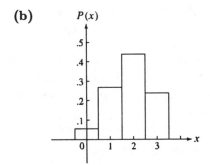

   (c) 1.86

STATISTICS

8.1 Frequency Distributions; Measures of Central Tendency

1. **(a)-(b)** Since 0-24 is to be the first interval and there are 25 numbers between 0 and 24 inclusive, we will let all six intervals be of size 25. The other five intervals are 25-49, 50-74, 75-99, 100-124, and 125-149. Making a tally of how many data values lie in each interval leads to the following frequency distribution.

Interval	Frequency
0-24	4
25-49	3
50-74	6
75-99	3
100-124	5
125-149	9

(c) Draw the histogram. It consists of 6 bars of equal width having heights as determined by the frequency of each interval. See the histogram in the answer section of the textbook.

(d) To construct the frequency polygon, join consecutive midpoints of the tops of the histogram bars with line segments. See the answer section of the textbook.

3. **(a)-(b)** There are eight intervals starting with 0-19. Making a tally of how many data values lie in each interval leads to the following frequency distribution.

Interval	Frequency
0-19	3
20-39	3
40-59	3
60-79	5
80-99	2
100-119	4
120-139	7
140-159	3

(c) Draw the histogram. It consists of 8 rectangles of equal width having heights as determined by the frequency of each interval. See the histogram in the answer section of the textbook.

(d) To construct the frequency polygon, join consecutive midpoints of the tops of the histogram bars with line segments. See the answer section of the textbook.

5. **(a)-(b)** Since 70-74 is to be the first interval, we let all the intervals be of size 5. The largest data value is 111, so the last interval that will be needed is 110-114. The frequency distribution is as follows.

Interval	Frequency
70-74	2
75-79	1
80-84	3
85-89	2
90-94	6
95-99	5
100-104	6
105-109	4
110-114	2

(c) Draw the histogram. It consists of 9 bars of equal width having heights as determined by the frequency of each interval. See the histogram in the answer section of the textbook.

(d) Construct the frequency polygon by joining consecutive midpoints of the tops of the histogram bars with line segments. See the answer section of the textbook.

9. The mean of the 5 numbers is

$$\bar{x} = \frac{\sum x}{n}$$
$$= \frac{8 + 10 + 16 + 21 + 25}{5}$$
$$= \frac{80}{5} = 16.$$

11. $\sum x = 21,900 + 22,850 + 24,930 + 29,710$
$\qquad + 28,340 + 40,000$
$\quad = 167,730$

The mean of the 6 numbers is

$$\bar{x} = \frac{\sum x}{n} = \frac{167,730}{6} = 27,955.$$

13. $\sum x = 9.4 + 11.3 + 10.5 + 7.4 + 9.1$
$\qquad + 8.4 + 9.7 + 5.2 + 1.1 + 4.7$
$\quad = 76.8$

The mean of the 10 numbers is

$$\bar{x} = \frac{\sum x}{n} = \frac{76.8}{10} = 7.68 \approx 7.7.$$

15. Add to the frequency distribution a new column, "Value $\times$ Frequency."

Value	Frequency	Value $\times$ Frequency
3	4	$3 \cdot 4 = 12$
5	2	$5 \cdot 2 = 10$
9	1	$9 \cdot 1 = 9$
12	3	$12 \cdot 3 = 36$
Totals:	10	67

The mean is

$$\bar{x} = \frac{67}{10} = 6.7.$$

17.

Value	Frequency	Value $\times$ Frequency
12	4	$12 \cdot 4 = 48$
13	2	$13 \cdot 2 = 26$
15	5	$15 \cdot 5 = 75$
19	3	$19 \cdot 3 = 57$
22	1	$22 \cdot 1 = 22$
23	5	$23 \cdot 5 = 115$
Totals:	20	343

The mean is

$$\bar{x} = \frac{343}{20} = 17.15 \approx 17.2.$$

19. 12, 18, 32, 51, 58, 92, 106

The median is the middle number, 51.

21. 100, 114, 125, 135, 150, 172

The median is the mean of the two middle numbers, which is

$$\frac{125 + 135}{2} = \frac{260}{2} = 130.$$

23. Arrange the numbers in numerical order, from smallest to largest.

$$3.4, 9.1, 27.6, 28.4, 29.8, 32.1, 47.6, 59.8$$

There are eight numbers here; the median is the mean of the two middle numbers, which is

$$\frac{28.4 + 29.8}{2} = \frac{58.2}{2} = 29.1.$$

25. Using a graphing calculator, $\bar{x} = 85.5$ and the median is 91.5.

27. 4, 9, 8, 6, 9, 2, 1, 3

The mode is the number that occurs most often. Here, the mode is 9.

29. 74, 68, 68, 68, 75, 75, 74, 74, 70

The mode is the number that occurs most often. Here, there are two modes, 68 and 74, since they both appear three times.

31. 6.8, 6.3, 6.3, 6.9, 6.7, 6.4, 6.1, 6.0

The mode is 6.3.

35.

Interval	Midpoint, x	Frequency, f	Product, xf
70-74	72	2	144
75-79	77	1	77
80-84	82	3	246
85-89	87	2	174
90-94	92	6	552
95-99	97	5	485
100-104	102	6	612
105-109	107	4	428
110-114	112	2	224
Totals:		31	2942

The mean of this collection of grouped data is

$$\bar{x} = \frac{\sum xf}{n} = \frac{2942}{31} \approx 94.9.$$

The intervals 90-94 and 100-104 each contain the most data values, 6, so they are the modal classes.

39. Find the mean of the numbers in the "Production" column.

$$\sum x = 2200 + 2000 + 1750 + 2200 + 2400$$
$$\qquad + 2800 + 2800 + 2450 + 2600 + 2750$$
$$\quad = 23,950$$

The mean production is

$$\bar{x} = \frac{\sum x}{n} = \frac{23,950}{10} = 2395 \text{ million bushels.}$$

To find the median, list the ten values in the "Production column" from smallest to largest.

$$1750, 2000, 2200, 2200, 2400,$$
$$2450, 2600, 2750, 2800, 2800$$

The median is the mean of the two middle entries.

$$\frac{2400 + 2450}{2} = 2425$$

The median production is 2425 million bushels.

41. From the histogram, we see that

(a) the height of the 10-19 histogram bar looks like about 17.5, so about 17.5% of the population was between 10 and 19 yr old;

(b) the 60-69 bar has a height of about 8, so about 8% of the population was between 60 and 69 yr old;

(c) the 20-29 bar appears to have the greatest height, so that age group had the largest percentage of the population.

43. (a)

Interval	Frequency
0-4	4
5-9	8
10-14	10
15-19	1
20-24	3
25-29	1

(b) Draw a histogram consisting of 6 rectangles of equal width having heights as determined by the frequency of each interval. To construct the frequency polygon, join consecutive midpoints of the tops of the histogram bars with line segments. See the histogram/frequency polygon in the answer section of the textbook.

(c) For the original data, $\sum x = 285$, $n = 27$, and

$$\bar{x} = \frac{\sum x}{n} = \frac{285}{27} \approx 10.56.$$

(d)

Interval	Midpoint, x	Frequency, f	Product, xf
0-4	2	4	8
5-9	7	8	56
10-14	12	10	120
15-19	17	1	17
20-24	22	3	66
25-29	27	1	27
Totals:		27	294

The mean of this collection of grouped data is

$$\bar{x} = \frac{\sum xf}{n} = \frac{294}{27} \approx 10.89.$$

(f) Arrange the data in increasing order.

$$0,\ 3,\ 4,\ 4,\ 5,\ 5,\ 6,\ 7,\ 7,$$
$$8,\ 8,\ 8,\ 10,\ 11,\ 11,\ 11, 11,\ 11,$$
$$12,\ 12,\ 14,\ 14,\ 16,\ 20,\ 21,\ 21,\ 25$$

There are 27 items. The middle item is the fourteenth item which is 11, so the median is 11.

The item with the greatest frequency is also 11, so the mode is 11.

45. (a) $\bar{x} = \dfrac{\sum x}{n}$

$$= \frac{37 + 37 + 37 + 37 + 49 + 57 + 65 + 67}{8}$$

$$= \frac{386}{8} = 48.25$$

The mean is 48.25.

(b) The median is the mean of the middle two numbers.

$$\frac{37 + 49}{2} = 43$$

(c) The mode is the most frequent number, which is 37.

8.2   Measures of Variation

1. The standard deviation of a sample of numbers is the square root of the variance of the sample.

3. The range is $74 - 29 = 53$, the difference of the highest and lowest numbers in the set. To find the standard deviation, first find the mean.

$$\bar{x} = \frac{\sum x}{n}$$

$$= \frac{42 + 38 + 29 + 74 + 82 + 71 + 35}{7}$$

$$= \frac{371}{7} = 53$$

To prepare for calculating the standard deviation, construct a table.

x	x^2
42	1764
38	1444
29	841
74	5476
82	6724
71	5041
35	1225
Total:	22,515

The total of the second column is $\sum x^2 = 22,515$. The variance is

$$s^2 = \frac{\sum x^2 - n\bar{x}^2}{n-1}$$

$$= \frac{22,515 - 7(53)^2}{7-1}$$

$$= \frac{22,515 - 19,663}{6}$$

$$= \frac{2852}{6} \approx 475.3,$$

and the standard deviation is

$$s = \sqrt{475.3} \approx 21.8.$$

5. The range is $287 - 241 = 46$. The mean is

$$\bar{x} = \frac{241 + 248 + 251 + 257 + 252 + 287}{6} = 256.$$

x	x^2
241	58,081
248	61,504
251	63,001
257	66,049
252	63,504
287	82,369
Total:	394,508

The standard deviation is

$$s = \sqrt{\frac{\sum x^2 - n\bar{x}^2}{n-1}}$$

$$= \sqrt{\frac{394,508 - 6(256)^2}{5}}$$

$$= \sqrt{258.4} \approx 16.1.$$

7. The range is $27 - 3 = 24$. The mean is

$$\bar{x} = \frac{\sum x}{n} = \frac{140}{10} = 14.$$

x	x^2
3	9
7	49
4	16
12	144
15	225
18	324
19	361
27	729
24	576
11	121
Total:	2554

The standard deviation is

$$s = \sqrt{\frac{\sum x^2 - n\bar{x}^2}{n-1}}$$

$$= \sqrt{\frac{2554 - 10(14)^2}{9}}$$

$$= \sqrt{66} \approx 8.1.$$

9. Expand the table to include columns for the midpoint x of each interval and for xf, x^2, and fx^2.

Interval	f	x	xf	x^2	fx^2
0-24	4	12	48	144	576
25-49	3	37	111	1369	4107
50-74	6	62	372	3844	23,064
75-99	3	87	261	7569	22,707
100-124	5	112	560	12,544	62,720
125-149	9	137	1233	18,769	168,821
Totals:	30		2585		282,095

The mean of the grouped data is

$$\bar{x} = \frac{\sum xf}{n} = \frac{2585}{30} \approx 86.2.$$

The standard deviation for the grouped data is

$$s = \sqrt{\frac{\sum fx^2 - n\bar{x}^2}{n-1}}$$

$$= \sqrt{\frac{282,095 - 30(86.2)^2}{30-1}}$$

$$\approx \sqrt{2046.7} \approx 45.2.$$

11. Start with the frequency distribution that was the answer to Exercise 5 of the previous section, and expand the table to include columns for the midpoint x of each interval and for xf, x^2, and fx^2.

Interval	f	x	xf	x^2	fx^2
70-74	2	72	144	5184	10,368
75-79	1	77	77	5929	5929
80-84	3	82	246	6724	20,172
85-89	2	87	174	7569	15,138
90-94	6	92	552	8464	50,784
95-99	5	97	485	9409	47,045
100-104	6	102	612	10,404	62,424
105-109	4	107	428	11,449	45,796
110-114	2	112	224	12,544	25,088
Totals:	31		2942		282,744

The mean of the grouped data is

$$\bar{x} = \frac{\sum fx}{n} = \frac{2942}{31} \approx 94.9.$$

The standard deviation for the grouped data is

$$s = \sqrt{\frac{\sum fx^2 - n\bar{x}^2}{n-1}}$$
$$= \sqrt{\frac{282{,}744 - 31(94.9)^2}{31-1}}$$
$$\approx \sqrt{118.0} \approx 10.9.$$

13. Use $k = 3$ in Chebyshev's theorem.

$$1 - \frac{1}{k^2} = 1 - \frac{1}{3^2} = \frac{8}{9},$$

so at least $\frac{8}{9}$ of the distribution is within 3 standard deviations of the mean.

15. Use $k = 5$ in Chebyshev's theorem.

$$1 - \frac{1}{k^2} = 1 - \frac{1}{5^2} = \frac{24}{25},$$

so at least $\frac{24}{25}$ of the distribution is within 5 standard deviations of the mean.

17. Here $32 = 50 - 3 \cdot 6 = \bar{x} - 3s$ and $68 = 50 + 3 \cdot 6 = \bar{x} + 3s$, so Chebyshev's theorem applies with $k = 3$; hence at least

$$1 - \frac{1}{k^2} = \frac{8}{9} \approx 88.9\%$$

of the numbers lie between 32 and 68.

19. The answer here is the complement of the answer to Exercise 17. It was found that at least 88.9% of the distribution of numbers are between 32 and 68, so at most $100\% - 88.9\% = 11.1\%$ of the numbers will be less than 32 or more than 68.

23. $15, 18, 19, 23, 25, 25, 28, 30, 34, 38$

(a) $\bar{x} = \dfrac{1}{10}(15 + 18 + 19 + 23 + 25 + 25 + 28$
$\qquad\qquad + 30 + 34 + 38)$

$\qquad = \dfrac{1}{10}(255) = 25.5$

The mean life of the sample of Brand X batteries is 25.5 hr.

x	x^2
15	225
18	324
19	361
23	529
25	625
25	625
28	784
30	900
34	1156
38	1444
Total:	6973

$$s = \sqrt{\frac{\sum x^2 - n\bar{x}^2}{n-1}}$$
$$= \sqrt{\frac{6973 - 10(25.5)^2}{9}}$$
$$\approx \sqrt{52.28} \approx 7.2$$

The standard deviation of the Brand X lives is 7.2 hr.

(b) Forever Power has a smaller standard deviation (4.1 hr, as opposed to 7.2 hr for Brand X), which indicates a more uniform life.

(c) Forever Power has a higher mean (26.2 hr, as opposed to 25.5 hr for Brand X), which indicates a longer average life.

25.

Sample Number	(a) $\bar{x}$	(b) s
1	$\frac{1}{3}$	2.1
2	2	2.6
3	$-\frac{1}{3}$	1.5
4	0	2.6
5	$\frac{5}{3}$	2.5
6	$\frac{7}{3}$	.6
7	1	1.0
8	$\frac{4}{3}$	2.1
9	$\frac{7}{3}$	.6
10	$\frac{2}{3}$	1.2

(c) $\overline{X} = \dfrac{\sum \bar{x}}{n} \approx \dfrac{11.3}{10} = 1.13$

(d) $\bar{s} = \dfrac{\sum s}{n} = \dfrac{16.8}{10} = 1.68$

(e) The upper control limit for the sample means is

$$\overline{X} + k_1 \bar{s} = 1.13 + 1.954(1.68) \approx 4.41.$$

The lower control limit for the sample means is

$$\overline{X} - k_1 \bar{s} = 1.13 - 1.954(1.68) \approx -2.15.$$

(f) The upper control limit for the sample standard deviations is

$$k_2 \bar{s} = 2.568(1.68) \approx 4.31.$$

The lower control limit for the sample standard deviations is

$$k_3 \bar{s} = 0(1.68) = 0.$$

27. This exercise should be solved using a calculator with a standard deviation key. The answers are $\bar{x} = 1.8158$ mm and $s = .4451$ mm.

29. (a) This exercise should be solved using a calculator with a standard deviation key. The answers are $\bar{x} = 7.3571$ and $s = .1326$.

(b) $\bar{x} + 2s = 7.3571 + 2(.1326) = 7.6223$
$\bar{x} - 2s = 7.3571 - 2(.1326) = 7.0919$

All the data or 100% are within these two values, that is, within 2 standard deviations of the mean.

8.3   The Normal Distribution

1. The peak in a normal curve occurs directly above *the mean.*

3. For normal distributions where $\mu \neq 0$ or $\sigma \neq 1$, z-scores are found by using the formula

$$z = \frac{x - \mu}{\sigma}.$$

5. Use the table, "Area Under a Normal Curve to the Left of z", in the Appendix. To find the percent of the area under a normal curve between the mean and 2.50 standard deviations from the mean, subtract the table entry for $z = 0$ (representing the mean) from the table entry for $z = 2.5$.

$$.9938 - .5000 = .4938$$

Therefore, 49.38% of the area lies between μ and $\mu + 2.5\sigma$.

7. Subtract the table entry for $z = -1.71$ from the table entry for $z = 0$.

$$.5000 - .0436 = .4564$$

45.64% of the area lies between μ and $\mu - 1.71\sigma$.

9. $P(1.41 \leq z \leq 2.83)$
$= P(z \leq 2.83) - P(z \leq 1.41)$
$=$ (area to the left of 2.83)
$\quad -$ (area to the left of 1.41)
$= .9977 - .9207$
$= .077$ or 7.7%

11. $P(-2.48 \leq z \leq -.05)$
$= P(z \leq -.05) - P(z \leq -2.48)$
$=$ (area to the left of $-.05$)
$\quad -$ (area to the left of -2.48)
$= .4801 - .0066$
$= .4735$ or 47.35%

13. $P(-3.11 \leq z \leq 1.44)$
$= P(z \leq 1.44) - P(z \leq -3.11)$
$=$ (area to the left of 1.44)
$\quad -$ (area to the left of -3.11)
$= .9251 - .0009$
$= .9242$ or 92.42%

15. 5% of the total area is to the left of z.

Use the table backwards. Look in the body of the table for an area of .05, and find the corresponding z using the left column and top column of the table.

The closest values to .05 in the body of the table are .0505, which corresponds to $z = -1.64$, and .0495, which corresponds to $z = -1.65$.

17. 15% of the total area is to the right of z.

If 15% of the area is to the right of z, then 85% of the area is to the left of z. The closest value to .85 in the body of the table is .8504, which corresponds to $z = 1.04$.

19. For any normal distribution, the value of $P(x \leq \mu)$ is .5 since half of the distribution is less than the mean. Similarly, $P(x \geq \mu)$ is .5 since half of the distribution is greater than the mean.

21. According to Chebyshev's theorem, the probability that a number will lie within 3 standard deviations of the mean of a probability distribution is at least

$$1 - \frac{1}{3^2} = 1 - \frac{1}{9} = \frac{8}{9} \approx .889.$$

Using the normal distribution, the probability that a number will lie within 3 standard deviations of the mean is .997.

These values are not contradictory since "at least .889" means .889 or more. For the normal distribution, the value is more.

In Exercises 23-27, let x represent the life of a light bulb.

23. Less than 500 hr

$$z = \frac{x - \mu}{\sigma} = \frac{500 - 500}{100} = 0, \text{ so}$$

$$P(x < 500) = P(z < 0)$$
$$= \text{ area to the left of } z = 0$$
$$= .5000.$$

Hence, $.5000(10,000) = 5000$ bulbs can be expected to last less than 500 hr.

25. Between 290 and 540 hr

For $x = 290$,

$$z = \frac{290 - 500}{100} = -2.1,$$

and for $x = 540$,

$$z = \frac{540 - 500}{100} = .4.$$

Then

$$P(290 < x < 540) = P(-2.1 < z < .4)$$
$$= \text{ area between } z = -2.1$$
$$\text{ and } z = .4$$
$$= .6554 - .0179$$
$$= .6375.$$

Hence, $.6375(10,000) = 6375$ bulbs should last between 290 and 540 hr.

27. More than 300 hr

For $x = 300$,

$$z = \frac{300 - 500}{100} = -2.$$

Then

$$P(x > 300) = P(z > -2)$$
$$= \text{ area to the right of } z = -2$$
$$= 1 - .0228$$
$$= .9772.$$

Hence, $.9772(10,000) = 9772$ bulbs should last more than 300 hr.

In Exercises 29 and 31, let x represent the weight of a package.

29. Here, $\mu = 16.5$, $\sigma = .5$.

For $x = 16$,

$$z = \frac{16 - 16.5}{.5} = -1.$$

$$P(x < 16) = P(z < -1) = .1587$$

The fraction of the boxes that are underweight is .1587.

31. Here, $\mu = 16.5$, $\sigma = .2$.

For $x = 16$,

$$z = \frac{16 - 16.5}{.2} = -2.5.$$

$$P(x < 16) = P(z < -2.5) = .0062$$

The fraction of the boxes that are underweight is .0062.

In Exercises 33-37, let x represent the weight of a chicken.

$$\mu = 1850, \ \sigma = 150$$

33. More than 1700 g means $x > 1700$.

For $x = 1700$,

$$z = \frac{1700 - 1850}{150} = -1.0.$$

$$\begin{aligned} P(x > 1700) &= 1 - P(x \leq 1700) \\ &= 1 - P(z \leq -1.0) \\ &= 1 - .1587 \\ &= .8413 \end{aligned}$$

Thus, 84.13% of the chickens will weigh more than 1700 g.

35. Between 1750 and 1900 g means $1750 \leq x \leq 1900$.

For $x = 1750$,

$$z = \frac{1750 - 1850}{150} = -.67.$$

For $x = 1900$,

$$z = \frac{1900 - 1850}{150} = .33.$$

$$\begin{aligned} P(1750 \leq x \leq 1900) &= P(-.67 \leq z \leq .33) \\ &= P(z \leq .33) - P(z \leq -.67) \\ &= .6293 - .2514 \\ &= .3779 \end{aligned}$$

Thus, 37.79% of the chickens will weigh between 1750 and 1900 g.

37. More than 2100 g or less than 1550 g

$$P(x < 1550 \text{ or } x > 2100) = 1 - P(1550 \leq x \leq 2100).$$

For $x = 1550$,

$$z = \frac{1550 - 1850}{150} = -2.00.$$

For $x = 2100$,

$$z = \frac{2100 - 1850}{150} = 1.67.$$

$$\begin{aligned} P&(x < 1550 \text{ or } x > 2100) \\ &= P(z \leq -2.00) + [1 - P(z \leq 1.67)] \\ &= .0228 + (1 - .9525) \\ &= .0228 + .0475 \\ &= .0703 \end{aligned}$$

Thus, 7.03% of chickens will weigh more than 2100 g or less than 1550 g.

39. Let x represent the bolt diameter.

$$\mu = .25, \ \sigma = .02$$

First, find the probability that a bolt has a diameter less than or equal to .3 in, that is, $P(x \leq .3)$. The z-score corresponding to $x = .3$ is

$$z = \frac{x - \mu}{\sigma} = \frac{.3 - .25}{.02} = 2.5.$$

Using the table, find the area to the left of $z = 2.5$. This gives

$$P(x \leq .3) = P(z \leq 2.5) = .9938.$$

Then

$$\begin{aligned} P(x > .3) &= 1 - P(x \leq .3) \\ &= 1 - .9938 \\ &= .0062. \end{aligned}$$

41. Let x represent a grocery bill.

$$\mu = 52.25, \ \sigma = 19.50$$

The middle 50% of the grocery bills have cutoffs at 25% below the mean and 25% above the mean. At 25% below the mean, the area to the left is .2500, which corresponds to about $z = -.67$. At 25% above the mean, the area to the left is .7500, which corresponds to about $z = .67$. Find the x-value that corresponds to each z-score.

For $z = -.67$,

$$-.67 = \frac{x - 52.25}{19.50} \quad \text{or} \quad x \approx 39.19.$$

For $z = .67$,

$$.67 = \frac{x - 52.25}{19.50} \quad \text{or} \quad x \approx 65.32.$$

The middle 50% of the customers spend between $39.19 and $65.32.

43. Let x represent the amount of vitamins a person needs. Then

$$\begin{aligned} P(x \leq \mu + 2.5\sigma) &= P(z \leq 2.5) \\ &= .9938. \end{aligned}$$

99.38% of the people will receive adequate amounts of vitamins.

45. The Recommended Daily Allowance is

$$\begin{aligned} \mu + 2.5\sigma &= 159 + 2.5(12) \\ &= 189 \text{ units.} \end{aligned}$$

47. Let x represent an individual's blood clotting time (in seconds).

$$\mu = 7.45, \ \sigma = 3.6$$

For $x = 7$,

$$z = \frac{x - \mu}{\sigma} = \frac{7 - 7.45}{3.6} \approx -.13,$$

and for $x = 8$,

$$z = \frac{8 - 7.45}{3.6} \approx .15.$$

Then

$$P(x < 7) + P(x > 8)$$
$$= P(z < -.13) + P(z > .15)$$
$$= \text{(area to the left of } z = -.13)$$
$$+ \text{(area to the right of } z = .15)$$
$$= .4483 + .4404$$
$$= .8887.$$

49. Let x represent a driving speed.

$$\mu = 50, \ \sigma = 10$$

At the 85th percentile, the area to the left is .8500, which corresponds to about $z = 1.04$. Find the x-value that corresponds to this z-score.

$$z = \frac{x - \mu}{\sigma}$$
$$1.04 = \frac{x - 50}{10}$$
$$10.4 = x - 50$$
$$60.4 = x$$

The 85th percentile speed for this road is 60.4 mph.

In Exercises 51 and 53, let x stand for a student's total points.

51. $P\left(x \geq \mu + \frac{3}{2}\sigma\right) = P(z \geq 1.5)$
$$= 1 - P(z \leq 1.5)$$
$$= 1 - .9332$$
$$= .0668$$

Thus, 6.68% of the students receive A's.

53. $P\left(\mu - \frac{1}{2}\sigma \leq x \leq \mu + \frac{1}{2}\sigma\right)$
$$= P(-.5 \leq z \leq .5)$$
$$= P(z \leq .5) - P(z \leq -.5)$$
$$= .6915 - .3085$$
$$= .383$$

Thus, 38.3% of the students receive C's.

In Exercises 55 and 57, let x represent a student's test score.

$$\mu = 74, \ \sigma = 6$$

55. Since the top 8% get A's, we want to find the number a for which

$$P(x \geq a) = .08,$$
$$\text{or} \quad P(x \leq a) = .92.$$

Read the table backwards to find the z-score for an area of .92, which is 1.41. Find the value of x that corresponds to $z = 1.41$.

$$z = \frac{x - \mu}{\sigma}$$
$$1.41 = \frac{x - 74}{6}$$
$$8.46 = x - 74$$
$$82.46 = x$$

The bottom cutoff score for an A is 82.

57. 23% of the students will receive D's and F's, so to find the bottom cutoff score for a C we need to find the number c for which

$$P(x \leq c) = .23.$$

Read the table backwards to find the z-score for an area of .23, which is $-.74$. Find the value of x that corresponds to $z = -.74$.

$$-.74 = \frac{x - 74}{6}$$
$$-4.44 = x - 74$$
$$69.56 = x$$

The bottom cutoff score for a C is 70.

59. Let x represent the height of a man.

$$\mu = 69.60, \ \sigma = 3.20$$

For $x = 66.27$,

$$z = \frac{66.27 - 69.60}{3.20} = -1.04.$$
$$P(x < 66.27) = P(x < -1.04) = .1492$$

8.4   Normal Approximation to the Binomial Distribution

1. In order to find the mean and standard deviation of a binomial distribution, you must know the number of trials and the probability of a success on each trial.

3. Let x represent the number of heads tossed. For this experiment, $n = 16$, $x = 8$, and $p = \frac{1}{2}$.

(a) $P(x = 8) = \binom{16}{8} \left(\frac{1}{2}\right)^8 \left(1 - \frac{1}{2}\right)^8$

$$\approx .1964$$

(b) $\mu = np = 16\left(\frac{1}{2}\right) = 8$

$$\sigma = \sqrt{np(1-p)} = \sqrt{16\left(\frac{1}{2}\right)\left(\frac{1}{2}\right)} = \sqrt{4} = 2$$

For $x = 7.5$,

$$z = \frac{7.5 - 8}{2} = -.25.$$

For $x = 8.5$,

$$z = \frac{8.5 - 8}{2} = .25.$$

$$P(z < .25) - P(z < -.25) = .5987 - .4013$$
$$= .1974$$

5. Let x represent the number of tails tossed. For this experiment, $n = 16$; $x = 13, 14, 15,$ or 16; and $p = \frac{1}{2}$.

(a)
$P(x = 13, 14, 15,$ or $16)$

$$= \binom{16}{13}\left(\frac{1}{2}\right)^{13}\left(1 - \frac{1}{2}\right)^3 + \binom{16}{14}\left(\frac{1}{2}\right)^{14}\left(1 - \frac{1}{2}\right)^2$$

$$+ \binom{16}{15}\left(\frac{1}{2}\right)^{15}\left(1 - \frac{1}{2}\right)^1 + \binom{16}{16}\left(\frac{1}{2}\right)^{16}\left(1 - \frac{1}{2}\right)^0$$

$$\approx .00854 + .00183 + .00024 + .00001$$
$$\approx .0106$$

(b) $\mu = np = 16\left(\frac{1}{2}\right) = 8$

$$\sigma = \sqrt{np(1-p)} = \sqrt{16\left(\frac{1}{2}\right)\left(\frac{1}{2}\right)} = \sqrt{4} = 2$$

For $x = 12.5$,

$$z = \frac{12.5 - 8}{2} = 2.25.$$

$$P(z > 2.25) = 1 - P(z \leq 2.25)$$
$$= 1 - .9878$$
$$= .0122$$

In Exercises 7 and 9, let x represent the number of heads tossed. Since $n = 1000$ and $p = \frac{1}{2}$,

$$\mu = np = 1000\left(\frac{1}{2}\right) = 500$$

and

$$\sigma = \sqrt{np(1-p)} = \sqrt{1000\left(\frac{1}{2}\right)\left(\frac{1}{2}\right)}$$
$$= \sqrt{250}$$
$$\approx 15.8.$$

7. To find $P(\text{exactly 500 heads})$, find the z-scores for $x = 499.5$ and $x = 500.5$.

For $x = 499.5$,

$$z = \frac{499.5 - 500}{15.8} \approx -.03.$$

For $x = 500.5$,

$$z = \frac{500.5 - 500}{15.8} \approx .03.$$

Using the table,

$$P(\text{exactly 500 heads}) = .5120 - .4880$$
$$= .0240.$$

9. Since we want 480 heads or more, we need to find the area to the right of $x = 479.5$. This will be $1 -$ the area to the left of $x = 479.5$. Find the z-score for $x = 479.5$.

$$z = \frac{479.5 - 500}{15.8} \approx -1.3$$

The area to the left of 479.5 is .0968, so

$$P(480 \text{ heads or more}) = 1 - .0968$$
$$= .9032.$$

11. Let x represent the number of 5's tossed.

$n = 120$, $p = \frac{1}{6}$

$$\mu = np = 120\left(\frac{1}{6}\right) = 20$$

$$\sigma = \sqrt{np(1-p)} = \sqrt{120\left(\frac{1}{6}\right)\left(\frac{5}{6}\right)} \approx 4.08$$

Since we want the probability of getting exactly twenty 5's, we need to find the area between $x = 19.5$ and $x = 20.5$. Find the corresponding z-scores.

For $x = 19.5$,

$$z = \frac{19.5 - 20}{4.08} \approx -.12.$$

For $x = 20.5$,

$$z = \frac{20.5 - 20}{4.08} \approx .12.$$

Using values from the table,

$$P(\text{exactly twenty 5's}) = .5478 - .4522$$
$$= .0956.$$

13. Let x represent the number of 3's tossed.

$n = 120$, $p = \frac{1}{6}$

$\mu = 20$, $\sigma \approx 4.08$

(These values for μ and σ are calculated in the solution for Exercise 11.)

Since

$P(\text{more than eighteen 3's})$
$\quad = 1 - P(\text{eighteen 3's or less})$,

find the z-score for $x = 18.5$.

$$z = \frac{18.5 - 20}{4.08} \approx -.37$$

Thus,

$$P(\text{more than eighteen 3's}) = 1 - .3557$$
$$= .6443.$$

15. Let x represent the number of times the chosen number appears.

$n = 130$; $x = 26, 27, 28, \ldots, 130$; and $p = \frac{1}{6}$

$$\mu = np = 130\left(\frac{1}{6}\right) = \frac{65}{3}$$

$$\sigma = \sqrt{np(1-p)} = \sqrt{130\left(\frac{1}{6}\right)\left(\frac{5}{6}\right)} = \frac{5}{6}\sqrt{26}$$

For $x = 25.5$,

$$z = \frac{25.5 - \frac{65}{3}}{\frac{5}{6}\sqrt{26}} \approx .90.$$

$$P(z > .90) = 1 - P(z \le .90)$$
$$= 1 - .8159$$
$$= .1841.$$

17. Let x represent the number of heaters that are defective.

$n = 10{,}000$, $p = .02$, $\mu = np = 200$

$$\sigma = \sqrt{np(1-p)} = \sqrt{10{,}000(.02)(.98)} = 14$$

We want the area to the right of $x = 222.5$. For $x = 222.5$,

$$z = \frac{222.5 - 200}{14} \approx 1.61.$$

$$P(\text{more than 222 defects}) = P(x \ge 222.5)$$
$$= P(z \ge 1.61)$$
$$= 1 - P(z \le 1.61)$$
$$= 1 - .9463$$
$$= .0537$$

19. Use a calculator or computer to complete this exercise. The answers are given.

(a) $P(\text{all 58 like it}) = 1.04 \times 10^{-9} \approx 0$

(b) $P(\text{exactly 28, 29, or 30 like it}) = .0018$

21. Let x be the number of nests escaping predation.

$n = 26, p = .3$

$\mu = np = 26(.3) = 7.8$

$\sigma = \sqrt{np(1-p)} = \sqrt{26(.3)(.7)}$
$\quad = \sqrt{5.46} \approx 2.337$

To find $P(\text{at least half escape predation})$, find the z-score for $x = 12.5$.

$$z = \frac{12.5 - 7.8}{2.337} \approx 2.01$$

$$P(z > 2.01) = 1 - .9778 = .0222$$

23. Let x represent the number of hospital patients struck by falling coconuts.

(a) $n = 20$; $x = 0$ or 1; and $p = .025$

$$P(x = 0 \text{ or } 1) = \binom{20}{0}(.025)^0(.975)^{20}$$
$$+ \binom{20}{1}(.025)^1(.975)^{19}$$
$$\approx .60269 + .30907$$
$$\approx .9118$$

(b) $n = 2000$; $x = 0, 1, 2, \ldots,$ or 70; $p = .025$

$\mu = np = 2000(.025) = 50$

$\sigma = \sqrt{np(1-p)} = \sqrt{2000(.025)(.975)} = \sqrt{48.75}$

To find $P(70$ or less$)$, find the z-score for $x = 70.5$.

$$z = \frac{70.5 - 50}{\sqrt{48.75}} \approx 2.94$$

$$P(x < 2.94) = .9984$$

25. Let x represent the number of people cured.

$$n = 25, p = .80$$

$$\mu = np = 25(.80) = 20$$

$$\sigma = \sqrt{np(1-p)} = \sqrt{25(.80)(.20)} = 2$$

To find

$$P(\text{exactly 20 cured}) = P(19.5 \leq x \leq 20.5),$$

find the z-scores for $x = 19.5$ and $x = 20.5$.

For $x = 19.5$,

$$z = \frac{19.5 - 20}{2} = -.25.$$

For $x = 20.5$,

$$z = \frac{20.5 - 20}{2} = .25.$$

Using the table,

$$P(\text{exactly 20 cured}) = .5987 - .4013$$
$$= .1974.$$

27. $P(x = 0) = \binom{25}{0}(.80)^0(.20)^{25}$

$$= (.20)^{25}$$
$$= 3.36 \times 10^{-18}$$
$$\approx 0$$

29. This exercise should be solved by calculator or computer methods. The answers, which may vary slightly, are

(a) .0001,

(b) .0002, and

(c) .0000.

31. Let x represent the number of students who carry a weapon.

$$n = 1200; \ x = 120, 121, 122, \ldots, \text{ or } 179; \ p = \tfrac{1}{8}$$

$$\mu = np = 1200\left(\frac{1}{8}\right) = 150$$

$$\sigma = \sqrt{np(1-p)} = \sqrt{1200\left(\frac{1}{8}\right)\left(\frac{7}{8}\right)}$$

$$= \frac{20}{8}\sqrt{21} = \frac{5}{2}\sqrt{21}$$

To find $P(\text{more than 120 but less than 180})$, find z-scores for $x = 120.5$ and $x = 179.5$.

For $x = 120.5$,

$$z = \frac{120.5 - 150}{\frac{5}{2}\sqrt{21}} \approx -2.57.$$

For $x = 179.5$,

$$z = \frac{179.5 - 150}{\frac{5}{2}\sqrt{21}} \approx 2.57.$$

$$P(-2.57 < z < 2.57)$$
$$= P(z < 2.57) - P(z < -2.57)$$
$$= .9949 - .0051$$
$$= .9898$$

33. Let x represent the number of questions.

$$n = 100; x = 60, 61, 62, \ldots, \text{ or } 100; \ p = \tfrac{1}{2}$$

$$\mu = np = 180\left(\frac{1}{2}\right) = 50$$

$$\sigma = \sqrt{np(1-p)} = \sqrt{100\left(\frac{1}{2}\right)\left(\frac{1}{2}\right)}$$

$$= \sqrt{25} = 5$$

To find $P(60$ or more correct$)$, find the z-score for $x = 59.5$.

$$z = \frac{59.5 - 50}{5} = 1.90$$

$$P(z > 1.90) = 1 - P(z \leq 1.90)$$
$$= 1 - .9713$$
$$= .0287$$

Chapter 8 Review Exercises

3. (a) Since 450-474 is to be the first interval, let all the intervals be of size 25. The largest data value is 566, so the last interval that will be needed is 550-574. The frequency distribution is as follows.

Interval	Frequency
450-474	5
475-499	6
500-524	5
525-549	2
550-574	2

(b) Draw the histogram. It consists of 5 bars of equal width having heights as determined by the frequency of each interval. See the histogram in the answer section of the textbook.

(c) Construct the frequency polygon by joining consecutive midpoints of the tops of the histogram bars with line segments. See the answer section of the textbook.

5. $\sum x = 41 + 60 + 67 + 68 + 72 + 74 + 78 + 83$
$\qquad + 90 + 97$
$\qquad = 730$

The mean of the 10 numbers is

$$\bar{x} = \frac{\sum x}{n} = \frac{730}{10} = 73.$$

7.

Interval	Midpoint, x	Frequency, f	Product, xf
10-19	14.5	6	87
20-29	24.5	12	294
30-39	34.5	14	483
40-49	44.5	10	445
50-59	54.5	8	436
Totals:		50	1745

The mean of this collection of grouped data is

$$\bar{x} = \frac{\sum xf}{n} = \frac{1745}{50} = 34.9.$$

11. Arrange the numbers in numerical order, from smallest to largest.

$$35, 36, 36, 38, 38, 42, 44, 48$$

There are 8 numbers here; the median is the mean of the two middle numbers, which is

$$\frac{38 + 38}{2} = \frac{76}{2} = 38.$$

The mode is the number that occurs most often. Here, there are two modes, 36 and 38, since they both appear twice.

13. The modal class for the distribution of Exercise 8 is the interval 55-59, since it contains more data values than any of the other intervals.

17. The range is $93 - 26 = 67$, the difference of the highest and lowest numbers in the distribution.

The mean is

$$\bar{x} = \frac{\sum x}{n} = \frac{520}{10} = 52.$$

Construct a table with the values of x and x^2.

x	x^2
26	676
43	1849
51	2601
29	841
37	1369
56	3136
29	841
82	6724
74	5476
93	8649
Total:	32,162

The standard deviation is

$$s = \sqrt{\frac{\sum x^2 - n\bar{x}^2}{n-1}}$$

$$= \sqrt{\frac{32,162 - 10(52)^2}{9}}$$

$$\approx \sqrt{569.1} \approx 23.9.$$

19. Start with the frequency distribution that was the answer to Exercise 8, and expand the table to include columns for the midpoint x of each interval and for xf, x^2, and fx^2.

Interval	f	x	xf	x^2	fx^2
40-44	2	42	84	1764	3528
45-49	5	47	235	2209	11,045
50-54	7	52	364	2704	18,928
55-59	10	57	570	3249	32,490
60-64	4	62	248	3844	15,376
65-69	1	67	67	4489	4489
Totals:	29		1568		85,856

The mean of the grouped data is

$$\bar{x} = \frac{\sum xf}{n} = \frac{1568}{29} \approx 54.07.$$

The standard deviation for the grouped data is

$$s = \sqrt{\frac{\sum fx^2 - n\bar{x}^2}{n-1}}$$

$$= \sqrt{\frac{85,856 - 29(54.07)^2}{29-1}}$$

$$\approx \sqrt{38.3} \approx 6.2.$$

21. A skewed distribution has the largest frequency at one end rather than in the middle.

23. To the left of $z = .41$

Using the standard normal curve table,

$$P(z < .41) = .6591.$$

25. Between $z = 1.53$ and $z = 2.82$

$P(1.53 \leq z \leq 2.82)$
$= P(z \leq 2.82) - P(z \leq 1.53)$
$= .9976 - .9370$
$= .0606$

27. The normal distribution is not a good approximation of a binomial distribution that has a value of p close to 0 or 1 because the histogram of such a binomial distribution is skewed and therefore not close to the shape of a normal distribution.

29.

Number of Heads, x	Frequency, f	xf	fx^2
0	1	0	0
1	5	5	5
2	7	14	28
3	5	15	45
4	2	8	32
Totals:	20	42	110

(a) $\bar{x} = \dfrac{\sum xf}{n} = \dfrac{42}{20} = 2.1$

$s = \sqrt{\dfrac{\sum fx^2 - n\bar{x}^2}{n-1}} = \sqrt{\dfrac{110 - 20(2.1)^2}{20-1}} \approx 1.07$

(b) For this binomial experiment,

$$\mu = np = 4\left(\frac{1}{2}\right) = 2,$$

and

$$\sigma = \sqrt{np(1-p)} = \sqrt{4\left(\frac{1}{2}\right)\left(\frac{1}{2}\right)} = \sqrt{1} = 1.$$

(c) The answers to parts (a) and (b) should be close.

31. (a) For Stock I,

$$\bar{x} = \frac{11 + (-1) + 14}{3} = 8,$$

so, the mean (average return) is 8%.

$s = \sqrt{\dfrac{\sum x^2 - n\bar{x}^2}{n-1}} = \sqrt{\dfrac{318 - 3(8)^2}{2}} = \sqrt{63} \approx 7.9$

so the standard deviation is 7.9%.

For Stock II,

$$\bar{x} = \frac{9 + 5 + 10}{3} = 8,$$

so the mean is also 8%.

$s = \sqrt{\dfrac{\sum x^2 - n\bar{x}^2}{n-1}} = \sqrt{\dfrac{206 - 3(8)^2}{2}} = \sqrt{7} \approx 2.6,$

so the standard deviation is 2.6%.

(b) Both stocks offer an average (mean) return of 8%. The smaller standard deviation for Stock II indicates a more stable return and thus greater security.

33. Let x represent the number of overstuffed frankfurters.

$$n = 500, p = .06$$

To approximate the binomial distribution, use a normal distribution with

$$\mu = np = 500(.06) = 30 \text{ and}$$
$$\sigma = \sqrt{np(1-p)} = \sqrt{28.2} \approx 5.31.$$

(a) 25 or fewer overstuffed corresponds to the area under the normal curve to the left of $x = 25.5$. The corresponding z-score is

$$z = \frac{25.5 - 30}{5.31} \approx -.85.$$

$P(25 \text{ or fewer overstuffed}) = P(x \leq 25.5)$
$= P(z < -.85)$
$= .1977$

(b) Exactly 30 overstuffed corresponds to the area between $x = 29.5$ and $x = 30.5$.

For $x = 29.5$,

$$z = \frac{29.5 - 30}{5.31} \approx -.09.$$

For $x = 30.5$,

$$z = \frac{30.5 - 30}{5.31} \approx .09.$$

$P(\text{exactly 30 overstuffed}) = P(29.5 \leq x \leq 30.5)$
$= P(-.09 \leq z \leq .09)$
$= .5359 - .4641$
$= .0718$

(c) More than 40 overstuffed corresponds to the area under the normal curve to the right of $x = 40.5$. The corresponding z-score is

$$z = \frac{40.5 - 30}{5.31} \approx 1.98.$$

$$P(\text{more than 40 overstuffed}) = P(x \geq 40.5)$$
$$= P(z \geq 1.98)$$
$$= 1 - .9761$$
$$= .0239$$

35. The table below records the mean and standard deviation for diet A and for diet B.

	$\bar{x}$	s
Diet A	2.7	2.26
Diet B	1.3	.95

(a) Diet A had the greater mean gain, since the mean for diet A is larger.

(b) Diet B had a more consistent gain, since diet B has a smaller standard deviation.

37. Let x represent the number of flies that are killed.

$n = 1000$; $x = 0, 1, 2, \ldots, 986$; $p = .98$

$\mu = np = 1000(.98) = 980$

$\sigma = \sqrt{np(1-p)} = \sqrt{1000(.98)(.02)} = \sqrt{19.6}$

To find $P(\text{no more than } 986)$, find the z-score for $x = 986.5$.

$$z = \frac{986.5 - 980}{\sqrt{19.6}} \approx 1.47$$

$P(z < 1.47) = .9292$

39. Again, let x represent the number of flies that are killed.

$n = 1000$; $x = 973, 974, 975, \ldots, 993$; $p = .98$

As in Exercise 37, $\mu = 980$ and $\sigma = \sqrt{19.6}$. To find $P(\text{between 973 and 993})$, find the z-scores for $x = 972.5$ and $x = 993.5$.

For $x = 972.5$,

$$z = \frac{972.5 - 980}{\sqrt{19.6}} \approx -1.69.$$

For $x = 993.5$,

$$z = \frac{993.5 - 980}{\sqrt{19.6}} \approx 3.05.$$

$$P(-1.69 \leq z \leq 3.05) = P(z \leq 3.05) - P(z \leq -1.69)$$
$$= .9989 - .0455$$
$$= .9534$$

41. No more than 35 min/day

$\mu = 42$, $\sigma = 12$

Find the z-score for $x = 35$.

$$z = \frac{35 - 42}{12} \approx -.58$$

$P(x \leq 35) = P(z \leq -.58) = .2810$

28.10% of the residents commute no more than 35 min/day.

43. Between 38 and 60 min/day

$\mu = 42$, $\sigma = 12$

Find the z-scores for $x = 38$ and $x = 60$.

For $x = 38$,

$$z = \frac{38 - 42}{12} \approx -.33.$$

For $x = 60$,

$$z = \frac{60 - 42}{12} = 1.5.$$

$$P(38 \leq x \leq 60) = P(-.33 \leq z \leq 1.5)$$
$$= P(z \leq 1.5) - P(z \leq -.33)$$
$$= .9332 - .3707$$
$$= .5625$$

56.25% of the residents commute between 38 and 60 min/day.

Chapter 8 Test

[8.1]

1. Refer to the following data, which give the number of pairs of socks sold at the Sock and Accessory Shop during the past 20 weeks.

125	155	148	110	162	128	132	119	150	129
132	168	124	115	153	143	148	143	128	138

   (a) Construct a grouped frequency distribution using six intervals, the first one being 110-119.

   (b) Construct a histogram.

   (c) Construct a frequency polygon.

2. Find the mean for the data shown in the following frequency distribution.

Value	Frequency
10	3
11	9
12	18
13	25
14	17
15	14
16	7

3. Find the mean, median, and mode (or modes) for the following list of numbers. Round the mean to the nearest tenth.

$$68, 72, 25, 49, 97, 58, 91, 25$$

[8.2]

4. Find the range, the mean, and the standard deviation for the following set of numbers.

$$3, 3, 4, 4, 4, 5, 6, 6, 6, 6$$

5. Find the standard deviation for the following grouped data. (Round your answer to the nearest tenth.)

Interval	Frequency
10-19	5
20-29	7
30-39	3
40-49	10
50-59	12

6. A distribution of 100 incomes has $\mu = \$10,500$ and $\sigma = \$1000$. Use Chebyshev's theorem to find:

(a) A range of incomes which would include at least 75 of the 100 incomes.

(b) The minimum number of incomes we would expect to find in the interval from $7500 to $13,500.

[8.3]

7. Using the normal curve table, find the following areas under the standard normal curve.

(a) The area between $z = -1.13$ and $z = 2.14$

(b) The area to the right of $z = 2.1$

8. If a normal distribution has mean 60 and standard deviation 6, find the following z-scores.

(a) The z-score for $x = 72$

(b) The z-score for $x = 51$

(c) The z-score for $x = 43.5$

9. A survey of students enrolled at a community college finds the students have an average age of 26.5 years, with a standard deviation of 5.5 years. The ages are closely approximated by a normal curve.

Using a normal curve table, find the percent of students whose ages are in the following ranges. (Round your answers to the nearest percent.)

(a) Younger than 24   (b) Between 24 and 29   (c) Over 35

[8.4]

10. Find the mean and standard deviation of a binomial distribution with $n = 56$ and $p = .4$.

11. Al's Quick Photo store develops an average roll of film in 2.3 minutes. The standard deviation in time is .6 minutes. Assume a normal distribution.

(a) Out of 1000 rolls of film, how many will be developed in less than 3.5 minutes?

(b) What percentage of film is developed in between 2 and 3 minutes?

12. A loaded coin with $P(\text{head}) = .6$ is tossed 100 times. Using the normal curve approximation to the binomial distribution, find the probability of getting each of the following:

(a) at least 55 heads;   (b) exactly 61 heads;

(c) between 55 and 65 heads (inclusive).

Chapter 8 Test Answers

1. (a)

Interval	Frequency
110-119	3
120-129	5
130-139	3
140-149	4
150-159	3
160-169	2

(b)-(c)

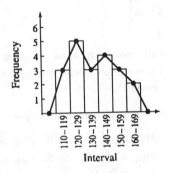

2. 13.2

3. Mean: 60.6; median: 63; mode: 25

4. Range: 3; mean: 4.7; standard deviation: 1.25

5. 14.6

6. (a) $8500 to $12,500 **(b)** 88

7. (a) .8546 **(b)** .0179

8. (a) 2 **(b)** −1.5 **(c)** −2.75

9. (a) 33% **(b)** 35% **(c)** 6%

10. $\mu = 22.4$, $\sigma \approx 3.67$

11. (a) 977 **(b)** 57%

12. (a) .8686 **(b)** .0819 **(c)** .7372

MARKOV CHAINS

9.1 Basic Properties of Markov Chains

1. $\begin{bmatrix} \frac{2}{3} & \frac{1}{2} \end{bmatrix}$ could not be a probability vector because the sum of the entries in the row is not equal to 1.

3. $\begin{bmatrix} 0 & 1 \end{bmatrix}$ could be a probability vector since it is a matrix of only one row, having nonnegative entries whose sum is 1.

5. $\begin{bmatrix} .4 & .2 & 0 \end{bmatrix}$ could not be a probability vector because the sum of the entries in the row is not equal to 1.

7. $\begin{bmatrix} .07 & .04 & .37 & .52 \end{bmatrix}$ could be a probability vector. It is a matrix of only one row, having nonnegative entries whose sum is 1.

9. $\begin{bmatrix} 0 & -.2 & .6 & .6 \end{bmatrix}$ could not be a probability vector because it has a negative entry.

11.
$$\begin{array}{cc} & \text{A} \quad \text{B} \\ \begin{array}{c} \text{A} \\ \text{B} \end{array} & \begin{bmatrix} \frac{2}{3} & \frac{1}{3} \\ 1 & 0 \end{bmatrix} \end{array}$$

This could be a transition matrix since it is a square matrix, all entries are between 0 and 1, inclusive, and the sum of the entries in each row is 1.

To draw the transition diagram, give names to the two states (such as A and B) and label the probabilities of going from one state to another. In this case,

$$P_{AA} = \frac{2}{3}, \qquad P_{AB} = \frac{1}{3},$$
$$P_{BA} = 1, \quad \text{and} \quad P_{BB} = 0.$$

See the transition diagram in the back of the textbook.

13. $\begin{bmatrix} \frac{1}{4} & \frac{3}{4} & 0 \\ 2 & 0 & 1 \\ 1 & \frac{2}{3} & 3 \end{bmatrix}$

This could not be a transition matrix because it has entries that are greater than 1.

15. $\begin{bmatrix} \frac{1}{3} & \frac{1}{2} & 1 \\ 0 & 1 & 0 \\ \frac{1}{2} & \frac{1}{2} & 1 \end{bmatrix}$

This could not be a transition matrix because the sum of the entries in the first row and in the third row is more than 1.

17. The transition diagram provides the information

$$P_{AA} = .9, \; P_{AB} = .1, \; P_{AC} = 0,$$
$$P_{BA} = .1, \; P_{BB} = .6, \; P_{BC} = .3,$$
$$P_{CA} = 0, \; P_{CB} = .3, \text{ and } P_{CC} = .7.$$

The transition matrix associated with this diagram is

$$\begin{array}{cc} & \begin{array}{ccc} \text{A} & \text{B} & \text{C} \end{array} \\ \begin{array}{c} \text{A} \\ \text{B} \\ \text{C} \end{array} & \begin{bmatrix} .9 & .1 & 0 \\ .1 & .6 & .3 \\ 0 & .3 & .7 \end{bmatrix} \end{array}.$$

19. $A = \begin{bmatrix} 1 & 0 \\ .8 & .2 \end{bmatrix}$

$$A^2 = \begin{bmatrix} 1 & 0 \\ .8 & .2 \end{bmatrix} \begin{bmatrix} 1 & 0 \\ .8 & .2 \end{bmatrix} = \begin{bmatrix} 1 & 0 \\ .96 & .04 \end{bmatrix}$$

$$A^3 = A \cdot A^2 = \begin{bmatrix} 1 & 0 \\ .8 & .2 \end{bmatrix} \begin{bmatrix} 1 & 0 \\ .96 & .04 \end{bmatrix} = \begin{bmatrix} 1 & 0 \\ .992 & .008 \end{bmatrix}$$

The entry in row 1, column 2 of A^3 gives the probability that state 1 changes to state 2 after 3 repetitions of the experiment. This probability is 0.

21. $C = \begin{bmatrix} .5 & .5 \\ .72 & .28 \end{bmatrix}$

$$C^2 = \begin{bmatrix} .5 & .5 \\ .72 & .28 \end{bmatrix} \begin{bmatrix} .5 & .5 \\ .72 & .28 \end{bmatrix} = \begin{bmatrix} .61 & .39 \\ .5616 & .4384 \end{bmatrix}$$

$$C^3 = \begin{bmatrix} .5 & .5 \\ .72 & .28 \end{bmatrix} \begin{bmatrix} .61 & .39 \\ .5616 & .4384 \end{bmatrix}$$

$$= \begin{bmatrix} .5858 & .4142 \\ .596448 & .403552 \end{bmatrix}$$

The probability that state 1 changes to state 2 after 3 repetitions is .4142, since that is the entry in row 1, column 2 of C^3.

23. $E = \begin{bmatrix} .8 & .1 & .1 \\ .3 & .6 & .1 \\ 0 & 1 & 0 \end{bmatrix}$

$E^2 = \begin{bmatrix} .8 & .1 & .1 \\ .3 & .6 & .1 \\ 0 & 1 & 0 \end{bmatrix} \begin{bmatrix} .8 & .1 & .1 \\ .3 & .6 & .1 \\ 0 & 1 & 0 \end{bmatrix}$

$= \begin{bmatrix} .67 & .24 & .09 \\ .42 & .49 & .09 \\ .3 & .6 & .1 \end{bmatrix}$

$E^3 = \begin{bmatrix} .8 & .1 & .1 \\ .3 & .6 & .1 \\ 0 & 1 & 0 \end{bmatrix} \begin{bmatrix} .67 & .24 & .09 \\ .42 & .49 & .09 \\ .3 & .6 & .1 \end{bmatrix}$

$= \begin{bmatrix} .608 & .301 & .091 \\ .483 & .426 & .091 \\ .42 & .49 & .09 \end{bmatrix}$

The probability that state 1 changes to state 2 after 3 repetitions is .301, since that is the entry in row 1, column 2 of E^3.

25. This exercise should be solved by graphing calculator methods. The solution may vary. The first five powers of the transition matrix are

$A = \begin{bmatrix} .1 & .2 & .2 & .3 & .2 \\ .2 & .1 & .1 & .2 & .4 \\ .2 & .1 & .4 & .2 & .1 \\ .3 & .1 & .1 & .2 & .3 \\ .1 & .3 & .1 & .1 & .4 \end{bmatrix}$,

$A^2 = \begin{bmatrix} .2 & .15 & .17 & .19 & .29 \\ .16 & .2 & .15 & .18 & .31 \\ .19 & .14 & .24 & .21 & .22 \\ .16 & .19 & .16 & .2 & .29 \\ .16 & .19 & .14 & .17 & .34 \end{bmatrix}$,

$A^3 = \begin{bmatrix} .17 & .178 & .171 & .191 & .29 \\ .171 & .178 & .161 & .185 & .305 \\ .18 & .163 & .191 & .197 & .269 \\ .175 & .174 & .164 & .187 & .3 \\ .167 & .184 & .158 & .182 & .309 \end{bmatrix}$,

$A^4 = \begin{bmatrix} .1731 & .175 & .1683 & .188 & .2956 \\ .1709 & .1781 & .1654 & .1866 & .299 \\ .1748 & .1718 & .1753 & .1911 & .287 \\ .1712 & .1775 & .1667 & .1875 & .2971 \\ .1706 & .1785 & .1641 & .1858 & .301 \end{bmatrix}$,

and

$A^5 = \begin{bmatrix} .17193 & .17643 & .1678 & .18775 & .29609 \\ .17167 & .17689 & .16671 & .18719 & .29754 \\ .17293 & .17488 & .17007 & .18878 & .29334 \\ .17192 & .17654 & .16713 & .18741 & .297 \\ .17142 & .17726 & .16629 & .18696 & .29807 \end{bmatrix}$.

The probability that state 2 changes to state 4 after 5 repetitions of the experiment is .18719, since that is the entry in row 2, column 4 of A^5.

27. (a) We are asked to show that

$$X_0(P^n) = (\cdots (((X_0 P)P)P) \cdots P)$$

is true for any natural number n, where the expression on the right side of the equation has a total of n factors of P. This may be proven by mathematical induction on n.

When $n = 1$, the statement becomes $X_0 = (X_0 P)$, which is obviously true. When $n = 2$, the statement becomes $(P^2) = (X_0 P)P$, or $X_0(PP) = (X_0 P)P$, which is true since matrix multiplication is associative. Next, assume the nth statement is true in order to show that the $(n+1)$st statement is true. That is, assume that

$$X_0(P^n) = (\cdots (((X_0 P)P)P) \cdots P)$$

is true. Associativity plays a role here also.

$$\begin{aligned} X_0(P^{n+1}) &= X_0(P^n \cdot P) \\ &= (X_0 P^n)P \\ &= (\cdots (((X_0 P)P)P) \cdots P)P \end{aligned}$$

Conclude that

$$X_0(P^n) = (\cdots (((X_0 P)P)P) \cdots P)$$

is true for any natural number n.

29. The probability vector is $\begin{bmatrix} .4 & .6 \end{bmatrix}$.

(a) $\begin{bmatrix} .4 & .6 \end{bmatrix} \begin{bmatrix} .8 & .2 \\ .35 & .65 \end{bmatrix} = \begin{bmatrix} .53 & .47 \end{bmatrix}$

Thus, after 1 wk Johnson has a 53% market share, and NorthClean has a 47% share.

(b) $C^2 = \begin{bmatrix} .8 & .2 \\ .35 & .65 \end{bmatrix} \begin{bmatrix} .8 & .2 \\ .35 & .65 \end{bmatrix} = \begin{bmatrix} .71 & .29 \\ .5075 & .4925 \end{bmatrix}$

$\begin{bmatrix} .4 & .6 \end{bmatrix} \begin{bmatrix} .71 & .29 \\ .5075 & .4925 \end{bmatrix} = \begin{bmatrix} .5885 & .4115 \end{bmatrix}$

After 2 wk, Johnson has a 58.85% market share, and NorthClean has a 41.15% share.

(c) $C^3 = \begin{bmatrix} .8 & .2 \\ .35 & .65 \end{bmatrix} \begin{bmatrix} .71 & .29 \\ .5075 & .4925 \end{bmatrix}$

$= \begin{bmatrix} .6695 & .3305 \\ .5784 & .4216 \end{bmatrix}$

$\begin{bmatrix} .4 & .6 \end{bmatrix} \begin{bmatrix} .6695 & .3305 \\ .5784 & .4216 \end{bmatrix} = \begin{bmatrix} .6148 & .3852 \end{bmatrix}$

After 3 wk, the shares are 61.48% and 38.52%, respectively.

(d) $C^4 = \begin{bmatrix} .8 & .2 \\ .35 & .65 \end{bmatrix} \begin{bmatrix} .6695 & .3305 \\ .5784 & .4216 \end{bmatrix}$

$= \begin{bmatrix} .6513 & .3487 \\ .6103 & .3897 \end{bmatrix}$

$\begin{bmatrix} .4 & .6 \end{bmatrix} \begin{bmatrix} .6513 & .3487 \\ .6103 & .3897 \end{bmatrix} = \begin{bmatrix} .62667 & .37333 \end{bmatrix}$

After 4 wk, the shares are 62.667% and 37.333%, respectively.

31. The transition matrix P is

$$\begin{array}{c} \\ G_0 \\ G_1 \\ G_2 \end{array} \begin{array}{ccc} G_0 & G_1 & G_2 \\ \begin{bmatrix} .85 & .1 & .05 \\ 0 & .8 & .2 \\ 0 & 0 & 1 \end{bmatrix} \end{array}.$$

We have 50,000 new policy holders, all in G_0. The probability vector for these people is

$$\begin{array}{ccc} G_0 & G_1 & G_2 \\ \begin{bmatrix} 1 & 0 & 0 \end{bmatrix} \end{array}.$$

(a) After 1 yr, the distribution of people in each group is

$\begin{bmatrix} 1 & 0 & 0 \end{bmatrix} \begin{bmatrix} .85 & .1 & .05 \\ 0 & .8 & .2 \\ 0 & 0 & 1 \end{bmatrix} = \begin{bmatrix} .85 & .1 & .05 \end{bmatrix}.$

There are

$.85(50,000) = 42,500$ people in G_0,
$.1(50,000) = 5000$ people in G_1, and
$.05(50,000) = 2500$ people in G_2.

(b) $P^2 = \begin{bmatrix} .85 & .1 & .05 \\ 0 & .8 & .2 \\ 0 & 0 & 1 \end{bmatrix} \begin{bmatrix} .85 & .1 & .05 \\ 0 & .8 & .2 \\ 0 & 0 & 1 \end{bmatrix}$

$= \begin{bmatrix} .7225 & .165 & .1125 \\ 0 & .64 & .36 \\ 0 & 0 & 1 \end{bmatrix}$

After 2 yr, the distribution of people in each group is

$\begin{bmatrix} 1 & 0 & 0 \end{bmatrix} \begin{bmatrix} .7225 & .165 & .1125 \\ 0 & .64 & .36 \\ 0 & 0 & 1 \end{bmatrix}$

$= \begin{bmatrix} .7225 & .165 & .1125 \end{bmatrix}.$

There are

$.7225(50,000) = 36,125$ in G_0,
$.165(50,000) = 8250$ in G_1, and
$.1125(50,000) = 5625$ in G_2.

(c) $P^3 = \begin{bmatrix} .85 & .1 & .05 \\ 0 & .8 & .2 \\ 0 & 0 & 1 \end{bmatrix} \begin{bmatrix} .7225 & .165 & .1125 \\ 0 & .64 & .36 \\ 0 & 0 & 1 \end{bmatrix}$

$= \begin{bmatrix} .61413 & .20425 & .18163 \\ 0 & .512 & .488 \\ 0 & 0 & 1 \end{bmatrix}$

After 3 yr, the distribution is

$\begin{bmatrix} 1 & 0 & 0 \end{bmatrix} \begin{bmatrix} .61413 & .20425 & .18163 \\ 0 & .512 & .488 \\ 0 & 0 & 1 \end{bmatrix}$

$= \begin{bmatrix} .61413 & .20425 & .18163 \end{bmatrix}.$

There are

$50,000(.61413) = 30,706$ in G_0,
$50,000(.20425) = 10,213$ in G_1, and
$50,000(.18163) = 9081$ in G_2.

(d)

$P^4 = \begin{bmatrix} .85 & .1 & .05 \\ 0 & .8 & .2 \\ 0 & 0 & 1 \end{bmatrix} \begin{bmatrix} .61413 & .20425 & .18163 \\ 0 & .512 & .488 \\ 0 & 0 & 1 \end{bmatrix}$

$= \begin{bmatrix} .52201 & .22481 & .25318 \\ 0 & .4096 & .5904 \\ 0 & 0 & 1 \end{bmatrix}$

The probabilities are

$\begin{bmatrix} 1 & 0 & 0 \end{bmatrix} \begin{bmatrix} .52201 & .22481 & .25318 \\ 0 & .4096 & .5904 \\ 0 & 0 & 1 \end{bmatrix}$

$= \begin{bmatrix} .52201 & .22481 & .25318 \end{bmatrix}.$

There are

$50,000(.52201) = 26,100$ in G_0,
$50,000(.22481) = 11,241$ in G_1, and
$50,000(.25318) = 12,659$ in G_2.

33. $P = \begin{bmatrix} .825 & .175 & 0 \\ .060 & .919 & .021 \\ .049 & 0 & .951 \end{bmatrix}$

The initial probability vector is

$\begin{bmatrix} .26 & .6 & .14 \end{bmatrix}$.

(a) The share held by each type after 1 yr is given by

$\begin{bmatrix} .26 & .6 & .14 \end{bmatrix} \begin{bmatrix} .825 & .175 & 0 \\ .060 & .919 & .021 \\ .049 & 0 & .951 \end{bmatrix}$

$= \begin{bmatrix} .257 & .597 & .146 \end{bmatrix}$.

(b)

$P^2 = \begin{bmatrix} .825 & .175 & 0 \\ .060 & .919 & .021 \\ .049 & 0 & .951 \end{bmatrix} \begin{bmatrix} .825 & .175 & 0 \\ .060 & .919 & .021 \\ .049 & 0 & .951 \end{bmatrix}$

$= \begin{bmatrix} .691 & .305 & .004 \\ .106 & .855 & .039 \\ .087 & .009 & .904 \end{bmatrix}$

The share held by each after 2 yr is

$\begin{bmatrix} .26 & .6 & .14 \end{bmatrix} \begin{bmatrix} .691 & .305 & .004 \\ .106 & .855 & .039 \\ .087 & .009 & .904 \end{bmatrix}$

$= \begin{bmatrix} .255 & .594 & .151 \end{bmatrix}$.

(c)

$P^3 = \begin{bmatrix} .825 & .175 & 0 \\ .060 & .919 & .021 \\ .049 & 0 & .951 \end{bmatrix} \begin{bmatrix} .691 & .305 & .004 \\ .106 & .855 & .039 \\ .087 & .009 & .904 \end{bmatrix}$

$= \begin{bmatrix} .589 & .401 & .01 \\ .141 & .804 & .055 \\ .117 & .023 & .860 \end{bmatrix}$

The share held by each after 3 yr is

$\begin{bmatrix} .26 & .6 & .14 \end{bmatrix} \begin{bmatrix} .589 & .401 & .01 \\ .141 & .804 & .055 \\ .117 & .023 & .860 \end{bmatrix}$

$= \begin{bmatrix} .254 & .590 & .156 \end{bmatrix}$.

35. (a) If there is no one in line, then after 1 min there will be either 0, 1, or 2 people in line with probabilities $p_{00} = \frac{1}{2}$, $p_{01} = \frac{1}{3}$, and $p_{02} = \frac{1}{6}$. If there is one person in line, then that person will be served and either 0, 1, or 2 new people will join the line, with probabilities $p_{10} = \frac{1}{2}$, $p_{11} = \frac{1}{3}$, and $p_{12} = \frac{1}{6}$. If there are two people in line, then one of them will be served and either 1 or 2 new people

will join the line, with probabilities $p_{21} = \frac{1}{2}$ and $p_{22} = \frac{1}{2}$; it is impossible for both people in line to be served, so $p_{20} = 0$. Therefore, the transition matrix is

$A = \begin{bmatrix} \frac{1}{2} & \frac{1}{3} & \frac{1}{6} \\ \frac{1}{2} & \frac{1}{3} & \frac{1}{6} \\ 0 & \frac{1}{2} & \frac{1}{2} \end{bmatrix}$.

(b) The transition matrix for a two-minute period is

$A^2 = \begin{bmatrix} \frac{1}{2} & \frac{1}{3} & \frac{1}{6} \\ \frac{1}{2} & \frac{1}{3} & \frac{1}{6} \\ 0 & \frac{1}{2} & \frac{1}{2} \end{bmatrix} \begin{bmatrix} \frac{1}{2} & \frac{1}{3} & \frac{1}{6} \\ \frac{1}{2} & \frac{1}{3} & \frac{1}{6} \\ 0 & \frac{1}{2} & \frac{1}{2} \end{bmatrix} = \begin{bmatrix} \frac{5}{12} & \frac{13}{36} & \frac{2}{9} \\ \frac{5}{12} & \frac{13}{36} & \frac{2}{9} \\ \frac{1}{4} & \frac{5}{12} & \frac{1}{3} \end{bmatrix}$.

(c) The probability that a queue with no one in line has two people in line 2 min later is $\frac{2}{9}$, since that is the entry in row 1, column 3 of A^2.

37. (a)

	Single	Multiple
Single	.90	.10
Multiple	.05	.95

(b) $\begin{bmatrix} .75 & .25 \end{bmatrix}$

(c) $\begin{bmatrix} .75 & .25 \end{bmatrix} \begin{bmatrix} .90 & .10 \\ .05 & .95 \end{bmatrix} = \begin{bmatrix} .688 & .313 \end{bmatrix}$

After 5 yr, 68.8% can be expected to live in single-family dwellings and 31.3% in multiple-family dwellings.

(d) $P^2 = \begin{bmatrix} .90 & .10 \\ .05 & .95 \end{bmatrix} \begin{bmatrix} .90 & .10 \\ .05 & .95 \end{bmatrix} = \begin{bmatrix} .815 & .185 \\ .0925 & .9075 \end{bmatrix}$

$\begin{bmatrix} .75 & .25 \end{bmatrix} \begin{bmatrix} .815 & .185 \\ .0925 & .9075 \end{bmatrix} = \begin{bmatrix} .634 & .366 \end{bmatrix}$

After 10 yr, 63.4% can be expected to live in single-family dwellings and 36.6% in multiple-family dwellings.

39. (a) $\begin{bmatrix} .443 & .364 & .193 \\ .277 & .436 & .287 \\ .266 & .304 & .430 \end{bmatrix} \begin{bmatrix} .443 & .364 & .193 \\ .277 & .436 & .287 \\ .266 & .304 & .430 \end{bmatrix}$

$= \begin{bmatrix} .348 & .379 & .273 \\ .320 & .378 & .302 \\ .316 & .360 & .323 \end{bmatrix}$

(b) The desired probability is found in row 1, column 1. The probability that if England won the last game, England will win the game after the next one is .348.

(c) The desired probability is found in row 2, column 1. The probability that if Australia won the last game, England will win the game after the next one is .320.

9.2 Regular Markov Chains

1. Let $A = \begin{bmatrix} .2 & .8 \\ .9 & .1 \end{bmatrix}$.

A is a regular transition matrix since $A^1 = A$ contains all positive entries.

3. Let $B = \begin{bmatrix} 1 & 0 \\ .6 & .4 \end{bmatrix}$.

$$B^2 = \begin{bmatrix} 1 & 0 \\ .6 & .4 \end{bmatrix} \begin{bmatrix} 1 & 0 \\ .6 & .4 \end{bmatrix} = \begin{bmatrix} 1 & 0 \\ .84 & .16 \end{bmatrix}$$

B is not regular since any power of B will have $\begin{bmatrix} 1 & 0 \end{bmatrix}$ as its first row and thus cannot have all positive entries.

5. Let $P = \begin{bmatrix} 0 & 1 & 0 \\ .4 & .2 & .4 \\ 1 & 0 & 0 \end{bmatrix}$.

$$P^2 = \begin{bmatrix} 0 & 1 & 0 \\ .4 & .2 & .4 \\ 1 & 0 & 0 \end{bmatrix} \begin{bmatrix} 0 & 1 & 0 \\ .4 & .2 & .4 \\ 1 & 0 & 0 \end{bmatrix}$$

$$= \begin{bmatrix} .4 & .2 & .4 \\ .48 & .44 & .08 \\ 0 & 1 & 0 \end{bmatrix}$$

$$P^3 = \begin{bmatrix} 0 & 1 & 0 \\ .4 & .2 & .4 \\ 1 & 0 & 0 \end{bmatrix} \begin{bmatrix} .4 & .2 & .4 \\ .48 & .44 & .08 \\ 0 & 1 & 0 \end{bmatrix}$$

$$= \begin{bmatrix} .48 & .44 & .08 \\ .256 & .568 & .176 \\ .4 & .2 & .4 \end{bmatrix}$$

P is a regular transition matrix since P^3 contains all positive entries.

7. Let $P = \begin{bmatrix} \frac{1}{4} & \frac{3}{4} \\ \frac{1}{2} & \frac{1}{2} \end{bmatrix}$, and let V be the probability vector $\begin{bmatrix} v_1 & v_2 \end{bmatrix}$. We want to find V such that

$$VP = V,$$

or $\begin{bmatrix} v_1 & v_2 \end{bmatrix} \begin{bmatrix} \frac{1}{4} & \frac{3}{4} \\ \frac{1}{2} & \frac{1}{2} \end{bmatrix} = \begin{bmatrix} v_1 & v_2 \end{bmatrix}$.

Use matrix multiplication on the left to obtain

$$\begin{bmatrix} \tfrac{1}{4}v_1 + \tfrac{1}{2}v_2 & \tfrac{3}{4}v_1 + \tfrac{1}{2}v_2 \end{bmatrix} = \begin{bmatrix} v_1 & v_2 \end{bmatrix}.$$

Set corresponding entries from the two matrices equal to get

$$\tfrac{1}{4}v_1 + \tfrac{1}{2}v_2 = v_1$$
$$\tfrac{3}{4}v_1 + \tfrac{1}{2}v_2 = v_2.$$

Multiply both equations by 4 to eliminate fractions.

$$v_1 + 2v_2 = 4v_1$$
$$3v_1 + 2v_2 = 4v_2$$

Simplify both equations.

$$-3v_1 + 2v_2 = 0$$
$$3v_1 - 2v_2 = 0$$

This is a dependent system. To find the values of v_1 and v_2, an additional equation is needed. Since $V = \begin{bmatrix} v_1 & v_2 \end{bmatrix}$ is a probability vector,

$$v_1 + v_2 = 1.$$

To find v_1 and v_2, solve the system

$$-3v_1 + 2v_2 = 0 \quad (1)$$
$$v_1 + v_2 = 1. \quad (2)$$

From equation (2), $v_1 = 1 - v_2$. Substitute $1 - v_2$ for v_1 in equation (1) to obtain

$$-3(1 - v_2) + 2v_2 = 0$$
$$-3 + 3v_2 + 2v_2 = 0$$
$$-3 + 5v_2 = 0$$
$$v_2 = \tfrac{3}{5}.$$

Since $v_1 = 1 - v_2$, $v_1 = \tfrac{2}{5}$, and the equilibrium vector is

$$V = \begin{bmatrix} \tfrac{2}{5} & \tfrac{3}{5} \end{bmatrix}.$$

9. Let $P = \begin{bmatrix} .3 & .7 \\ .4 & .6 \end{bmatrix}$, and let V be the probability vector $\begin{bmatrix} v_1 & v_2 \end{bmatrix}$. We want to find V such that

$$VP = V,$$

or $\begin{bmatrix} v_1 & v_2 \end{bmatrix} \begin{bmatrix} .3 & .7 \\ .4 & .6 \end{bmatrix} = \begin{bmatrix} v_1 & v_2 \end{bmatrix}$.

By matrix multiplication and equality of matrices,

$$.3v_1 + .4v_2 = v_1$$
$$.7v_1 + .6v_2 = v_2.$$

Simplify these equations to get the dependent system

$$-.7v_1 + .4v_2 = 0$$
$$.7v_1 - .4v_2 = 0.$$

Since V is a probability vector,

$$v_1 + v_2 = 1.$$

To find v_1 and v_2, solve the system

$$.7v_1 - .4v_2 = 0$$
$$v_1 + v_2 = 1$$

by the substitution method. Observe that $v_2 = 1 - v_1$.

$$.7v_1 - .4(1 - v_1) = 0$$
$$1.1v_1 - .4 = 0$$
$$v_1 = \frac{.4}{1.1} = \frac{4}{11}$$
$$v_2 = 1 - \frac{4}{11} = \frac{7}{11}$$

The equilibrium vector is

$$\begin{bmatrix} \frac{4}{11} & \frac{7}{11} \end{bmatrix}.$$

11. Let V be the probability vector $\begin{bmatrix} v_1 & v_2 & v_3 \end{bmatrix}$.

$$\begin{bmatrix} v_1 & v_2 & v_3 \end{bmatrix} \begin{bmatrix} .1 & .1 & .8 \\ .4 & .4 & .2 \\ .1 & .2 & .7 \end{bmatrix} = \begin{bmatrix} v_1 & v_2 & v_3 \end{bmatrix}$$

$$.1v_1 + .4v_2 + .1v_3 = v_1$$
$$.1v_1 + .4v_2 + .2v_3 = v_2$$
$$.8v_1 + .2v_2 + .7v_3 = v_3$$

Simplify these equations to get the dependent system

$$-.9v_1 + .4v_2 + .1v_3 = 0$$
$$.1v_1 - .6v_2 + .2v_3 = 0$$
$$.8v_1 + .2v_2 - .3v_3 = 0.$$

Since V is a probability vector,

$$v_1 + v_2 + v_3 = 1.$$

Solving the above system of four equations using the Gauss-Jordan method, we obtain

$$v_1 = \frac{14}{83}, \ v_2 = \frac{19}{83}, \ v_3 = \frac{50}{83}.$$

Thus, the equilibrium vector is

$$\begin{bmatrix} \frac{14}{83} & \frac{19}{83} & \frac{50}{83} \end{bmatrix}.$$

13. Let V be the probability vector $\begin{bmatrix} v_1 & v_2 & v_3 \end{bmatrix}$.

$$\begin{bmatrix} v_1 & v_2 & v_3 \end{bmatrix} \begin{bmatrix} .25 & .35 & .4 \\ .1 & .3 & .6 \\ .55 & .4 & .05 \end{bmatrix} = \begin{bmatrix} v_1 & v_2 & v_3 \end{bmatrix}$$

$$.25v_1 + .1v_2 + .55v_3 = v_1$$
$$.35v_1 + .3v_2 + .4v_3 = v_2$$
$$.4v_1 + .6v_2 + .05v_3 = v_3$$

Simplify these equations to get the dependent system

$$-.75v_1 + .1v_2 + .55v_3 = 0$$
$$.35v_1 - .7v_2 + .4v_3 = 0$$
$$.4v_1 + .6v_2 - .95v_3 = 0.$$

Since V is a probability vector,

$$v_1 + v_2 + v_3 = 1.$$

Solving this system we obtain

$$v_1 = \frac{170}{563}, \ v_2 = \frac{197}{563}, \ v_3 = \frac{196}{563}.$$

Thus, the equilibrium vector is

$$\begin{bmatrix} \frac{170}{563} & \frac{197}{563} & \frac{196}{563} \end{bmatrix}.$$

15. $\begin{bmatrix} v_1 & v_2 & v_3 \end{bmatrix} \begin{bmatrix} .85 & .10 & .05 \\ 0 & .80 & .20 \\ 0 & 0 & 1 \end{bmatrix}$

$= \begin{bmatrix} v_1 & v_2 & v_3 \end{bmatrix}$

$$.85v_1 = v_1$$
$$.10v_1 + .80v_2 = v_2$$
$$.05v_1 + .20v_2 + v_3 = v_3$$

We also have $v_1 + v_2 + v_3 = 1$.
Solving this system, we obtain

$$v_1 = 0, \ v_2 = 0, \ v_3 = 1.$$

The equilibrium vector is

$$\begin{bmatrix} 0 & 0 & 1 \end{bmatrix}.$$

17. $\begin{bmatrix} v_1 & v_2 & v_3 \end{bmatrix} \begin{bmatrix} .825 & .175 & 0 \\ .060 & .919 & .021 \\ .049 & 0 & .951 \end{bmatrix}$

$= \begin{bmatrix} v_1 & v_2 & v_3 \end{bmatrix}$

$$.825v_1 + .060v_2 + .049v_3 = v_1$$
$$.175v_1 + .919v_2 \qquad\quad = v_2$$
$$.021v_2 + .951v_3 = v_3$$

Also, $v_1 + v_2 + v_3 = 1$.
Solving this system, we obtain

$$v_1 = \tfrac{81}{331}, \ v_2 = \tfrac{175}{331}, \ v_3 = \tfrac{75}{331}.$$

The equilibrium vector is

$$\begin{bmatrix} \tfrac{81}{331} & \tfrac{175}{331} & \tfrac{75}{331} \end{bmatrix}.$$

19. $\begin{bmatrix} v_1 & v_2 & v_3 \end{bmatrix} \begin{bmatrix} .80 & .15 & .05 \\ .20 & .70 & .10 \\ .20 & .20 & .60 \end{bmatrix}$

$= \begin{bmatrix} v_1 & v_2 & v_3 \end{bmatrix}$

$$.80v_1 + .20v_2 + .20v_3 = v_1$$
$$.15v_1 + .70v_2 + .20v_3 = v_2$$
$$.05v_1 + .10v_2 + .60v_3 = v_3$$

Solving this system, we obtain

$$v_1 = \tfrac{1}{2}, \ v_2 = \tfrac{7}{20}, \ v_3 = \tfrac{3}{20}.$$

The equilibrium vector is

$$\begin{bmatrix} \tfrac{1}{2} & \tfrac{7}{20} & \tfrac{3}{20} \end{bmatrix}.$$

21. This exercise should be solved by graphing calculator methods. The solution may vary. The answer is

$$V = \begin{bmatrix} .171898 & .176519 & .167414 & .187526 & .296644 \end{bmatrix}.$$

23. Let V be the probability vector $\begin{bmatrix} x_1 & x_2 \end{bmatrix}$. We want to find V such that

$$V \begin{bmatrix} p & 1-p \\ 1-q & q \end{bmatrix} = V.$$

The system of equations is

$$px_1 + (1-q)x_2 = x_1$$
$$(1-p)x_1 + qx_2 = x_2.$$

Collecting like terms and simplifying leads to

$$(p-1)x_1 + (1-q)x^2 = 0,$$

so $\qquad\qquad x_1 = \dfrac{1-q}{1-p}x_2.$

Substituting this into $x_1 + x_2 = 1$, we obtain

$$\frac{1-q}{1-p}x_2 + x_2 = 1$$

or $\qquad \dfrac{2-p-q}{1-p}x^2 = 1;$

therefore,

$$x_2 = \frac{1-p}{2-p-q}$$

and $\quad x_1 = \dfrac{1-q}{2-p-q},$

so $\quad V = \begin{bmatrix} \dfrac{1-q}{2-p-q} & \dfrac{1-p}{2-p-q} \end{bmatrix}.$

Since $0 < p < 1$ and $0 < q < 1$, the matrix is always regular.

25. Let V be the probability vector $\begin{bmatrix} x_1 & x_2 \end{bmatrix}$.

We have $P = \begin{bmatrix} a_{11} & a_{12} \\ a_{21} & a_{22} \end{bmatrix}$,

where $\quad a_{11} + a_{21} = 1$
and $\qquad a_{12} + a_{22} = 1$.

The resulting equations are

$$a_{11}x_1 + a_{21}x_2 = x_1$$
$$a_{12}x_1 + a_{22}x_2 = x_2,$$

which we simplify to

$$(a_{11} - 1)x_1 + a_{21}x_2 = 0$$
$$a_{12}x_1 + (a_{22} - 1)x_2 = 0.$$

Hence, $x_1 = \dfrac{a_{21}}{1-a_{11}}x_2 = \dfrac{a_{21}}{a_{21}}x_2 = x_2,$

which we substitute into $x_1 + x_2 = 1$, obtaining

$$x_2 + x_2 = 1$$
$$2x_2 = 1$$
$$x_2 = \tfrac{1}{2}$$

and, therefore,

$$x_1 = 1 - \tfrac{1}{2} = \tfrac{1}{2}.$$

The equilibrium vector is

$$\begin{bmatrix} \tfrac{1}{2} & \tfrac{1}{2} \end{bmatrix}.$$

27. The transition matrix for the given information is

$$
\begin{array}{cc}
 & \begin{array}{cc} \text{Works} & \begin{array}{c}\text{Doesn't}\\\text{Work}\end{array} \end{array}\\
\begin{array}{c}\text{Works}\\\text{Doesn't Work}\end{array} & \begin{bmatrix} .9 & .1 \\ .7 & .3 \end{bmatrix}.
\end{array}
$$

Let V be the probability vector $\begin{bmatrix} v_1 & v_2 \end{bmatrix}$.

$$\begin{bmatrix} v_1 & v_2 \end{bmatrix}\begin{bmatrix} .9 & .1 \\ .7 & .3 \end{bmatrix} = \begin{bmatrix} v_1 & v_2 \end{bmatrix}$$

$$.9v_1 + .7v_2 = v_1$$
$$.1v_1 + .3v_2 = v_2$$

Simplify these equations to get the dependent system

$$-.1v_1 + .7v_2 = 0$$
$$.1v_1 - .7v_2 = 0.$$

Also, $v_1 + v_2 = 1$, so $v_1 = 1 - v_2$.

$$.1(1 - v_2) - .7v_2 = 0$$
$$.1 - .8v_2 = 0$$

$$v_2 = \tfrac{1}{8}, \; v_1 = \tfrac{7}{8}$$

The equilibrium vector is

$$\begin{bmatrix} \tfrac{7}{8} & \tfrac{1}{8} \end{bmatrix}.$$

The long-range probability that the line will work correctly is $\tfrac{7}{8}$.

29. The transition matrix is

$$
\begin{array}{c}
 & \begin{array}{ccc} \text{30-yr} & \text{15-yr} & \text{Adjustable} \end{array}\\
\begin{array}{c}\text{30-yr}\\\text{15-yr}\\\text{Adjustable}\end{array} & \begin{bmatrix} .444 & .479 & .077 \\ .150 & .802 & .048 \\ .463 & .367 & .170 \end{bmatrix}.
\end{array}
$$

To find the long-range trend, use the system

$$v_1 + v_2 + v_3 = 1$$
$$.444v_1 + .150v_2 + .463v_3 = v_1$$
$$.479v_1 + .802v_2 + .367v_3 = v_2$$
$$.077v_1 + .048v_2 + .170v_3 = v_3.$$

Simplify these equations to obtain the system

$$v_1 + v_2 + v_3 = 1$$
$$-.556v_1 + .150v_2 + .463v_3 = 0$$
$$.479v_1 - .198v_2 + .367v_3 = 0$$
$$.077v_1 + .048v_2 - .830v_3 = 0.$$

Solve this system by the Gauss-Jordan method to obtain $v_1 = .240$, $v_2 = .697$, and $v_3 = .063$.

The long range trend is 24.0% 30-yr fixed-rate, 69.7% 15-yr fixed-rate, and 6.3% adjustable.

31. The transition matrix is

$$
\begin{array}{c}
 & \begin{array}{ccc} \text{A} & \text{B} & \text{C} \end{array}\\
\begin{array}{c}\text{A}\\\text{B}\\\text{C}\end{array} & \begin{bmatrix} .3 & .3 & .4 \\ .15 & .3 & .55 \\ .3 & .6 & .1 \end{bmatrix}.
\end{array}
$$

To find the long-range distribution, use the system

$$v_1 + v_2 + v_3 = 1$$
$$.3v_1 + .15v_2 + .3v_3 = v_1$$
$$.3v_1 + .3v_2 + .6v_3 = v_2$$
$$.4v_1 + .55v_2 + .1v_3 = v_3.$$

Simplify these equations to obtain the system

$$v_1 + v_2 + v_3 = 1$$
$$-.7v_1 + .15v_2 + .3v_3 = 0$$
$$.3v_1 - .7v_2 + .6v_3 = 0$$
$$.4v_1 + .55v_2 - .9v_3 = 0.$$

Solve this system by the Gauss-Jordan method to obtain $v_1 = \frac{60}{251}$, $v_2 = \frac{102}{251}$, and $v_3 = \frac{89}{251}$. The long-range prediction is

$$\begin{bmatrix} \tfrac{60}{251} & \tfrac{102}{251} & \tfrac{89}{251} \end{bmatrix}.$$

33. (a) For the 1967-1979 trends, the transition matrix is

$$
\begin{array}{c}
 & \begin{array}{ccc} \text{Poor} & \begin{array}{c}\text{Middle}\\\text{Class}\end{array} & \text{Affluent} \end{array}\\
\begin{array}{c}\text{Poor}\\\text{Middle Class}\\\text{Affluent}\end{array} & \begin{bmatrix} .645 & .355 & 0 \\ .062 & .875 & .063 \\ 0 & .311 & .689 \end{bmatrix}.
\end{array}
$$

To find the long-range trend, use the system

$$v_1 + v_2 + v_3 = 1$$
$$.645v_1 + .062v_2 = v_1$$
$$.355v_1 + .875v_2 + .311v_3 = v_2$$
$$.063v_2 + .689v_3 = v_3.$$

Simplify these equations to obtain the system

$$v_1 + v_2 + v_3 = 1$$
$$-.355v_1 + .062v_2 = 0$$
$$.355v_1 - .125v_2 + .311v_3 = 0$$
$$.063v_2 - .311v_3 = 0.$$

Solve this system by the Gauss-Jordan method to obtain $v_1 = .127$, $v_2 = .726$, and $v_3 = .147$.

The long-range trend is 12.7% poor, 72.6% middle class, and 14.7% affluent.

(b) For the 1980-1991 trends, the transition matrix is

	Poor	Middle Class	Affluent
Poor	.696	.304	0
Middle Class	.085	.840	.075
Affluent	0	.271	.729

To find the long-range trend, use the system

$$v_1 + v_2 + v_3 = 1$$
$$.696v_1 + .085v_2 = v_1$$
$$.304v_1 + .840v_2 + .271v_3 = v_2$$
$$.075v_2 + .729v_3 = v_3.$$

Simplify these equations to obtain the system

$$v_1 + v_2 + v_3 = 1$$
$$-.304v_1 + .085v_2 = 0$$
$$.304v_1 - .160v_2 + .271v_3 = 0$$
$$.075v_2 - .271v_3 = 0.$$

Solve this system by the Gauss-Jordan method to obtain $v_1 = .180$, $v_2 = .643$, and $v_3 = .178$.

The long-range trend is 18.0% poor, 64.3% middle class, and 17.8% affluent.

35. The transition matrix is

$$\begin{bmatrix} p & 1-p \\ 1-p & p \end{bmatrix}.$$

The columns sum to $p + (1-p) = 1$, so by Exercise 25 the equilibrium vector is $\begin{bmatrix} \frac{1}{2} & \frac{1}{2} \end{bmatrix}$.

The long-range prediction for the fraction of the people who will hear the decision correctly is $\frac{1}{2}$.

37. The transition matrix is

$$P = \begin{bmatrix} .12 & .88 \\ .54 & .46 \end{bmatrix}.$$

Let V be the probability vector $\begin{bmatrix} v_1 & v_2 \end{bmatrix}$.

$$\begin{bmatrix} v_1 & v_2 \end{bmatrix} \begin{bmatrix} .12 & .88 \\ .54 & .46 \end{bmatrix} = \begin{bmatrix} v_1 & v_2 \end{bmatrix}$$
$$.12v_1 + .54v_2 = v_1$$
$$.88v_1 + .46v_2 = v_2$$

Simplify these equations to get the dependent system

$$-.88v_1 + .54v_2 = 0$$
$$.88v_1 - .54v_2 = 0.$$

Also, $v_1 + v_2 = 1$.

Solving this system, we obtain

$$v_1 = \frac{27}{71} \text{ and } v_2 = \frac{44}{71},$$

and note that

$$\frac{27}{71} \approx .38 = 38\%.$$

About 38% of letters in English text are expected to be vowels.

9.3 Absorbing Markov Chains

1.

	1	2	3
1	.15	.05	.8
2	0	1	0
3	.4	.6	0

Since $p_{22} = 1$, state 2 is absorbing. There is a probability of .05 of going from state 1 to state 2 and a probability of .6 of going from state 3 to state 2, so it is possible to go from each nonabsorbing state to the absorbing state. Thus, this is the transition matrix of an absorbing Markov chain.

3.

	1	2	3
1	.4	0	.6
2	0	1	0
3	.9	0	.1

Since $p_{22} = 1$, state 2 is absorbing. Since $p_{12} = 0$ and $p_{32} = 0$, it is not possible to go from either of the nonabsorbing states (state 1 and state 3) to the absorbing state (state 2). Thus, this is not the transition matrix of an absorbing Markov chain.

5.

	1	2	3	4
1	.2	.5	.1	.2
2	0	1	0	0
3	.9	.02	.04	.04
4	0	0	0	1

Since $p_{22} = 1$ and $p_{44} = 1$, states 2 and 4 are absorbing. It is possible to get from state 1 to states 2 and 4, and from state 3 to states 2 and 4. Thus, this is the transition matrix of an absorbing Markov chain.

7.

$$
\begin{array}{cc}
& \begin{array}{cccc} 1 & 2 & 3 & 4 \end{array} \\
\begin{array}{c} 1 \\ 2 \\ 3 \\ 4 \end{array} &
\left[\begin{array}{cccc}
.1 & .8 & 0 & .1 \\
0 & 1 & 0 & 0 \\
1 & 0 & 0 & 0 \\
0 & 0 & 0 & 1
\end{array} \right]
\end{array}
$$

Since $p_{22} = 1$ and $p_{44} = 1$, states 2 and 4 are absorbing. It is possible to go from state 1 to states 2 and 4 and from state 3 to states 2 and 4 through state 1. Thus, this is the transition matrix of an absorbing Markov chain.

9. $P = \left[\begin{array}{cc|c} 1 & 0 & 0 \\ 0 & 1 & 0 \\ \hline .2 & .3 & .5 \end{array} \right]$

Here $R = [.2 \quad .3]$ and $Q = [.5]$.
Find the fundamental matrix F.

$$
F = [I_1 - Q]^{-1} = [1 - .5]^{-1}
$$

$$
= [.5]^{-1} = \left[\tfrac{1}{.5} \right] = [2]
$$

The product FR is

$$
FR = [2] [.2 \quad .3] = [.4 \quad .6].
$$

11.

$$
\begin{array}{cc}
& \begin{array}{ccc} 1 & 2 & 3 \end{array} \\
\begin{array}{c} 1 \\ 2 \\ 3 \end{array} &
\left[\begin{array}{ccc}
.8 & .15 & .05 \\
0 & 1 & 0 \\
0 & 0 & 1
\end{array} \right]
\end{array} = P
$$

Rearrange the rows and columns of P so that the absorbing states come first.

$$
\begin{array}{cc}
& \begin{array}{ccc} 2 & 3 & 1 \end{array} \\
\begin{array}{c} 2 \\ 3 \\ 1 \end{array} &
\left[\begin{array}{cc|c}
1 & 0 & 0 \\
0 & 1 & 0 \\ \hline
.15 & .05 & .8
\end{array} \right]
\end{array}
$$

$$
R = [.15 \quad .05], \ Q = [.8]
$$

$$
F = [I_1 - Q]^{-1} = [1 - .8]^{-1} = [.2]^{-1}
$$

$$
= \left[\tfrac{1}{.2} \right] = [5]
$$

$$
FR = [5] [.15 \quad .05] = [.75 \quad .25]
$$

13. $\left[\begin{array}{cc|c} 1 & 0 & 0 \\ 0 & 1 & 0 \\ \hline \frac{1}{3} & \frac{1}{3} & \frac{1}{3} \end{array} \right] = P$

$$
R = \left[\tfrac{1}{3} \quad \tfrac{1}{3} \right], \ Q = \left[\tfrac{1}{3} \right]
$$

$$
F = [I_1 - Q]^{-1} = \left[1 - \tfrac{1}{3} \right]^{-1} = \left[\tfrac{2}{3} \right]^{-1} = \left[\tfrac{3}{2} \right]
$$

$$
FR = \left[\tfrac{3}{2} \right] \left[\tfrac{1}{3} \quad \tfrac{1}{3} \right] = \left[\tfrac{1}{2} \quad \tfrac{1}{2} \right]
$$

15.

$$
\begin{array}{cc}
& \begin{array}{cccc} 1 & 2 & 3 & 4 \end{array} \\
\begin{array}{c} 1 \\ 2 \\ 3 \\ 4 \end{array} &
\left[\begin{array}{cccc}
1 & 0 & 0 & 0 \\
\frac{1}{3} & 0 & \frac{2}{3} & 0 \\
0 & 0 & 1 & 0 \\
\frac{1}{4} & \frac{1}{4} & \frac{1}{4} & \frac{1}{4}
\end{array} \right]
\end{array} = P
$$

Rearrange the rows and columns of P so that the absorbing states come first.

$$
\begin{array}{cc}
& \begin{array}{cccc} 1 & 3 & 2 & 4 \end{array} \\
\begin{array}{c} 1 \\ 3 \\ 2 \\ 4 \end{array} &
\left[\begin{array}{cc|cc}
1 & 0 & 0 & 0 \\
0 & 1 & 0 & 0 \\ \hline
\frac{1}{3} & \frac{2}{3} & 0 & 0 \\
\frac{1}{4} & \frac{1}{4} & \frac{1}{4} & \frac{1}{4}
\end{array} \right]
\end{array}
$$

$$
R = \left[\begin{array}{cc} \frac{1}{3} & \frac{2}{3} \\ \frac{1}{4} & \frac{1}{4} \end{array} \right]; Q = \left[\begin{array}{cc} 0 & 0 \\ \frac{1}{4} & \frac{1}{4} \end{array} \right]
$$

$$
F = [I_2 - Q]^{-1} = \left(\left[\begin{array}{cc} 1 & 0 \\ 0 & 1 \end{array} \right] - \left[\begin{array}{cc} 0 & 0 \\ \frac{1}{4} & \frac{1}{4} \end{array} \right] \right)^{-1}
$$

$$
= \left[\begin{array}{cc} 1 & 0 \\ -\frac{1}{4} & \frac{3}{4} \end{array} \right]^{-1} = \left[\begin{array}{cc} 1 & 0 \\ \frac{1}{3} & \frac{4}{3} \end{array} \right]
$$

$$
FR = \left[\begin{array}{cc} 1 & 0 \\ \frac{1}{3} & \frac{4}{3} \end{array} \right] \left[\begin{array}{cc} \frac{1}{3} & \frac{2}{3} \\ \frac{1}{4} & \frac{1}{4} \end{array} \right] = \left[\begin{array}{cc} \frac{1}{3} & \frac{2}{3} \\ \frac{4}{9} & \frac{5}{9} \end{array} \right]
$$

17.

$$
\begin{array}{cc}
& \begin{array}{ccccc} 1 & 2 & 3 & 4 & 5 \end{array} \\
\begin{array}{c} 1 \\ 2 \\ 3 \\ 4 \\ 5 \end{array} &
\left[\begin{array}{ccccc}
1 & 0 & 0 & 0 & 0 \\
0 & 1 & 0 & 0 & 0 \\
.1 & .2 & .3 & .2 & .2 \\
.3 & .5 & .1 & 0 & .1 \\
0 & 0 & 0 & 0 & 1
\end{array} \right]
\end{array} = P
$$

Rearranging, we obtain the matrix

$$
\begin{array}{cc}
& \begin{array}{ccccc} 1 & 2 & 5 & 3 & 4 \end{array} \\
\begin{array}{c} 1 \\ 2 \\ 5 \\ 3 \\ 4 \end{array} &
\left[\begin{array}{ccc|cc}
1 & 0 & 0 & 0 & 0 \\
0 & 1 & 0 & 0 & 0 \\
0 & 0 & 1 & 0 & 0 \\ \hline
.1 & .2 & .2 & .3 & .2 \\
.3 & .5 & .1 & .1 & 0
\end{array} \right]
\end{array} .
$$

$$
R = \left[\begin{array}{ccc} .1 & .2 & .2 \\ .3 & .5 & .1 \end{array} \right]; Q = \left[\begin{array}{cc} .3 & .2 \\ .1 & 0 \end{array} \right]
$$

$$F = [I_2 - Q]^{-1} = \left(\begin{bmatrix} 1 & 0 \\ 0 & 1 \end{bmatrix} - \begin{bmatrix} .3 & .2 \\ .1 & 0 \end{bmatrix} \right)^{-1}$$

$$= \begin{bmatrix} .7 & -.2 \\ -.1 & 1 \end{bmatrix}^{-1} = \begin{bmatrix} \frac{25}{17} & \frac{5}{17} \\ \frac{5}{34} & \frac{35}{34} \end{bmatrix}$$

$$FR = \begin{bmatrix} \frac{25}{17} & \frac{5}{17} \\ \frac{5}{34} & \frac{35}{34} \end{bmatrix} \begin{bmatrix} \frac{1}{10} & \frac{2}{10} & \frac{2}{10} \\ \frac{3}{10} & \frac{5}{10} & \frac{1}{10} \end{bmatrix}$$

$$= \begin{bmatrix} \frac{4}{17} & \frac{15}{34} & \frac{11}{34} \\ \frac{11}{34} & \frac{37}{68} & \frac{9}{68} \end{bmatrix}$$

19. (a) The transition matrix is

$$\begin{array}{c} \\ 0 \\ 1 \\ 2 \\ 3 \\ 4 \end{array} \begin{array}{ccccc} 0 & 1 & 2 & 3 & 4 \\ \begin{bmatrix} 1 & 0 & 0 & 0 & 0 \\ \frac{1}{2} & 0 & \frac{1}{2} & 0 & 0 \\ 0 & \frac{1}{2} & 0 & \frac{1}{2} & 0 \\ 0 & 0 & \frac{1}{2} & 0 & \frac{1}{2} \\ 0 & 0 & 0 & 0 & 1 \end{bmatrix} \end{array}.$$

Rearranging, we have

$$\begin{array}{c} \\ 0 \\ 4 \\ 1 \\ 2 \\ 3 \end{array} \begin{array}{ccccc} 0 & 4 & 1 & 2 & 3 \\ \left[\begin{array}{cc|ccc} 1 & 0 & 0 & 0 & 0 \\ 0 & 1 & 0 & 0 & 0 \\ \hline \frac{1}{2} & 0 & 0 & \frac{1}{2} & 0 \\ 0 & 0 & \frac{1}{2} & 0 & \frac{1}{2} \\ 0 & \frac{1}{2} & 0 & \frac{1}{2} & 0 \end{array} \right] \end{array}.$$

$$R = \begin{bmatrix} \frac{1}{2} & 0 \\ 0 & 0 \\ 0 & \frac{1}{2} \end{bmatrix}; \quad Q = \begin{bmatrix} 0 & \frac{1}{2} & 0 \\ \frac{1}{2} & 0 & \frac{1}{2} \\ 0 & \frac{1}{2} & 0 \end{bmatrix}$$

$$F = [I_3 - Q]^{-1} = \left(\begin{bmatrix} 1 & 0 & 0 \\ 0 & 1 & 0 \\ 0 & 0 & 1 \end{bmatrix} - \begin{bmatrix} 0 & \frac{1}{2} & 0 \\ \frac{1}{2} & 0 & \frac{1}{2} \\ 0 & \frac{1}{2} & 0 \end{bmatrix} \right)^{-1}$$

$$= \begin{bmatrix} 1 & -\frac{1}{2} & 0 \\ -\frac{1}{2} & 1 & -\frac{1}{2} \\ 0 & -\frac{1}{2} & 1 \end{bmatrix}^{-1} = \begin{bmatrix} \frac{3}{2} & 1 & \frac{1}{2} \\ 1 & 2 & 1 \\ \frac{1}{2} & 1 & \frac{3}{2} \end{bmatrix}$$

$$FR = \begin{bmatrix} \frac{3}{2} & 1 & \frac{1}{2} \\ 1 & 2 & 1 \\ \frac{1}{2} & 1 & \frac{3}{2} \end{bmatrix} \begin{bmatrix} \frac{1}{2} & 0 \\ 0 & 0 \\ 0 & \frac{1}{2} \end{bmatrix} = \begin{bmatrix} \frac{3}{4} & \frac{1}{4} \\ \frac{1}{2} & \frac{1}{2} \\ \frac{1}{4} & \frac{3}{4} \end{bmatrix}$$

(b) If player A starts with $1, the probability of ruin for A is $\frac{3}{4}$, since that is the entry in row 1, column 1 of FR. The $\frac{3}{4}$ is the probability that the nonabsorbing state of starting with $1 will lead to the absorbing state of ruin.

(c) If player A starts with $3, the probability of ruin for A is $\frac{1}{4}$, since that is the entry in row 3, column 1 of FR.

21. Use the formulas given in the textbook to calculate r and then x_a if $a = 10, b = 30$, and $p = .49$.

$$r = \frac{1-p}{p} = \frac{1-.49}{.49} \approx 1.0408$$

The probability that A will be ruined in this situation is

$$x_a = \frac{r^a - r^{a+b}}{1 - r^{a+b}}$$

$$= \frac{(1.0408)^{10} - (1.0408)^{40}}{1 - (1.0408)^{40}}$$

$$\approx .8756.$$

23. $a = 10$, $b = 10$

Complete the chart by using the formulas given in Exercise 21 for r and x_a.

p	r	x_a
.1	9	.9999999997
.2	4	.99999905
.3	$\frac{7}{3}$	.99979
.4	1.5	.98295
.5	1	.5
.6	$\frac{2}{3}$	.017046
.7	$\frac{3}{7}$	.000209
.8	.25	.00000095
.9	$\frac{1}{9}$	.0000000003

25. If an absorbing Markov chain has only one absorbing state, then the equilibrium vector has 1 in the position corresponding to the absorbing state and 0 in all of the other positions. Regardless of the initial state, the long-term trend will be for all states to end up in the single absorbing state.

27. (a) The transition matrix P is

$$\begin{array}{c} \\ 1 \\ 2 \\ 3 \\ 4 \end{array} \begin{array}{cccc} 1 & 2 & 3 & 4 \\ \begin{bmatrix} .15 & .6 & .25 & 0 \\ 0 & .15 & .25 & .6 \\ 0 & 0 & 1 & 0 \\ 0 & 0 & 0 & 1 \end{bmatrix} \end{array}.$$

Note that states 3 and 4 are absorbing. We arrange the matrix to obtain

$$\begin{array}{c} \\ 3 \\ 4 \\ 1 \\ 2 \end{array} \begin{array}{cccc} 3 & 4 & 1 & 2 \\ \left[\begin{array}{cc|cc} 1 & 0 & 0 & 0 \\ 0 & 1 & 0 & 0 \\ \hline .25 & 0 & .15 & .6 \\ .25 & .6 & 0 & .15 \end{array} \right] \end{array}.$$

Then

$$F = \begin{bmatrix} .85 & -.6 \\ 0 & .85 \end{bmatrix}^{-1} = \begin{bmatrix} \frac{20}{17} & \frac{240}{289} \\ 0 & \frac{20}{17} \end{bmatrix}, \text{ so}$$

$$FR = \begin{matrix} 1 \\ 2 \end{matrix} \begin{bmatrix} \overset{3}{\frac{145}{289}} & \overset{4}{\frac{144}{289}} \\ \frac{5}{17} & \frac{12}{17} \end{bmatrix}.$$

(b) The probability of a freshman's graduating is the probability of state 1 eventually becoming state 4. From FR we see this probability is $\frac{144}{289}$.

(c) To find the expected number of years that a freshman will be in college before graduating or flunking out, add the entries in row 1 of F to obtain

$$\frac{20}{17} + \frac{240}{289} = \frac{580}{289} \approx 2.007 \text{ yr.}$$

29. (a)

$$\begin{array}{c} \\ 1 \\ 2 \\ 3 \\ 4 \\ 5 \end{array} \begin{array}{ccccc} 1 & 2 & 3 & 4 & 5 \\ \begin{bmatrix} .4 & .3 & .2 & .1 & 0 \\ .2 & .1 & 0 & .6 & .1 \\ 0 & 0 & 1 & 0 & 0 \\ .1 & .1 & .4 & .1 & .3 \\ 0 & 0 & 0 & 0 & 1 \end{bmatrix} \end{array}$$

is the transition matrix. States 3 and 5 are absorbing. Upon rearranging, we obtain

$$\begin{array}{c} \\ 3 \\ 5 \\ 1 \\ 2 \\ 4 \end{array} \begin{array}{ccccc} 3 & 5 & 1 & 2 & 4 \\ \left[\begin{array}{cc|ccc} 1 & 0 & 0 & 0 & 0 \\ 0 & 1 & 0 & 0 & 0 \\ \hline .2 & 0 & .4 & .3 & .1 \\ 0 & .1 & .2 & .1 & .6 \\ .4 & .3 & .1 & .1 & .1 \end{array} \right] \end{array}.$$

Then

$$F = \begin{bmatrix} .6 & -.3 & -.1 \\ -.2 & .9 & -.6 \\ -.1 & -.1 & .9 \end{bmatrix}^{-1}$$

$$= \begin{bmatrix} 2.0436 & .7629 & .7357 \\ .6540 & 1.4441 & 1.0354 \\ .2997 & .2452 & 1.3079 \end{bmatrix},$$

and

$$FR = \begin{matrix} 1 \\ 2 \\ 4 \end{matrix} \begin{bmatrix} \overset{3}{.703} & \overset{5}{.297} \\ .545 & .455 \\ .583 & .417 \end{bmatrix}.$$

The second column of FR gives the probability of ending up in compartment 5 given the initial compartment.

(b) From compartment 1, the probability is .297.

(c) From compartment 2, the probability is .455.

(d) From compartment 3, the probability is 0 since state 3 is absorbing.

(e) From compartment 4, the probability is .417.

(f) To find the expected number of times that a rat in compartment 1 will be in compartment 1 before ending up in compartment 3 or 5, look at the entry in row 1, column 1 of F, $2.0436 \approx 2.04$.

(g) To find the expected number of times that a rat in compartment 4 will be in compartment 4 before ending up in compartment 3 or 5, look at the entry in row 3, column 3 of F, $1.3079 \approx 1.31$.

31. This exercise should be solved by graphing calculator methods. The solution may vary. The answers are as follows.

(a) .6

(b) .5

Chapter 9 Review Exercises

3. $\begin{bmatrix} .4 & .6 \\ 1 & 0 \end{bmatrix}$

This could be a transition matrix since it is a square matrix, all entries are between 0 and 1, inclusive, and the sum of the entries in each row is 1.

5. $\begin{bmatrix} .8 & .2 & 0 \\ 0 & 1 & 0 \\ .1 & .4 & .5 \end{bmatrix}$

This could be a transition matrix for the same reasons stated in Exercise 3.

7. (a) $C = \begin{bmatrix} .6 & .4 \\ 1 & 0 \end{bmatrix}$

$$C^2 = \begin{bmatrix} .6 & .4 \\ 1 & 0 \end{bmatrix} \begin{bmatrix} .6 & .4 \\ 1 & 0 \end{bmatrix} = \begin{bmatrix} .76 & .24 \\ .6 & .4 \end{bmatrix}$$

$$C^3 = \begin{bmatrix} .6 & .4 \\ 1 & 0 \end{bmatrix} \begin{bmatrix} .76 & .24 \\ .6 & .4 \end{bmatrix}$$

$$= \begin{bmatrix} .696 & .304 \\ .76 & .24 \end{bmatrix}$$

(b) The probability that state 2 changes to state

1 after 3 repetitions is .76, since that is the entry in row 2, column 1 of C^3.

9. (a) $E = \begin{bmatrix} .2 & .5 & .3 \\ .1 & .8 & .1 \\ 0 & 1 & 0 \end{bmatrix}$

$E^2 = \begin{bmatrix} .2 & .5 & .3 \\ .1 & .8 & .1 \\ 0 & 1 & 0 \end{bmatrix} \begin{bmatrix} .2 & .5 & .3 \\ .1 & .8 & .1 \\ 0 & 1 & 0 \end{bmatrix}$

$= \begin{bmatrix} .09 & .8 & .11 \\ .1 & .79 & .11 \\ .1 & .8 & .1 \end{bmatrix}$

$E^3 = \begin{bmatrix} .2 & .5 & .3 \\ .1 & .8 & .1 \\ 0 & 1 & 0 \end{bmatrix} \begin{bmatrix} .09 & .8 & .11 \\ .1 & .79 & .11 \\ .1 & .8 & .1 \end{bmatrix}$

$= \begin{bmatrix} .098 & .795 & .107 \\ .099 & .792 & .109 \\ .1 & .79 & .11 \end{bmatrix}$

(b) The probability that state 2 changes to state 1 after 3 repetitions is .099, since that is the entry in row 2, column 1 of E^3.

11. $T^2 = \begin{bmatrix} .4 & .6 \\ .5 & .5 \end{bmatrix} \begin{bmatrix} .4 & .6 \\ .5 & .5 \end{bmatrix} = \begin{bmatrix} .46 & .54 \\ .45 & .55 \end{bmatrix}$

The distribution after 2 repetitions is

$[.3 \quad .7] \begin{bmatrix} .46 & .54 \\ .45 & .55 \end{bmatrix} = [.453 \quad .547]$.

To predict the long-range distribution, let V be the probability vector $[v_1 \quad v_2]$.

$[v_1 \quad v_2] \begin{bmatrix} .46 & .54 \\ .45 & .55 \end{bmatrix} = [v_1 \quad v_2]$

$.46v_1 + .45v_2 = v_1$
$.54v_1 + .55v_2 = v_2$
$v_1 + \quad v_2 = 1$

$-.54v_1 + .45v_2 = 0$
$.54v_1 - .45v_2 = 0$
$v_2 = 1 - v_1$

$.54v_1 + .45(1 - v_1) = 0$
$.99v_1 = .45$

$v_1 = \dfrac{45}{99} = \dfrac{5}{11} \approx .455$

$v_2 = \dfrac{54}{99} = \dfrac{6}{11} \approx .545$

The long-range distribution is

$\left[\frac{5}{11} \quad \frac{6}{11}\right]$ or $[.455 \quad .545]$.

13. $T^2 = \begin{bmatrix} .6 & .2 & .2 \\ .3 & .3 & .4 \\ .5 & .4 & .1 \end{bmatrix} \begin{bmatrix} .6 & .2 & .2 \\ .3 & .3 & .4 \\ .5 & .4 & .1 \end{bmatrix}$

$= \begin{bmatrix} .52 & .26 & .22 \\ .47 & .31 & .22 \\ .47 & .26 & .27 \end{bmatrix}$

The distribution after 2 repetitions is

$[.2 \quad .4 \quad .4] \begin{bmatrix} .52 & .26 & .22 \\ .47 & .31 & .22 \\ .47 & .26 & .27 \end{bmatrix}$

$= [.48 \quad .28 \quad .24]$.

To predict the long-range distribution, let V be the probability vector $[v_1 \quad v_2 \quad v_3]$.

$[v_1 \quad v_2 \quad v_3] \begin{bmatrix} .6 & .2 & .2 \\ .3 & .3 & .4 \\ .5 & .4 & .1 \end{bmatrix}$

$= [v_1 \quad v_2 \quad v_3]$

$.6v_1 + .3v_2 + .5v_3 = v_1$
$.2v_1 + .3v_2 + .4v_3 = v_2$
$.2v_1 + .4v_2 + .1v_3 = v_3$

Also, $v_1 + v_2 + v_3 = 1$.
Solving this system by the Gauss-Jordan method gives

$v_1 = \dfrac{47}{95} \approx .495,$

$v_2 = \dfrac{26}{95} \approx .274,$

$v_3 = \dfrac{22}{95} \approx .232.$

The long-range distribution is

$\left[\frac{47}{95} \quad \frac{26}{95} \quad \frac{22}{95}\right]$ or $[.495 \quad .274 \quad .232]$.

17. $A = \begin{bmatrix} .4 & .2 & .4 \\ 0 & 1 & 0 \\ .6 & .3 & .1 \end{bmatrix}$

$A^2 = \begin{bmatrix} .4 & .4 & .2 \\ 0 & 1 & 0 \\ .3 & .45 & .25 \end{bmatrix}$

$A^3 = \begin{bmatrix} .28 & .54 & .18 \\ 0 & 1 & 0 \\ .27 & .585 & .145 \end{bmatrix}$

Note that the second row will always have zeros; hence, the matrix is not regular.

23.

$$\begin{array}{c} \\ 1 \\ 2 \\ 3 \end{array} \begin{array}{ccc} 1 & 2 & 3 \\ \end{array} \\ \begin{bmatrix} .2 & 0 & .8 \\ 0 & 1 & 0 \\ .7 & 0 & .3 \end{bmatrix}$$

Since $p_{22} = 1$, state 2 is absorbing. Since $p_{12} = 0$ and $p_{32} = 0$, it is not possible to go from either of the nonabsorbing states to the absorbing state. Thus, this is not the transition matrix of an absorbing Markov chain.

25.

$$\begin{array}{c} \\ 1 \\ 2 \\ 3 \end{array} \begin{array}{ccc} 1 & 2 & 3 \\ \end{array} \\ \begin{bmatrix} .2 & .5 & .3 \\ 0 & 1 & 0 \\ 0 & 0 & 1 \end{bmatrix} = P$$

Rearranging, we have

$$\begin{array}{c} \\ 2 \\ 3 \\ 1 \end{array} \begin{array}{ccc} 2 & 3 & 1 \\ \end{array} \\ \left[\begin{array}{cc|c} 1 & 0 & 0 \\ 0 & 1 & 0 \\ \hline .5 & .3 & .2 \end{array} \right].$$

$$R = [.5 \quad .3], \quad Q = [.2]$$

$$F = [I_1 - Q]^{-1} = [1 - .2]^{-1} = [.8]^{-1}$$

$$= \left[\tfrac{10}{8} \right] = \left[\tfrac{5}{4} \right] \quad \text{or} \quad [1.25]$$

$$FR = [1.25] \, [.5 \quad .3] = [.625 \quad .375]$$

$$\text{or} \quad \left[\tfrac{5}{8} \quad \tfrac{3}{8} \right]$$

27.

$$\begin{array}{c} \\ 1 \\ 2 \\ 3 \\ 4 \end{array} \begin{array}{cccc} 1 & 2 & 3 & 4 \\ \end{array} \\ \begin{bmatrix} \tfrac{1}{5} & \tfrac{1}{5} & \tfrac{2}{5} & \tfrac{1}{5} \\ 0 & 1 & 0 & 0 \\ \tfrac{1}{2} & \tfrac{1}{4} & \tfrac{1}{8} & \tfrac{1}{8} \\ 0 & 0 & 0 & 1 \end{bmatrix} = P$$

Rearranging, we have

$$\begin{array}{c} \\ 2 \\ 4 \\ 1 \\ 3 \end{array} \begin{array}{cccc} 2 & 4 & 1 & 3 \\ \end{array} \\ \left[\begin{array}{cc|cc} 1 & 0 & 0 & 0 \\ 0 & 1 & 0 & 0 \\ \hline \tfrac{1}{5} & \tfrac{1}{5} & \tfrac{1}{5} & \tfrac{2}{5} \\ \tfrac{1}{4} & \tfrac{1}{8} & \tfrac{1}{2} & \tfrac{1}{8} \end{array} \right].$$

$$R = \begin{bmatrix} \tfrac{1}{5} & \tfrac{1}{5} \\ \tfrac{1}{4} & \tfrac{1}{8} \end{bmatrix} \text{ and } Q = \begin{bmatrix} \tfrac{1}{5} & \tfrac{2}{5} \\ \tfrac{1}{2} & \tfrac{1}{8} \end{bmatrix}$$

$$F = [I_2 - Q]^{-1} = \left(\begin{bmatrix} 1 & 0 \\ 0 & 1 \end{bmatrix} - \begin{bmatrix} \tfrac{1}{5} & \tfrac{2}{5} \\ \tfrac{1}{2} & \tfrac{1}{8} \end{bmatrix} \right)^{-1}$$

$$= \begin{bmatrix} \tfrac{4}{5} & -\tfrac{2}{5} \\ -\tfrac{1}{2} & \tfrac{7}{8} \end{bmatrix}^{-1} = \begin{bmatrix} \tfrac{7}{4} & \tfrac{4}{5} \\ 1 & \tfrac{8}{5} \end{bmatrix}$$

$$FR = \begin{bmatrix} \tfrac{7}{4} & \tfrac{4}{5} \\ 1 & \tfrac{8}{5} \end{bmatrix} \begin{bmatrix} \tfrac{1}{5} & \tfrac{1}{5} \\ \tfrac{1}{4} & \tfrac{1}{8} \end{bmatrix} = \begin{bmatrix} \tfrac{11}{20} & \tfrac{9}{20} \\ \tfrac{3}{5} & \tfrac{2}{5} \end{bmatrix} \text{ or } \begin{bmatrix} .55 & .45 \\ .6 & .4 \end{bmatrix}$$

29. (a) The distribution after the campaign is

$$[.35 \quad .65] \begin{bmatrix} .8 & .2 \\ .4 & .6 \end{bmatrix} = [.54 \quad .46].$$

(b) $P^3 = \begin{bmatrix} .688 & .312 \\ .624 & .376 \end{bmatrix}$

The distribution after 3 campaigns is

$$[.35 \quad .65] \begin{bmatrix} .688 & .312 \\ .624 & .376 \end{bmatrix} = [.6464 \quad .3536].$$

31. The distribution after 1 mo is

$$[.4 \quad .4 \quad .2] \begin{bmatrix} .8 & .15 & .05 \\ .25 & .55 & .2 \\ .04 & .21 & .75 \end{bmatrix} = [.428 \quad .332 \quad .25].$$

33. $P^2 = \begin{bmatrix} .8 & .15 & .05 \\ .25 & .55 & .2 \\ .04 & .21 & .75 \end{bmatrix} \begin{bmatrix} .8 & .15 & .05 \\ .25 & .55 & .2 \\ .04 & .21 & .75 \end{bmatrix}$

$$= \begin{bmatrix} .6795 & .213 & .1075 \\ .3455 & .382 & .2725 \\ .1145 & .279 & .6065 \end{bmatrix}$$

$$P^3 = \begin{bmatrix} .8 & .15 & .05 \\ .25 & .55 & .2 \\ .04 & .21 & .75 \end{bmatrix} \begin{bmatrix} .6795 & .213 & .1075 \\ .3455 & .382 & .2725 \\ .1145 & .279 & .6065 \end{bmatrix}$$

$$= \begin{bmatrix} .60115 & .24165 & .1572 \\ .3828 & .31915 & .29805 \\ .18561 & .29799 & .5164 \end{bmatrix}$$

The distribution after 3 mo is

$$[.4 \quad .4 \quad .2] \begin{bmatrix} .60115 & .24165 & .1572 \\ .3828 & .31915 & .29805 \\ .18561 & .29799 & .5164 \end{bmatrix}$$

$$= [.431 \quad .284 \quad .285].$$

In Exercises 35-43, the original transition matrix is

$$P = \begin{bmatrix} .3 & .5 & .2 \\ .2 & .6 & .2 \\ .1 & .5 & .4 \end{bmatrix}.$$

35. The probability that a man of normal weight will have a thin son is given by the entry in row 2, column 1 of P, which is .2.

37. $P^2 = \begin{bmatrix} .3 & .5 & .2 \\ .2 & .6 & .2 \\ .1 & .5 & .4 \end{bmatrix} \begin{bmatrix} .3 & .5 & .2 \\ .2 & .6 & .2 \\ .1 & .5 & .4 \end{bmatrix} = \begin{bmatrix} .21 & .55 & .24 \\ .2 & .56 & .24 \\ .17 & .55 & .28 \end{bmatrix}$

$P^3 = \begin{bmatrix} .3 & .5 & .2 \\ .2 & .6 & .2 \\ .1 & .5 & .4 \end{bmatrix} \begin{bmatrix} .21 & .55 & .24 \\ .2 & .56 & .24 \\ .17 & .55 & .28 \end{bmatrix}$

$= \begin{bmatrix} .197 & .555 & .248 \\ .196 & .556 & .248 \\ .189 & .555 & .256 \end{bmatrix}$

The probability that a man of normal weight will have a thin great-grandson is given by the entry in row 2, column 1 of P^3, which is .196.

39. $P^2 = \begin{bmatrix} .21 & .55 & .24 \\ .2 & .56 & .24 \\ .17 & .55 & .28 \end{bmatrix}$

The probability that an overweight man will have an overweight grandson is given by the entry in row 3, column 3 of P^2, which is .28.

41. The distribution of men by weight after 1 generation is

$\begin{bmatrix} .2 & .55 & .25 \end{bmatrix} \begin{bmatrix} .3 & .5 & .2 \\ .2 & .6 & .2 \\ .1 & .5 & .4 \end{bmatrix}$

$= \begin{bmatrix} .195 & .555 & .25 \end{bmatrix}.$

43. The distribution of men by weight after 3 generations is

$\begin{bmatrix} .2 & .55 & .25 \end{bmatrix} \begin{bmatrix} .197 & .555 & .248 \\ .196 & .556 & .248 \\ .189 & .555 & .256 \end{bmatrix}$

$= \begin{bmatrix} .194 & .556 & .25 \end{bmatrix}.$

45. If the offspring both carry genes AA, then so must their offspring; hence, state 1 ends up in state 1 with probability 1. If the offspring both carry genes aa, then so must their offspring; hence, state 6 ends up in state 6 with probability 1. If AA mates with aa, then the offspring will carry genes

Aa; hence, state 3 ends up in state 4 with probability 1. If AA mates with Aa, then there are four possible outcomes for a pair of offspring: AA and AA is one of the outcomes, so state 2 ends up in state 1 with probability $\frac{1}{4}$; AA and Aa can happen two ways, so state 2 ends up in state 2 with probability $\frac{2}{4}$ or $\frac{1}{2}$; and Aa and Aa is the last possible outcome, so state 2 ends up in state 4 with probability $\frac{1}{4}$. If Aa mates with Aa, then there are sixteen possible outcomes for a pair of offspring: state 4 ends up in states 1, 2, 3, 4, 5, 6 with respective probabilities $\frac{1}{16}$, $\frac{1}{4}$, $\frac{1}{8}$, $\frac{1}{4}$, $\frac{1}{4}$, and $\frac{1}{16}$. If Aa mates with aa, then there are four possible outcomes for a pair of offspring, corresponding to three of the possible states: state 5 ends up in states 4, 5, 6 with respective probabilities $\frac{1}{4}$, $\frac{1}{2}$, and $\frac{1}{4}$. This verifies that the transition matrix for this mating experiment is

	1	2	3	4	5	6
1	1	0	0	0	0	0
2	$\frac{1}{4}$	$\frac{1}{2}$	0	$\frac{1}{4}$	0	0
3	0	0	0	1	0	0
4	$\frac{1}{16}$	$\frac{1}{4}$	$\frac{1}{8}$	$\frac{1}{4}$	$\frac{1}{4}$	$\frac{1}{16}$
5	0	0	0	$\frac{1}{4}$	$\frac{1}{2}$	$\frac{1}{4}$
6	0	0	0	0	0	1

47. Rearrange the rows and columns of the transition matrix so that the absorbing states come first.

	1	6	2	3	4	5
1	1	0	0	0	0	0
6	0	1	0	0	0	0
2	$\frac{1}{4}$	0	$\frac{1}{2}$	0	$\frac{1}{4}$	0
3	0	0	0	0	1	0
4	$\frac{1}{16}$	$\frac{1}{16}$	$\frac{1}{4}$	$\frac{1}{8}$	$\frac{1}{4}$	$\frac{1}{4}$
5	0	$\frac{1}{4}$	0	0	$\frac{1}{4}$	$\frac{1}{2}$

From this rearranged matrix, observe that

$$Q = \begin{bmatrix} \frac{1}{2} & 0 & \frac{1}{4} & 0 \\ 0 & 0 & 1 & 0 \\ \frac{1}{4} & \frac{1}{8} & \frac{1}{4} & \frac{1}{4} \\ 0 & 0 & \frac{1}{4} & \frac{1}{2} \end{bmatrix}.$$

49. In Exercises 48, it was shown that the fundamental matrix for this absorbing Markov chain is

$$F = \begin{bmatrix} \frac{8}{3} & \frac{1}{6} & \frac{4}{3} & \frac{2}{3} \\ \frac{4}{3} & \frac{4}{3} & \frac{8}{3} & \frac{4}{3} \\ \frac{4}{3} & \frac{1}{3} & \frac{8}{3} & \frac{4}{3} \\ \frac{2}{3} & \frac{1}{6} & \frac{4}{3} & \frac{8}{3} \end{bmatrix}.$$

If Aa mates with Aa (which corresponds to state 4, which in turn corresponds to row 3 of F), $\frac{8}{3}$ pairs of offspring with these genes can be expected before ending up in one of the two absorbing states. This is because $\frac{8}{3}$ is the entry in row 3, column 3 of F.

51. (a) After the duplication, there are $2n$ genes and n of them are being selected; this can be done in $\binom{2n}{n}$ different ways. Suppose there are i mutant genes before the duplication and j mutant genes in the next generation. After the duplication, there will be $2i$ mutant genes, of which j will be selected; this can be done in $\binom{2i}{j}$ different ways. Also, there are $2n - 2i$ nonmutant genes, of which $n - j$ will be selected; this can be done in $\binom{2n-2i}{n-j}$ different ways.

Therefore, the probability of a generation with i mutant genes being followed by a generation with j mutant genes, which is the transition probability from state i to state j, is

$$p_{ij} = \frac{\binom{2i}{j}\binom{2n-2i}{n-j}}{\binom{2n}{n}}.$$

(b) The absorbing states are state 0 and state n. If a generation has no mutant genes, then after duplication there will still be none, and if a generation consists entirely of mutant genes, its successor will also.

(c) Use

$$p_{ij} = \frac{\binom{2i}{j}\binom{2n-2i}{n-j}}{\binom{2n}{n}}$$

with $n = 3$ and $i = 0, 1, 2, 3$ and $j = 0, 1, 2, 3$ to calculate the entries of the transition matrix. Let $\binom{n}{r} = 0$ when $n < r$.

$$p_{00} = \frac{\binom{0}{0}\binom{6}{3}}{\binom{6}{3}} = 1, \; p_{01} = \frac{\binom{0}{1}\binom{6}{2}}{\binom{6}{3}} = 0,$$

$$p_{02} = 0, \; p_{03} = 0,$$

$$p_{10} = \frac{\binom{2}{0}\binom{4}{3}}{\binom{6}{3}} = \frac{1}{5}, \; p_{11} = \frac{\binom{2}{1}\binom{4}{2}}{\binom{6}{3}} = \frac{3}{5},$$

$$p_{12} = \frac{\binom{2}{2}\binom{4}{1}}{\binom{6}{3}} = \frac{1}{5}, \; p_{13} = \frac{\binom{2}{3}\binom{4}{0}}{\binom{6}{3}} = 0,$$

$$p_{20} = \frac{\binom{4}{0}\binom{2}{3}}{\binom{6}{3}} = 0, \; p_{21} = \frac{\binom{4}{1}\binom{2}{2}}{\binom{6}{3}} = \frac{1}{5},$$

$$p_{22} = \frac{\binom{4}{2}\binom{2}{1}}{\binom{6}{3}} = \frac{3}{5}, \; p_{23} = \frac{\binom{4}{3}\binom{2}{0}}{\binom{6}{3}} = \frac{1}{5},$$

$$p_{30} = \frac{\binom{6}{0}\binom{0}{3}}{\binom{6}{3}} = 0, \; p_{31} = 0, \; p_{32} = 0,$$

$$p_{33} = \frac{\binom{6}{3}\binom{0}{0}}{\binom{6}{3}} = 1$$

The transition matrix is

$$\begin{array}{c} \\ 0 \\ 1 \\ 2 \\ 3 \end{array} \begin{array}{cccc} 0 & 1 & 2 & 3 \end{array} \\ \begin{bmatrix} 1 & 0 & 0 & 0 \\ \frac{1}{5} & \frac{3}{5} & \frac{1}{5} & 0 \\ 0 & \frac{1}{5} & \frac{3}{5} & \frac{1}{5} \\ 0 & 0 & 0 & 1 \end{bmatrix} = P.$$

(d) Rearrange the rows and columns of P.

$$\begin{array}{c} \\ 0 \\ 3 \\ 1 \\ 2 \end{array} \begin{array}{cccc} 0 & 3 & 1 & 2 \end{array} \\ \left[\begin{array}{cc|cc} 1 & 0 & 0 & 0 \\ 0 & 1 & 0 & 0 \\ \hline \frac{1}{5} & 0 & \frac{3}{5} & \frac{1}{5} \\ 0 & \frac{1}{5} & \frac{1}{5} & \frac{3}{5} \end{array} \right]$$

$$R = \begin{bmatrix} \frac{1}{5} & 0 \\ 0 & \frac{1}{5} \end{bmatrix}, \; Q = \begin{bmatrix} \frac{3}{5} & \frac{1}{5} \\ \frac{1}{5} & \frac{3}{5} \end{bmatrix}$$

$$F = [I_2 - Q]^{-1}$$

$$= \left(\begin{bmatrix} 1 & 0 \\ 0 & 1 \end{bmatrix} - \begin{bmatrix} \frac{3}{5} & \frac{1}{5} \\ \frac{1}{5} & \frac{3}{5} \end{bmatrix} \right)^{-1}$$

$$= \begin{bmatrix} \frac{2}{5} & -\frac{1}{5} \\ -\frac{1}{5} & \frac{2}{5} \end{bmatrix}^{-1} = \begin{bmatrix} \frac{10}{3} & \frac{5}{3} \\ \frac{5}{3} & \frac{10}{3} \end{bmatrix}$$

$$FR = \begin{bmatrix} \frac{10}{3} & \frac{5}{3} \\ \frac{5}{3} & \frac{10}{3} \end{bmatrix} \begin{bmatrix} \frac{1}{5} & 0 \\ 0 & \frac{1}{5} \end{bmatrix} = \begin{bmatrix} \frac{2}{3} & \frac{1}{3} \\ \frac{1}{3} & \frac{2}{3} \end{bmatrix}$$

(e) If a set of 3 genes has 1 mutant gene, the probability that the mutant gene will disappear is $\frac{2}{3}$, since that is the entry in row 1, column 1 of FR.

(f) If a set of 3 genes has 1 mutant gene, 5 generations would be expected to have 1 mutant gene before either the mutant genes or the nonmutant genes disappear, since that is the sum of the entries in row 1 of F.

Extended Application: Cavities and Restoration

1. This verification should be accomplished by graphing calculator or computer methods.

2. The entry in row 1, column 1 of the fundamental matrix indicates that there will be about 6.6 visits until there is some tooth decay. Each visit represents 6 mo, so 6.6 visits = 39.6 mo = 3.3 yr.

3. Use a graphing calculator or computer to show that

$$FR = \begin{bmatrix} .768332 & .231668 \\ .814382 & .185618 \\ .775304 & .224696 \\ .807692 & .192308 \\ .5 & .5 \\ .5 & .5 \\ .5 & .5 \\ .5 & .5 \end{bmatrix}.$$

4. The probability that a healthy tooth is eventually lost is $.231668 \approx .23$, since that is the entry in row 1, column 2 of FR.

5. The four .5 entries in column 2 of FR each correspond to a five-digit number whose first digit is 1, which corresponds to decay on the occlusal surface.

Chapter 9 Test

[9.1]

1. Decide which of the following matrices could be transition matrices. Justify your answers.

   (a) $\begin{bmatrix} .1 & .9 \\ 0 & 1 \end{bmatrix}$
   (b) $\begin{bmatrix} .1 & .2 & .5 & .2 \\ 0 & .8 & .3 & -.1 \\ .1 & .8 & .1 & 0 \\ 0 & .2 & 0 & 1 \end{bmatrix}$
   (c) $\begin{bmatrix} .21 & 0 & .88 \\ 0 & .45 & .55 \\ .79 & .03 & .18 \end{bmatrix}$

2. Let A be the transition matrix.

$$\begin{bmatrix} \frac{1}{2} & \frac{1}{2} \\ \frac{1}{3} & \frac{2}{3} \end{bmatrix}$$

   (a) Find the first three powers of A.

   (b) Find the probability that state 1 changes to state 2 after two repetitions of the experiment.

   (c) Find the probability that state 2 changes to state 1 after three repetitions of the experiment.

3. A survey conducted for General Motors revealed that 40% of GM buyers would buy GM again while 75% of Toyota buyers would buy Toyotas again. Suppose the initial distribution vector is $\begin{bmatrix} 100 & 50 \end{bmatrix}$ with GM coming first.

   (a) Write the transition matrix for this problem.

   (b) Find the probability vector after one time period.

   (c) Find the probability vector after two time periods.

[9.2]

4. Which of the following transition matrices are regular?

   (a) $\begin{bmatrix} 1 & 0 \\ .6 & .4 \end{bmatrix}$
   (b) $\begin{bmatrix} 1 & 0 & 0 \\ 0 & 1 & 0 \\ .1 & .2 & .7 \end{bmatrix}$
   (c) $\begin{bmatrix} .15 & .55 & .30 \\ 1 & 0 & 0 \\ 0 & 1 & 0 \end{bmatrix}$

5. Find the equilibrium vector for the regular transition matrix $\begin{bmatrix} .2 & .8 \\ .4 & .6 \end{bmatrix}$.

[9.1-9.2]

6. In the small Illinois town of Red Bud, there are three political parties: the Corn Party, the Wheat Party, and the Tea Party. Each year 50% of the Corn Party members switch to the Wheat Party, while the rest remain loyal. Among the Wheat Party members, 25% switch to Corn, and another 25% switch to Tea, and the rest remain loyal. 50% of the Tea Party members switch to Wheat and the rest remain loyal.

(a) Draw the transition diagram for this problem.

(b) Write the transition matrix.

(c) Show that the matrix is regular.

(d) Find the long-range prediction for the proportion of voters in each party.

[9.3]

7. Let $P = \begin{bmatrix} 1 & 0 & 0 \\ 0 & 1 & 0 \\ \frac{1}{4} & \frac{1}{2} & \frac{1}{4} \end{bmatrix}$ be the transition matrix for an absorbing Markov chain.

(a) Find all absorbing states.

(b) Find the fundamental matrix F.

(c) Find the matrix FR.

(d) What is the long-range probability of the chain terminating in state 1?

Chapter 9 Test Answers

1. **(a)** This could be a transition matrix since it is a square matrix with all entries between 0 and 1, inclusive, and the sum of the entries in each row is 1.

   (b) This could not be a transition matrix because it has a negative entry.
   (c) This could not be a transition matrix because the sum of the entries in the first row is not 1.

2. **(a)** $A = \begin{bmatrix} \frac{1}{2} & \frac{1}{2} \\ \frac{1}{3} & \frac{2}{3} \end{bmatrix}$, $A^2 = \begin{bmatrix} \frac{5}{12} & \frac{7}{12} \\ \frac{7}{18} & \frac{11}{18} \end{bmatrix}$, $A^3 = \begin{bmatrix} \frac{29}{72} & \frac{43}{72} \\ \frac{43}{108} & \frac{65}{108} \end{bmatrix}$ **(b)** $\frac{7}{12}$ **(c)** $\frac{43}{108}$

3. **(a)** $A = \begin{bmatrix} .4 & .6 \\ .25 & .75 \end{bmatrix}$ **(b)** $\begin{bmatrix} 52.5 & 97.5 \end{bmatrix}$ **(c)** $\begin{bmatrix} 45.375 & 104.625 \end{bmatrix}$

4. **(a)** Not regular **(b)** Not regular **(c)** Regular

5. $\begin{bmatrix} \frac{1}{3} & \frac{2}{3} \end{bmatrix}$

6. **(a)**

   (b)

	Corn	Wheat	Tea
Corn	.5	.5	0
Wheat	.25	.5	.25
Tea	0	.5	.5

   (c) $A^2 = \begin{bmatrix} .375 & .5 & .125 \\ .25 & .5 & .25 \\ .125 & .25 & .375 \end{bmatrix}$, which has all positive entries.

   (d) $\begin{bmatrix} .25 & .5 & .25 \end{bmatrix}$

7. **(a)** States 1 and 2 **(b)** $F = \begin{bmatrix} \frac{4}{3} \end{bmatrix}$ **(c)** $FR = \begin{bmatrix} \frac{1}{3} & \frac{2}{3} \end{bmatrix}$ **(d)** $\frac{1}{3}$

GAME THEORY

10.1 Strictly Determined Games

For Exercises 1-7, use the following game.

$$
\begin{array}{c}
 & & \text{B} \\
 & & 1 \quad\; 2 \quad\; 3 \\
\text{A} \begin{array}{c} 1 \\ 2 \\ 3 \end{array} &
\left[\begin{array}{rrr}
6 & -4 & 0 \\
3 & -2 & 6 \\
-1 & 5 & 11
\end{array}\right]
\end{array}
$$

1. Consider the strategy $(1,1)$. Player A chooses row 1, and player B chooses column 1. A positive number represents a payoff from B to A. The first row, first column entry is 6, indicating a payoff of $6 from B to A.

3. Consider the strategy $(2,2)$. Player A chooses row 2, and player B chooses column 2. A negative number represents a payoff from A to B. The second row, second column entry is -2, indicating a payoff of $2 from A to B.

5. Consider the strategy $(3,1)$. Player A chooses row 3, and player B chooses column 1. The third row, first column entry is -1, indicating a payoff of $1 from A to B.

7. Yes, each entry in column 2 is smaller than the corresponding entry in column 3, so column 2 dominates column 3.

9. $\begin{bmatrix} 0 & -2 & 8 \\ 3 & -1 & -9 \end{bmatrix}$

   Column 2 dominates column 1, so remove column 1 to obtain
   $$\begin{bmatrix} -2 & 8 \\ -1 & -9 \end{bmatrix}.$$

11. $\begin{bmatrix} 1 & 4 \\ 4 & -1 \\ 3 & 5 \\ -4 & 0 \end{bmatrix}$

    Row 3 dominates rows 1 and 4, so remove rows 1 and 4 to obtain
    $$\begin{bmatrix} 4 & -1 \\ 3 & 5 \end{bmatrix}.$$

13. $\begin{bmatrix} 8 & 12 & -7 \\ -2 & 1 & 4 \end{bmatrix}$

    Column 1 dominates column 2, so remove column 2 to obtain
    $$\begin{bmatrix} 8 & -7 \\ -2 & 4 \end{bmatrix}.$$

15. $\begin{bmatrix} \boxed{\underline{3}} & \boxed{5} \\ 2 & \underline{-5} \end{bmatrix}$

    Underline the smallest number in each row, and draw a box around the largest number in each column. The 3 at $(1,1)$ is the smallest number in its row and the largest number in its column, so the saddle point is 3 at $(1,1)$. This game is strictly determined, and its value is 3.

17. $\begin{bmatrix} 3 & \underline{-4} & \boxed{1} \\ \boxed{5} & \boxed{3} & \underline{-2} \end{bmatrix}$

    Underline the smallest number in each row, and box the largest number in each column; in this matrix, the two categorizations do not overlap. There is no saddle point. This game is not strictly determined.

19. $\begin{bmatrix} \underline{-6} & 2 \\ -1 & \underline{-10} \\ \boxed{\underline{3}} & \boxed{5} \end{bmatrix}$

    The 3 at $(3,1)$ is the smallest number in its row and the largest number in its column, so the saddle point is 3 at $(3,1)$. This game is strictly determined, and its value is 3.

21. $\begin{bmatrix} 2 & 3 & \boxed{1} \\ -1 & \boxed{4} & \underline{-7} \\ 5 & 2 & \underline{0} \\ \boxed{8} & \underline{-4} & -1 \end{bmatrix}$

    The 1 at $(1,3)$ is the smallest number in its row and the largest number in its column, so the saddle point is 1 at $(1,3)$. This game is strictly determined, and its value is 1.

23. $\begin{bmatrix} \underline{-6} & 1 & \boxed{4} & \boxed{2} \\ \boxed{9} & \boxed{3} & \underline{-8} & -7 \end{bmatrix}$

There is no saddle point. This game is not strictly determined.

25. Focus on any single column of the payoff matrix. In that column, suppose the row 1 entry is x_1, the row 2 entry is x_2, and the row 3 entry is x_3.

Since row 1 dominates row 2, $x_1 > x_2$, and since row 2 dominates row 3, $x_2 > x_3$. By transitivity, it follows that $x_1 > x_3$. The same phenomenon involving these three rows occurs in every column, so row 1 dominates row 3.

We have shown that, whenever row 1 dominates row 2 and row 2 dominates row 3, it must also be true that row 1 dominates row 3.

27.

	.01	.10	.20
Repair	$130	$130	$130
No Repair	$25	$200	$500

(a) An optimist should make no repairs; minimum cost is $25.

(b) A pessimist should make repairs; a worst case of $130 is better than a possible cost of $500 if no repairs are made.

(c) Find the expected cost of each strategy.

Make repairs:

$$.7(130) + .2(130) + .1(130) = \$130$$

Make no repairs:

$$.7(25) + .2(200) + .1(500) = \$107.50$$

He should make no repairs. The expected cost to the company if this strategy is chosen is $107.50.

29. (a)

	Fails	Doesn't Fail
Overhaul	−$8600	−$2600
Don't Overhaul	−$6000	$0

(b) Find the expected cost under each strategy.

Overhaul:

$$.1(-8600) + .9(-2600) = -\$3200$$

Don't overhaul:

$$.3(-6000) + .7(0) = -\$1800$$

To minimize his expected costs, the businessman should not overhaul the machine before shipping.

(c) Column 1 dominates column 2, and row 2 dominates row 1. The saddle point is −$6000.

31.

		B		
		City 1	City 2	City 3
	City 1	5	−2	6
A	City 2	7	5	9
	City 3	3	−3	5

To get the entries in the above matrix, look, for example, at the entry in row 2, column 1. If merchant A locates in city 2 and merchant B in city 1, then merchant A will get 80% of the business in city 2, 20% in city 3, and 60% in city 1. Taking into account the fraction of the population living in each city, we get

$$.80(.45) + .20(.30) + .60(.25) = .57.$$

Thus, merchant A gets 57% of the total business. Now 57% is 7 percentage points above 50%, so the entry in row 2, column 1 is +7.

Likewise for row 3, column 1, we get

$$.80(.25) + .20(.30) + .60(.45) = .53 = 53\%,$$

which is 3 percentage points above 50%. The other entries are found in a similar manner. (Note that all diagonal entries are 5 since 55% is 5 percentage points above 50%.)

The 5 at $(2,2)$ is the smallest entry in its row and the largest in its column, so the saddle point is the 5 at $(2,2)$. The value of the game is 5.

33. $\begin{bmatrix} \boxed{3} & -8 & \boxed{\underline{-9}} \\ 0 & \boxed{6} & \underline{-12} \\ -8 & 4 & \underline{-10} \end{bmatrix}$

Underline the smallest number in each row, and box the largest number in each column. −9 is the smallest entry in its row and the largest in its column. The saddle point is −9 at $(1,3)$, and the value of the game is −9.

35. The payoff matrix is as follows.

	Rock	Scissors	Paper
Rock	0	1	−1
Scissors	−1	0	1
Paper	1	−1	0

Underline the smallest number in each row, and box the largest number in each column; in this matrix, the two categorizations do not overlap. The game is not strictly determined since it does not have a saddle point.

10.2 Mixed Strategies

1. (a) $AMB = \begin{bmatrix} .5 & .5 \end{bmatrix} \begin{bmatrix} 3 & -4 \\ -5 & 2 \end{bmatrix} \begin{bmatrix} .3 \\ .7 \end{bmatrix}$

$= \begin{bmatrix} .5 & .5 \end{bmatrix} \begin{bmatrix} -1.9 \\ -.1 \end{bmatrix}$

$= \begin{bmatrix} -.95 - .05 \end{bmatrix} = \begin{bmatrix} -1 \end{bmatrix}$

The expected value is -1.

(b) $AMB = \begin{bmatrix} .1 & .9 \end{bmatrix} \begin{bmatrix} 3 & -4 \\ -5 & 2 \end{bmatrix} \begin{bmatrix} .3 \\ .7 \end{bmatrix}$

$= \begin{bmatrix} .1 & .9 \end{bmatrix} \begin{bmatrix} -1.9 \\ -.1 \end{bmatrix} = \begin{bmatrix} -.28 \end{bmatrix}$

The expected value is $-.28$.

(c) $AMB = \begin{bmatrix} .8 & .2 \end{bmatrix} \begin{bmatrix} 3 & -4 \\ -5 & 2 \end{bmatrix} \begin{bmatrix} .3 \\ .7 \end{bmatrix}$

$= \begin{bmatrix} .8 & .2 \end{bmatrix} \begin{bmatrix} -1.9 \\ -.1 \end{bmatrix} = \begin{bmatrix} -1.54 \end{bmatrix}$

The expected value is -1.54.

(d) $AMB = \begin{bmatrix} .2 & .8 \end{bmatrix} \begin{bmatrix} 3 & -4 \\ -5 & 2 \end{bmatrix} \begin{bmatrix} .3 \\ .7 \end{bmatrix}$

$= \begin{bmatrix} .2 & .8 \end{bmatrix} \begin{bmatrix} -1.9 \\ -.1 \end{bmatrix} = \begin{bmatrix} -.46 \end{bmatrix}$

The expected value is $-.46$.

3. $\begin{bmatrix} 5 & 1 \\ 3 & 4 \end{bmatrix}$

There are no saddle points so the game is not strictly determined, and mixed strategies must be used. Here $a_{11} = 5, a_{12} = 1, a_{21} = 3$, and $a_{22} = 4$. For player A, the optimum strategy is

$$p_1 = \frac{a_{22} - a_{21}}{a_{11} - a_{21} - a_{12} + a_{22}}$$

$$= \frac{4 - 3}{5 - 3 - 1 + 4} = \frac{1}{5},$$

$$p_2 = 1 - p_1 = 1 - \frac{1}{5} = \frac{4}{5}.$$

For player B, the optimum strategy is

$$q_1 = \frac{a_{22} - a_{12}}{a_{11} - a_{21} - a_{12} + a_{22}}$$

$$= \frac{4 - 1}{5 - 3 - 1 + 4} = \frac{3}{5},$$

$$q_2 = 1 - q_1 = 1 - \frac{3}{5} = \frac{2}{5}.$$

The value of the game is

$$\frac{a_{11}a_{22} - a_{12}a_{21}}{a_{11} - a_{21} - a_{12} + a_{22}} = \frac{5(4) - 1(3)}{5 - 3 - 1 + 4}$$

$$= \frac{20 - 3}{5} = \frac{17}{5}.$$

5. $\begin{bmatrix} -2 & 0 \\ 3 & -4 \end{bmatrix}$

There are no saddle points so the game is not strictly determined, and mixed strategies must be used. Here $a_{11} = -2, a_{12} = 0, a_{21} = 3$, and $a_{22} = -4$. For player A, the optimum strategy is

$$p_1 = \frac{-4 - 3}{-2 - 3 - 0 + (-4)} = \frac{-7}{-9} = \frac{7}{9},$$

$$p_2 = 1 - \frac{7}{9} = \frac{2}{9}.$$

For player B, the optimum strategy is

$$q_1 = \frac{-4 - 0}{-2 - 3 - 0 + (-4)} = \frac{-4}{-9} = \frac{4}{9},$$

$$q_2 = 1 - \frac{4}{9} = \frac{5}{9}.$$

The value of the game is

$$\frac{-2(-4) - 0(3)}{-2 - 3 - 0 + (-4)} = -\frac{8}{9}.$$

7. $\begin{bmatrix} 4 & -3 \\ -1 & 7 \end{bmatrix}$

There are no saddle points. For player A, the optimum strategy is

$$p_1 = \frac{7 - (-1)}{4 - (-1) - (-3) + 7} = \frac{8}{15},$$

$$p_2 = 1 - \frac{8}{15} = \frac{7}{15}.$$

For player B, the optimum strategy is

$$q_1 = \frac{7 - (-3)}{4 - (-1) - (-3) + 7} = \frac{10}{15} = \frac{2}{3},$$

$$q_2 = 1 - \frac{2}{3} = \frac{1}{3}.$$

The value of the game is

$$\frac{4(7) - (-3)(-1)}{4 - (-1) - (-3) + 7} = \frac{25}{15} = \frac{5}{3}.$$

9. $\begin{bmatrix} -2 & \frac{1}{2} \\ 0 & -3 \end{bmatrix}$

There are no saddle points. For player A, the optimum strategy is

$$p_1 = \frac{-3 - 0}{-2 - 0 - \frac{1}{2} + (-3)} = \frac{-3}{\frac{-11}{2}} = \frac{6}{11},$$

$$p_2 = 1 - \frac{6}{11} = \frac{5}{11}.$$

For player B, the optimum strategy is

$$q_1 = \frac{-3 - \frac{1}{2}}{-2 - 0 - \frac{1}{2} + (-3)} = \frac{-\frac{7}{2}}{\frac{-11}{2}} = \frac{7}{11},$$

$$q_2 = 1 - \frac{7}{11} = \frac{4}{11}.$$

The value of the game is

$$\frac{-2(-3) - \frac{1}{2}(0)}{-2 - 0 - \frac{1}{2} + (-3)} = \frac{6}{\frac{-11}{2}} = -\frac{12}{11}.$$

11. $\begin{bmatrix} \frac{8}{3} & -\frac{1}{2} \\ \frac{3}{4} & -\frac{5}{12} \end{bmatrix}$

The game is strictly determined since it has a saddle point at $(2, 2)$. The value of the game is $-\frac{5}{12}$.

13. $\begin{bmatrix} -1 & 2 \\ 3 & 1 \end{bmatrix}$

There are no saddle points. For player A, the optimum strategy is

$$p_1 = \frac{1 - 3}{-1 - 3 - 2 + 1} = \frac{-2}{-5} = \frac{2}{5},$$

$$p_2 = 1 - \frac{2}{5} = \frac{3}{5}.$$

For player B, the optimum strategy is

$$q_1 = \frac{1 - 2}{-1 - 3 - 2 + 1} = \frac{-1}{-5} = \frac{1}{5},$$

$$q_2 = 1 - \frac{1}{5} = \frac{4}{5}.$$

The value of the game is

$$\frac{-1(1) - 2(3)}{-1 - 3 - 2 + 1} = \frac{-7}{-5} = \frac{7}{5}.$$

15. $\begin{bmatrix} -4 & 9 \\ 3 & -5 \\ 8 & 7 \end{bmatrix}$

Row 3 dominates row 2, so remove row 2. This gives the matrix

$$\begin{bmatrix} -4 & 9 \\ 8 & 7 \end{bmatrix}.$$

For player A, the optimum strategy is

$$p_1 = \frac{7 - 8}{-4 - 8 - 9 + 7} = \frac{-1}{-14} = \frac{1}{14},$$

$$p_2 = 0 \text{ (row 2 was removed)},$$

$$p_3 = 1 - (p_1 + p_2) = 1 - \frac{1}{14} = \frac{13}{14}.$$

For player B, the optimum strategy is

$$q_1 = \frac{7 - 9}{-4 - 8 - 9 + 7} = \frac{-2}{-14} = \frac{1}{7},$$

$$q_2 = 1 - \frac{1}{7} = \frac{6}{7}.$$

The value of the game is

$$\frac{-4(7) - 9(8)}{-4 - 8 - 9 + 7} = \frac{-100}{-14} = \frac{50}{7}.$$

17. $\begin{bmatrix} 8 & 6 & 3 \\ -1 & -2 & 4 \end{bmatrix}$

Column 2 dominates column 1, so remove column 1. This gives the matrix

$$\begin{bmatrix} 6 & 3 \\ -2 & 4 \end{bmatrix}.$$

For player A, the optimum strategy is

$$p_1 = \frac{4 - (-2)}{6 - (-2) - 3 + 4} = \frac{6}{9} = \frac{2}{3},$$

$$p_2 = 1 - \frac{2}{3} = \frac{1}{3}.$$

For player B, the optimum strategy is

$$q_1 = 0 \text{ (column 1 was removed)},$$

$$q_2 = \frac{4 - 3}{6 - (-2) - 3 + 4} = \frac{1}{9},$$

$$q_3 = 1 - (q_1 + q_2) = 1 - \frac{1}{9} = \frac{8}{9}.$$

The value of the game is

$$\frac{6(4) - 3(-2)}{6 - (-2) - 3 + 4} = \frac{30}{9} = \frac{10}{3}.$$

19.
$$\begin{bmatrix} 9 & -1 & 6 \\ 13 & 11 & 8 \\ 6 & 0 & 9 \end{bmatrix}$$

Row 2 dominates row 1, so remove row 1. This gives the matrix

$$\begin{bmatrix} 13 & 11 & 8 \\ 6 & 0 & 9 \end{bmatrix}.$$

Now, column 2 dominates column 1, so remove column 1. This gives the matrix

$$\begin{bmatrix} 11 & 8 \\ 0 & 9 \end{bmatrix}.$$

For player A, the optimum strategy is

$$p_1 = 0 \text{ (row 1 was removed)},$$

$$p_2 = \frac{9 - 0}{11 - 0 - 8 + 9} = \frac{9}{12} = \frac{3}{4},$$

$$p_3 = 1 - \frac{3}{4} = \frac{1}{4}.$$

For player B, the optimum strategy is

$$q_1 = 0 \text{ (column 1 was removed)},$$

$$q_2 = \frac{9 - 8}{11 - 0 - 8 + 9} = \frac{1}{12},$$

$$q_3 = 1 - \frac{1}{12} = \frac{11}{12}.$$

The value of the game is

$$\frac{11(9) - 8(0)}{11 - 0 - 8 + 9} = \frac{99}{12} = \frac{33}{4}.$$

21. In a non-strictly-determined game, there is no saddle point. Let

$$M = \begin{bmatrix} a_{11} & a_{12} \\ a_{21} & a_{22} \end{bmatrix}$$

be the payoff matrix of the game. Assume that player B chooses column 1 with probability p_1. The expected value for B, assuming A plays row 1, is E_1, where

$$E_1 = a_{11}p_1 + a_{12}(1 - p_1).$$

The expected value for B if A plays row 2 is E_2, where

$$E_2 = a_{21}p_1 + a_{22}(1 - p_1).$$

The optimum strategy for player B is found by letting $E_1 = E_2$.

$$a_{11}p_1 + a_{12}(1 - p_1) = a_{21}p_1 + a_{22}(1 - p_1)$$
$$a_{11}p_1 + a_{12} - a_{12}p_1 = a_{21}p_1 + a_{22} - a_{22}p_1$$
$$a_{11}p_1 - a_{21}p_1 - a_{12}p_1 + a_{22}p_1 = a_{22} - a_{12}$$
$$p_1(a_{11} - a_{21} - a_{12} + a_{22}) = a_{22} - a_{12}$$

$$p_1 = \frac{a_{22} - a_{12}}{a_{11} - a_{21} - a_{12} + a_{22}}$$

Since $p_2 = 1 - p_1$,

$$p_2 = 1 - \frac{a_{22} - a_{12}}{a_{11} - a_{21} - a_{12} + a_{22}}$$

$$= \frac{a_{11} - a_{21} - a_{12} + a_{22} - (a_{22} - a_{12})}{a_{11} - a_{21} - a_{12} + a_{22}}$$

$$= \frac{a_{11} - a_{21}}{a_{11} - a_{21} - a_{21} + a_{22}}.$$

These are the formulas given in the text for p_1 and p_2.

25.

		Bates	
		T.V.	Radio
Allied	T.V.	$\begin{bmatrix} 1.0 \end{bmatrix}$	$-.7$
	Radio	$-.5$	$.5$

The optimum strategy for Allied is

$$p_1 = \frac{.5 - (-.5)}{1 - (-.5) - (-.7) + .5} = \frac{1}{2.7} = \frac{10}{27},$$

$$p_2 = 1 - \frac{10}{27} = \frac{17}{27}.$$

Allied should use T.V. with probability $\frac{10}{27}$ and radio with probability $\frac{17}{27}$.

The value of the game is

$$\frac{1(.5) - (-.7)(-.5)}{2.7} = \frac{1.5}{2.7} = \frac{1}{18},$$

which represents increased sales of

$$1,000,000\left(\frac{1}{18}\right) \approx 1,000,000(.055556)$$
$$= \$55,556.$$

27.

	Selling during:	
	Rain	Shine
Buying for: Rain	250	−150
Shine	−150	350

The best mixed strategy for Merrill is

$$p_1 = \frac{350 - (-150)}{250 - (-150) - (-150) + 350}$$

$$= \frac{500}{900} = \frac{5}{9},$$

$$p_2 = 1 - \frac{5}{9} = \frac{4}{9}.$$

He should invest in rainy day goods $\frac{5}{9}$ of the time and in sunny day goods $\frac{4}{9}$ of the time.

The value of the game is

$$\frac{250(350) - (-150)^2}{900} = \frac{65,000}{900} = \frac{650}{9}$$

$$\approx \$72.22.$$

Thus, his profit is $72.22.

29. The payoff matrix is as follows.

	Jamie	
	Pounce	Freeze
Euclid Pounce	3	1
Freeze	−2	2

The optimum strategy for Euclid is

$$p_1 = \frac{2 - (-2)}{3 - (-2) - 1 + 2} = \frac{4}{6} = \frac{2}{3},$$

$$p_2 = 1 - \frac{2}{3} = \frac{1}{3}.$$

Euclid should pounce $\frac{2}{3}$ of the time and freeze $\frac{1}{3}$ of the time.

The optimum strategy for Jamie is

$$q_1 = \frac{2 - 1}{3 - (-2) - 1 + 2} = \frac{1}{6},$$

$$q_2 = 1 - \frac{1}{6} = \frac{5}{6}.$$

Jamie should pounce $\frac{1}{6}$ of the time and freeze $\frac{5}{6}$ of the time.

The value of the game is

$$\frac{3(2) - 1(-2)}{3 - (-2) - 1 + 2} = \frac{8}{6} = \frac{4}{3}.$$

31. (a) The payoff matrix is as follows.

	Number of Fingers B Shows	
	1	2
Number of Fingers A Shows 1	2	−3
2	−3	4

(b) For player A, the optimum strategy is

$$p_1 = \frac{4 - (-3)}{2 - (-3) - (-3) + 4} = \frac{7}{12},$$

$$p_2 = 1 - \frac{7}{12} = \frac{5}{12}.$$

For player B, the optimum strategy is

$$q_1 = \frac{4 - (-3)}{2 - (-3) - (-3) + 4} = \frac{7}{12},$$

$$q_2 = 1 - \frac{7}{12} = \frac{5}{12}.$$

This means that each player should show 1 finger $\frac{7}{12}$ of the time and 2 fingers $\frac{5}{12}$ of the time.

The value of the game is

$$\frac{2(4) - (-3)(-3)}{2 - (-3) - (-3) + 4} = -\frac{1}{12}.$$

10.3 Game Theory and Linear Programming

1. $\begin{bmatrix} 1 & 2 \\ 3 & 1 \end{bmatrix}$

To find the optimum strategy for player A, use the following linear programming problem.

Minimize $w = x + y$
subject to: $x + 3y \geq 1$
                    $2x + y \geq 1$
with $x \geq 0, y \geq 0.$

Solve this linear programming problem by the graphical method. Sketch the feasible region.

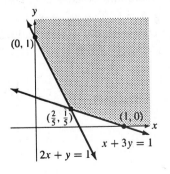

The region is unbounded, with corner points $(0,1)$, $\left(\frac{2}{5}, \frac{1}{5}\right)$, and $(1,0)$.

Corner Point	Value of $w = x + y$
$(0,1)$	$0 + 1 = 1$
$\left(\frac{2}{5}, \frac{1}{5}\right)$	$\frac{2}{5} + \frac{1}{5} = \frac{3}{5}$
$(1,0)$	$1 + 0 = 1$

The minimum value is $w = \frac{3}{5}$ at the point where $x = \frac{2}{5}, y = \frac{1}{5}$. Thus, the value of the game is $g = \frac{1}{w} = \frac{5}{3}$, and the optimum strategy for A is

$$p_1 = gx = \frac{5}{3}\left(\frac{2}{5}\right) = \frac{2}{3},$$

$$p_2 = gy = \frac{5}{3}\left(\frac{1}{5}\right) = \frac{1}{3}.$$

To find the optimum strategy for player B, use the following linear programming problem.

Maximize $\quad z = x + y$
subject to: $\quad x + 2y \leq 1$
$\qquad\qquad 3x + \ y \leq 1$
with $\qquad\quad x \geq 0, y \geq 0.$

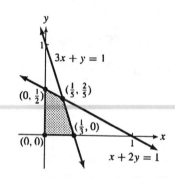

Corner Point	Value of $z = x + y$
$\left(0, \frac{1}{2}\right)$	$0 + \frac{1}{2} = \frac{1}{2}$
$\left(\frac{1}{5}, \frac{2}{5}\right)$	$\frac{1}{5} + \frac{2}{5} = \frac{3}{5}$
$\left(\frac{1}{3}, 0\right)$	$\frac{1}{3} + 0 = \frac{1}{3}$
$(0,0)$	$0 + 0 = 0$

The maximum value is $z = \frac{3}{5}$ at the point where $x = \frac{1}{5}, y = \frac{2}{5}$. The value of the game is $g = \frac{1}{z} = \frac{5}{3}$ (agreeing with our earlier findings), and the optimum strategy for B is

$$q_1 = gx = \frac{5}{3}\left(\frac{1}{5}\right) = \frac{1}{3},$$

$$q_2 = gy = \frac{5}{3}\left(\frac{2}{5}\right) = \frac{2}{3}.$$

3. $\begin{bmatrix} 4 & -2 \\ -1 & 6 \end{bmatrix}$

Get rid of negative numbers by adding 2 to each entry to obtain

$$\begin{bmatrix} 6 & 0 \\ 1 & 8 \end{bmatrix}.$$

(The 2 will have to be subtracted later, after the calculations have been performed.)

To find the optimum strategy for player A,

Minimize $\quad w = x + y$
subject to: $\quad 6x + \ y \geq 1$
$\qquad\qquad\qquad 8y \geq 1$
with $\qquad\quad x \geq 0, y \geq 0.$

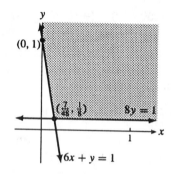

Corner Point	Value of $w = x + y$
$(0,1)$	$0 + 1 = 1$
$\left(\frac{7}{48}, \frac{1}{8}\right)$	$\frac{7}{48} + \frac{1}{8} = \frac{13}{48}$

The minimum value is $w = \frac{13}{48}$ at $\left(\frac{7}{48}, \frac{1}{8}\right)$. Thus, $g = \frac{1}{w} = \frac{48}{13}$, and the optimum strategy for A is

$$p_1 = gx = \frac{48}{13}\left(\frac{7}{48}\right) = \frac{7}{13},$$

$$p_2 = gy = \frac{48}{13}\left(\frac{1}{8}\right) = \frac{6}{13}.$$

The value of the game is

$$\frac{48}{13} - 2 = \frac{22}{13}.$$

To find the optimum strategy for player B,

Maximize $\quad z = x + y$
subject to: $\quad 6x \qquad\ \leq 1$
$\qquad\qquad x + 8y \leq 1$
with $\qquad\quad x \geq 0, y \geq 0.$

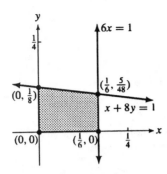

Corner Point	Value of $z = x + y$
$(0, \frac{1}{8})$	$0 + \frac{1}{8} = \frac{1}{8}$
$(\frac{1}{6}, \frac{5}{48})$	$\frac{1}{6} + \frac{5}{48} = \frac{13}{48}$
$(\frac{1}{6}, 0)$	$\frac{1}{6} + 0 = \frac{1}{6}$
$(0, 0)$	$0 + 0 = 0$

The maximum value is $z = \frac{13}{48}$ at $(\frac{1}{6}, \frac{5}{48})$. So $g = \frac{1}{z} = \frac{48}{13}$, and the optimum strategy for B is

$$q_1 = gx = \frac{48}{13}\left(\frac{1}{6}\right) = \frac{8}{13},$$

$$q_2 = gy = \frac{48}{13}\left(\frac{5}{48}\right) = \frac{5}{13}.$$

5. $\begin{bmatrix} 8 & -7 \\ -2 & 4 \end{bmatrix}$

Add 7 to each entry to obtain

$$\begin{bmatrix} 15 & 0 \\ 5 & 11 \end{bmatrix}.$$

To find the optimum strategy for player A,

Minimize $w = x + y$
subject to: $15x + 5y \geq 1$
$\qquad\qquad\quad 11y \geq 1$
with $x \geq 0, y \geq 0.$

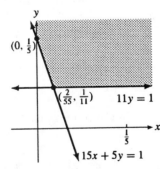

Corner Point	Value of $w = x + y$
$(0, \frac{1}{5})$	$0 + \frac{1}{5} = \frac{1}{5}$
$(\frac{2}{55}, \frac{1}{11})$	$\frac{2}{55} + \frac{1}{11} = \frac{7}{55}$

The minimum value is $w = \frac{7}{55}$ at $(\frac{2}{55}, \frac{1}{11})$. Thus, $g = \frac{1}{w} = \frac{55}{7}$, and the optimum strategy for A is

$$p_1 = gx = \frac{55}{7}\left(\frac{2}{55}\right) = \frac{2}{7},$$

$$p_2 = gy = \frac{55}{7}\left(\frac{1}{11}\right) = \frac{5}{7}.$$

The value of the game is

$$\frac{55}{7} - 7 = \frac{6}{7}.$$

To find the optimum strategy for player B,

Maximize $z = x + y$
subject to: $15x \qquad\quad \leq 1$
$\qquad\qquad\; 5x + 11y \leq 1$
with $x \geq 0, y \geq 0.$

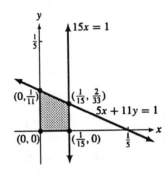

Corner Point	Value of $z = x + y$
$(0, \frac{1}{11})$	$0 + \frac{1}{11} = \frac{1}{11}$
$(\frac{1}{15}, \frac{2}{33})$	$\frac{1}{15} + \frac{2}{33} = \frac{21}{165} = \frac{7}{55}$
$(\frac{1}{15}, 0)$	$\frac{1}{15} + 0 = \frac{1}{15}$
$(0, 0)$	$0 + 0 = 0$

The maximum value is $z = \frac{7}{55}$ at $(\frac{1}{15}, \frac{2}{33})$. The optimum strategy for B is

$$q_1 = gx = \frac{55}{7}\left(\frac{1}{15}\right) = \frac{11}{21},$$

$$q_2 = gy = \frac{55}{7}\left(\frac{2}{33}\right) = \frac{10}{21}.$$

7. $\begin{bmatrix} 3 & -4 & 1 \\ 5 & 3 & -2 \end{bmatrix}$

Column 1 is dominated by the other two columns, so remove it.

$$\begin{bmatrix} -4 & 1 \\ 3 & -2 \end{bmatrix}$$

Add 4 to each entry to obtain

$$\begin{bmatrix} 0 & 5 \\ 7 & 2 \end{bmatrix}.$$

The linear programming problem to be solved is as follows.

Maximize $z = x_2 + x_3$
subject to: $5x_3 \leq 1$
$7x_2 + 2x_3 \leq 1$
with $x_2 \geq 0, x_3 \geq 0$.

Use the simplex method to solve the problem. The initial tableau is

x_2	x_3	s_1	s_2	z	
0	5	1	0	0	1
[7]	2	0	1	0	1
-1	-1	0	0	1	0

Pivot on the indicated entry.

$R_2 + 7R_3 \to R_3$

x_2	x_3	s_1	s_2	z	
0	[5]	1	0	0	1
7	2	0	1	0	1
0	-5	0	1	7	1

Pivot again.

$-2R_1 + 5R_2 \to R_2$
$R_1 + R_3 \to R_3$

x_2	x_3	s_1	s_2	z	
0	5	1	0	0	1
35	0	-2	5	0	3
0	0	1	1	7	2

Create a 1 in the columns corresponding to x_2, x_3, and z.

	x_2	x_3	s_1	s_2	z	
$\frac{1}{5}R_1 \to R_1$	0	1	$\frac{1}{5}$	0	0	$\frac{1}{5}$
$\frac{1}{35}R_2 \to R_2$	1	0	$-\frac{2}{35}$	$\frac{1}{7}$	0	$\frac{3}{35}$
$\frac{1}{7}R_3 \to R_3$	0	0	$\frac{1}{7}$	$\frac{1}{7}$	1	$\frac{2}{7}$

From this final tableau, we have $x_2 = \frac{3}{35}, x_3 = \frac{1}{5}, y_1 = \frac{1}{7}, y_2 = \frac{1}{7}, z = \frac{2}{7}$. Note that $g = \frac{1}{z} = \frac{7}{2}$. The optimum strategy for player A is

$$p_1 = gy_1 = \frac{7}{2}\left(\frac{1}{7}\right) = \frac{1}{2},$$

$$p_2 = gy_2 = \frac{7}{2}\left(\frac{1}{7}\right) = \frac{1}{2}.$$

The optimum strategy for player B is

$q_1 = 0$ (column 1 was removed),

$$q_2 = gx_2 = \frac{7}{2}\left(\frac{3}{35}\right) = \frac{3}{10},$$

$$q_3 = gx_3 = \frac{7}{2}\left(\frac{1}{5}\right) = \frac{7}{10}.$$

The value of the game is

$$\frac{7}{2} - 4 = -\frac{1}{2}.$$

9. $\begin{bmatrix} -1 & 2 & 4 \\ 3 & -2 & 0 \end{bmatrix}$

Remove column 3, since it is dominated by column 2.

$$\begin{bmatrix} -1 & 2 \\ 3 & -2 \end{bmatrix}$$

Add 2 to each entry to obtain

$$\begin{bmatrix} 1 & 4 \\ 5 & 0 \end{bmatrix}.$$

The linear programming problem to be solved is as follows.

Maximize $z = x_1 + x_2$
subject to: $x_1 + 4x_2 \leq 1$
$5x_1 \qquad \leq 1$
with $x_1 \geq 0, x_2 \geq 0$.

x_1	x_2	s_1	s_2	z	
1	4	1	0	0	1
[5]	0	0	1	0	1
-1	-1	0	0	1	0

Pivot on each indicated entry.

$-R_2 + 5R_1 \to R_1$

$R_2 + 5R_3 \to R_3$

x_1	x_2	s_1	s_2	z	
0	[20]	5	-1	0	4
5	0	0	1	0	1
0	-5	0	1	5	1

	x_1	x_2	s_1	s_2	z	
	0	20	5	-1	0	4
	5	0	0	1	0	1
$R_1 + 4R_3 \to R_3$	0	0	5	3	20	8

Create a 1 in the columns corresponding to x_1, x_2, and z.

	x_1	x_2	s_1	s_2	z	
$\frac{1}{20}R_1 \to R_1$	0	1	$\frac{1}{4}$	$-\frac{1}{20}$	0	$\frac{1}{5}$
$\frac{1}{5}R_2 \to R_2$	1	0	0	$\frac{1}{5}$	0	$\frac{1}{5}$
$\frac{1}{20}R_3 \to R_3$	0	0	$\frac{1}{4}$	$\frac{3}{20}$	1	$\frac{2}{5}$

From this final tableau, we have $x_1 = \frac{1}{5}, x_2 = \frac{1}{5}, y_1 = \frac{1}{4}, y_2 = \frac{3}{20}$, and $z = \frac{2}{5}$. Note that $g = \frac{1}{z} = \frac{5}{2}$. The optimum strategy for player A is

$$p_1 = gy_1 = \frac{5}{2}\left(\frac{1}{4}\right) = \frac{5}{8},$$

$$p_2 = gy_2 = \frac{5}{2}\left(\frac{3}{20}\right) = \frac{3}{8}.$$

The optimum strategy for player B is

$$q_1 = gx_1 = \frac{5}{2}\left(\frac{1}{5}\right) = \frac{1}{2},$$

$$q_2 = gx_2 = \frac{5}{2}\left(\frac{1}{5}\right) = \frac{1}{2},$$

$$q_3 = 0 \text{ (column 3 was removed)}.$$

The value of the game is

$$\frac{5}{2} - 2 = \frac{1}{2}.$$

11. $\begin{bmatrix} 1 & 0 & -1 \\ -1 & 0 & 1 \\ 2 & -1 & 2 \end{bmatrix}$

Add 1 to each entry to obtain

$$\begin{bmatrix} 2 & 1 & 0 \\ 0 & 1 & 2 \\ 3 & 0 & 3 \end{bmatrix}.$$

The linear programming problem to be solved is as follows.

Maximize $z = x_1 + x_2 + x_3$
subject to: $2x_1 + x_2 \leq 1$
$\qquad\qquad x_2 + 2x_3 \leq 1$
$\qquad 3x_1 \qquad + 3x_3 \leq 1$
with $\qquad x_1 \geq 0, x_2 \geq 0, x_3 \geq 0.$

The initial tableau is

x_1	x_2	x_3	s_1	s_2	s_3	z	
2	1	0	1	0	0	0	1
0	1	2	0	1	0	0	1
3	0	3	0	0	1	0	1
-1	-1	-1	0	0	0	1	0

Pivot on each indicated entry.

	x_1	x_2	x_3	s_1	s_2	s_3	z	
$-2R_3+3R_1 \to R_1$	0	3	-6	3	0	-2	0	1
	0	1	2	0	1	0	0	1
	3	0	3	0	0	1	0	1
$R_3+3R_4 \to R_4$	0	-3	0	0	0	1	3	1

	x_1	x_2	x_3	s_1	s_2	s_3	z	
	0	3	-6	3	0	-2	0	1
$-R_1+3R_2 \to R_2$	0	0	12	-3	3	2	0	2
	3	0	3	0	0	1	0	1
$R_1+R_4 \to R_4$	0	0	-6	3	0	-1	3	2

	x_1	x_2	x_3	s_1	s_2	s_3	z	
$R_2+2R_1 \to R_1$	0	6	0	3	3	-2	0	4
	0	0	12	-3	3	2	0	2
$-R_2+4R_3 \to R_3$	12	0	0	3	-3	2	0	2
$R_2+2R_4 \to R_4$	0	0	0	3	3	0	6	6

Create a 1 in the columns corresponding to x_1, x_2, x_3, and z.

	x_1	x_2	x_3	s_1	s_2	s_3	z	
$\frac{1}{6}R_1 \to R_1$	0	1	0	$\frac{1}{2}$	$\frac{1}{2}$	$-\frac{1}{3}$	0	$\frac{2}{3}$
$\frac{1}{12}R_2 \to R_2$	0	0	1	$-\frac{1}{4}$	$\frac{1}{4}$	$\frac{1}{6}$	0	$\frac{1}{6}$
$\frac{1}{12}R_3 \to R_3$	1	0	0	$\frac{1}{4}$	$-\frac{1}{4}$	$\frac{1}{6}$	0	$\frac{1}{6}$
$\frac{1}{6}R_4 \to R_4$	0	0	0	$\frac{1}{2}$	$\frac{1}{2}$	0	1	1

From this final tableau, we have $x_1 = \frac{1}{6}, x_2 = \frac{2}{3}, x_3 = \frac{1}{6}, y_1 = \frac{1}{2}, y_2 = \frac{1}{2}, y_3 = 0$, and $z = 1$. Note that $g = \frac{1}{z} = 1$. The optimum strategy for player A is

$$p_1 = gy_1 = 1\left(\frac{1}{2}\right) = \frac{1}{2}.$$

$$p_2 = gy_2 = 1\left(\frac{1}{2}\right) = \frac{1}{2},$$

$$p_3 = gy_3 = 1(0) = 0.$$

The optimum strategy for player B is

$$q_1 = gx_1 = 1\left(\frac{1}{6}\right) = \frac{1}{6},$$

$$q_2 = gx_2 = 1\left(\frac{2}{3}\right) = \frac{2}{3},$$

$$q_3 = gx_3 = 1\left(\frac{1}{6}\right) = \frac{1}{6}.$$

The value of the game is $1 - 1 = 0$.

13. The payoff matrix is as follows.

	Strike	No Strike
Bid $30,000	−$5500	$4500
Bid $40,000	$4500	$0

Add $5500 to each entry to obtain

$$\begin{bmatrix} 0 & 10{,}000 \\ 10{,}000 & 5500 \end{bmatrix}.$$

To find the optimum strategy for the contractor,

Minimize $\quad w = x + y$

subject to: $\qquad 10{,}000y \geq 1$

$\qquad\qquad 10{,}000x + 5500y \geq 1$

with $\qquad x \geq 0, y \geq 0.$

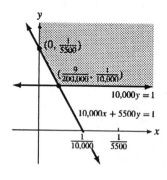

Corner Point	Value of $w = x + y$
$\left(0, \frac{1}{5500}\right)$	$0 + \frac{1}{5500} = \frac{1}{5500}$
$\left(\frac{9}{200{,}000}, \frac{1}{10{,}000}\right)$	$\frac{9}{200{,}000} + \frac{1}{10{,}000} = \frac{29}{200{,}000}$

The minimum value is $w = \frac{29}{200{,}000}$ at $\left(\frac{9}{200{,}000}, \frac{1}{10{,}000}\right)$.
Thus, $g = \frac{1}{w} = \frac{200{,}000}{29}$, and the optimum strategy
for the contractor is

$$p_1 = gx = \frac{200{,}000}{29}\left(\frac{9}{200{,}000}\right) = \frac{9}{29},$$

$$p_2 = gy = \frac{200{,}000}{29}\left(\frac{1}{10{,}000}\right) = \frac{20}{29}.$$

That is, the contractor should bid $30,000 with
probability $\frac{9}{29}$ and $40,000 with probability $\frac{20}{29}$.
The value of the game is

$$\frac{200{,}000}{29} - 5500 \approx \$1396.55.$$

15. (a) The payoff matrix is as follows.

			OI	
		A	B	C
GI	A	5000	10,000	10,000
	B	8000	4000	8000
	C	6000	6000	3000

Note that $5000 = \frac{1}{2}(10{,}000), 4000 = \frac{1}{2}(8000)$, and
$3000 = \frac{1}{2}(6000)$ are the reduced profits for General Items when the two companies run ads in the
same city.

(b) The linear programming problem to be solved
is as follows.

Maximize $\quad z = x_1 + x_2 + x_3$

subject to: $\quad 5000x_1 + 10{,}000x_2 + 10{,}000x_3 \leq 1$

$\qquad\qquad 8000x_1 + 4000x_2 + 8000x_3 \leq 1$

$\qquad\qquad 6000x_1 + 6000x_2 + 3000x_3 \leq 1$

with $\qquad x_1 \geq 0, x_2 \geq 0, x_3 \geq 0.$

The initial tableau is

x_1	x_2	x_3	s_1	s_2	s_3	z	
5000	10,000	10,000	1	0	0	0	1
8000	4000	8000	0	1	0	0	1
6000	6000	3000	0	0	1	0	1
−1	−1	−1	0	0	0	1	0

(Solution continues on the next page.)

Pivot on each indicated entry.

$$
\begin{array}{c}
\\
-5R_2 + 8R_1 \rightarrow R_1 \\
\\
\\
-3R_2 + 4R_3 \rightarrow R_3 \\
R_2 + 8000R_4 \rightarrow R_4
\end{array}
\begin{array}{c}
\begin{array}{ccccccc}
x_1 & x_2 & x_3 & s_1 & s_2 & s_3 & z \\
\end{array}\\
\left[\begin{array}{ccccccc|c}
0 & \boxed{60{,}000} & 40{,}000 & 8 & -5 & 0 & 0 & 3 \\
8000 & 4000 & 8000 & 0 & 1 & 0 & 0 & 1 \\
0 & 12{,}000 & -12{,}000 & 0 & -3 & 4 & 0 & 1 \\
0 & -4000 & 0 & 0 & 1 & 1 & 8000 & 1
\end{array}\right]
\end{array}
$$

$$
\begin{array}{c}
\\
-R_1 + 15R_2 \rightarrow R_2 \\
-R_1 + 5R_3 \rightarrow R_3 \\
R_1 + 15R_4 \rightarrow R_4
\end{array}
\begin{array}{c}
\begin{array}{ccccccc}
x_1 & x_2 & x_3 & s_1 & s_2 & s_3 & z \\
\end{array}\\
\left[\begin{array}{ccccccc|c}
0 & 60{,}000 & 40{,}000 & 8 & -5 & 0 & 0 & 3 \\
120{,}000 & 0 & 80{,}000 & -8 & 20 & 0 & 0 & 12 \\
0 & 0 & -100{,}000 & -8 & -10 & 20 & 0 & 2 \\
0 & 0 & 40{,}000 & 8 & 10 & 0 & 120{,}000 & 18
\end{array}\right]
\end{array}
$$

Create a 1 in the columns corresponding to x_1, x_2, s_3, and z.

$$
\begin{array}{c}
\frac{1}{60{,}000}R_1 \rightarrow R_1 \\
\frac{1}{120{,}000}R_2 \rightarrow R_2 \\
\frac{1}{20}R_3 \rightarrow R_3 \\
\frac{1}{120{,}000}R_4 \rightarrow R_4
\end{array}
\begin{array}{c}
\begin{array}{cccccccc}
x_1 & x_2 & x_3 & s_1 & s_2 & s_3 & z \\
\end{array}\\
\left[\begin{array}{ccccccc|c}
0 & 1 & \frac{2}{3} & \frac{1}{7500} & -\frac{1}{12{,}000} & 0 & 0 & \frac{1}{20{,}000} \\
1 & 0 & \frac{2}{3} & -\frac{1}{15{,}000} & \frac{1}{6000} & 0 & 0 & \frac{1}{10{,}000} \\
0 & 0 & -5000 & -\frac{2}{5} & -\frac{1}{2} & 1 & 0 & \frac{1}{10} \\
0 & 0 & \frac{1}{3} & \frac{1}{15{,}000} & \frac{1}{12{,}000} & 0 & 1 & \frac{3}{20{,}000}
\end{array}\right]
\end{array}
$$

From this final tableau, we have

$$
x_1 = \frac{1}{10{,}000}, x_2 = \frac{1}{20{,}000}, x_3 = 0, y_1 = \frac{1}{15{,}000}, y_2 = \frac{1}{12{,}000}, y_3 = 0, \text{ and } z = \frac{3}{20{,}000}.
$$

Note that

$$
g = \frac{1}{z} = \frac{20{,}000}{3} \approx 6666.67,
$$

so the value of the game is \$6666.67. The optimum strategy for General Items is

$$
p_1 = gy_1 = \frac{20{,}000}{3}\left(\frac{1}{15{,}000}\right) = \frac{4}{9},
$$

$$
p_2 = gy_2 = \frac{20{,}000}{3}\left(\frac{1}{12{,}000}\right) = \frac{5}{9},
$$

$$
p_3 = gy_3 = \frac{20{,}000}{3}(0) = 0.
$$

That is, General Items should advertise in Atlanta with probability $\frac{4}{9}$, in Boston with probability $\frac{5}{9}$, and never in Cleveland.

The optimum strategy for Original Imitators is

$$
q_1 = gx_1 = \frac{20{,}000}{3}\left(\frac{1}{10{,}000}\right) = \frac{2}{3},
$$

$$
q_2 = gx_2 = \frac{20{,}000}{3}\left(\frac{1}{20{,}000}\right) = \frac{1}{3},
$$

$$
q_3 = gx_3 = \frac{20{,}000}{3}(0) = 0.
$$

That is, Original Imitators should advertise in Atlanta with probability $\frac{2}{3}$, in Boston with probability $\frac{1}{3}$, and never in Cleveland.

17. This exercise should be solved by graphing calculator or computer methods. The solution may vary slightly. The answer is that merchant A should locate in cities 1, 2, and 3 with probabilities $\frac{27}{101}$, $\frac{129}{202}$, and $\frac{19}{202}$, respectively; merchant B should locate in cities 1, 2, and 3 with probabilities $\frac{39}{101}$, $\frac{9}{101}$, and $\frac{53}{101}$, respectively. The value of the game is $\frac{885}{101} \approx 8.76$ percentage points.

19. $\begin{bmatrix} 50 & 0 \\ 10 & 40 \end{bmatrix}$

To find the student's optimum strategy,

Minimize $w = x + y$
subject to: $50x + 10y \geq 1$
            $40y \geq 1$
with $x \geq 0, y \geq 0.$

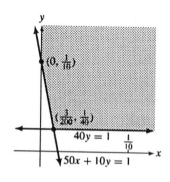

Corner Point	Value of $w = x + y$
$(0, \frac{1}{10})$	$0 + \frac{1}{10} = \frac{1}{10}$
$(\frac{3}{200}, \frac{1}{40})$	$\frac{3}{200} + \frac{1}{40} = \frac{1}{25}$

The minimum value is $w = \frac{1}{25}$ at $(\frac{3}{200}, \frac{1}{40})$. Thus, the value of the game is $g = \frac{1}{w} = 25$ points, and the optimum strategy for the student is

$$p_1 = gx = 25 \left(\frac{3}{200}\right) = \frac{3}{8},$$

$$p_2 = gy = 25 \left(\frac{1}{40}\right) = \frac{5}{8}.$$

That is, the student should choose the calculator with probability $\frac{3}{8}$ and the book with probability $\frac{5}{8}$.

21. (a) For player A, choice 1 is to believe B when B says "ace," and choice 2 is to ask B to show his card when B says "ace." For player B, there are four choices. Choice 1 is to always tell the truth.

Choice 2 is to lie only if the card is a queen. Choice 3 is to lie only if the card is a king. Choice 4 is to lie if the card is a queen or king. These lead to the following payoff matrix:

$$A \begin{matrix} & & B & \\ \end{matrix}$$
$$A \begin{bmatrix} 0 & -\frac{2}{3} & -\frac{1}{3} & -1 \\ -\frac{1}{3} & 0 & \frac{1}{3} & \frac{2}{3} \end{bmatrix}.$$

(b) Because of the negative entries, we will add 1 to all the entries, giving the matrix

$$\begin{bmatrix} 1 & \frac{1}{3} & \frac{2}{3} & 0 \\ \frac{2}{3} & 1 & \frac{4}{3} & \frac{5}{3} \end{bmatrix}.$$

The linear programming problem to be solved is:

Maximize $z = x_1 + x_2 + x_3 + x_4$
subject to: $x_1 + \frac{1}{3}x_2 + \frac{2}{3}x_3 \leq 1$

$$\frac{2}{3}x_1 + x_2 + \frac{4}{3}x_3 + \frac{5}{3}x_4 \leq 1$$

with $x_1 \geq 0, x_2 \geq 0, x_3 \geq 0, x_4 \geq 0.$

The initial tableau is

$$\begin{matrix} x_1 & x_2 & x_3 & x_4 & s_1 & s_2 & z \end{matrix}$$
$$\begin{bmatrix} \boxed{1} & \frac{1}{3} & \frac{2}{3} & 0 & 1 & 0 & 0 & | & 1 \\ \frac{2}{3} & 1 & \frac{4}{3} & \frac{5}{3} & 0 & 1 & 0 & | & 1 \\ -1 & -1 & -1 & -1 & 0 & 0 & 1 & | & 0 \end{bmatrix}.$$

We arbitrarily choose the first column. The smallest ratio is formed by the 1 in row 1. We use this as the first pivot and arrive at the following matrix.

$$\begin{matrix} & x_1 & x_2 & x_3 & x_4 & s_1 & s_2 & z \end{matrix}$$
$$\begin{matrix} & \\ -\frac{2}{3}R_1 + R_2 \to R_2 \\ R_1 + R_3 \to R_3 \end{matrix} \begin{bmatrix} 1 & \frac{1}{3} & \frac{2}{3} & 0 & 1 & 0 & 0 & | & 1 \\ 0 & \frac{7}{9} & \frac{8}{9} & \boxed{\frac{5}{3}} & -\frac{2}{3} & 1 & 0 & | & \frac{1}{3} \\ 0 & -\frac{2}{3} & -\frac{1}{3} & -1 & 1 & 0 & 1 & | & 1 \end{bmatrix}$$

The next pivot is the $\frac{5}{3}$ in row 2, column 4.

$$\begin{matrix} & x_1 & x_2 & x_3 & x_4 & s_1 & s_2 & z \end{matrix}$$
$$\begin{matrix} & \\ & \\ R_2 + \frac{5}{3}R_3 \to R_3 \end{matrix} \begin{bmatrix} 1 & \frac{1}{3} & \frac{2}{3} & 0 & 1 & 0 & 0 & | & 1 \\ 0 & \boxed{\frac{7}{9}} & \frac{8}{9} & \frac{5}{3} & -\frac{2}{3} & 1 & 0 & | & \frac{1}{3} \\ 0 & -\frac{1}{3} & \frac{1}{3} & 0 & 1 & 1 & \frac{5}{3} & | & 2 \end{bmatrix}$$

The next pivot is the $\frac{7}{9}$ in row 2, column 2.

$$
\begin{array}{c}
-R_2 + \frac{7}{3}R_1 \to R_1 \\[2mm]
\\
R_2 + \frac{7}{3}R_3 \to R_3
\end{array}
\begin{array}{cccccc}
x_1 & x_2 & x_3 & x_4 & s_1 & s_2 & z \\
\left[\begin{array}{cccccc|c}
\frac{7}{3} & 0 & \frac{2}{3} & -\frac{5}{3} & 3 & -1 & 0 \\[1mm]
0 & \frac{7}{9} & \frac{8}{9} & \frac{5}{3} & -\frac{2}{3} & 1 & 0 \\[1mm]
\hline
0 & 0 & \frac{5}{3} & \frac{5}{3} & \frac{5}{3} & \frac{10}{3} & \frac{35}{9}
\end{array}\right.
& & & & & & \left.\begin{array}{c} 2 \\[1mm] \frac{1}{3} \\[1mm] \hline 5 \end{array}\right]
\end{array}
$$

Dividing the bottom row by $\frac{35}{9}$ gives a z-value of $\frac{45}{35} = \frac{9}{7}$, so $g = \frac{1}{z} = \frac{7}{9}$. The values of y_1 and y_2 are read from the bottom of the columns for the two slack variables after dividing the bottom row by $\frac{35}{9}$: $y_1 = \frac{3}{7}, y_2 = \frac{6}{7}$. We find the values of p_1 and p_2 by multiplying these values by g:

$$p_1 = \frac{7}{9}\left(\frac{3}{7}\right) = \frac{1}{3},$$

$$p_2 = \frac{7}{9}\left(\frac{6}{7}\right) = \frac{2}{3}.$$

Next, we find the values of $x_1, x_2, x_3,$ and x_4 by using the first four columns combined with the last column: $x_1 = \frac{6}{7}, x_2 = \frac{3}{7}, x_3 = 0, x_4 = 0$. We find the values of $q_1, q_2, q_3,$ and q_4 by multiplying the values of $x_1, x_2, x_3,$ and x_4 by g:

$$p_1 = \frac{7}{9}\left(\frac{6}{7}\right) = \frac{2}{3},$$

$$p_2 = \frac{7}{9}\left(\frac{3}{7}\right) = \frac{1}{3},$$

$$p_3 = 0, p_4 = 0.$$

Finally, the value of the game is found by subtracting from g the 1 that was added at the beginning, yielding $\frac{7}{9} - 1 = -\frac{2}{9}$.

Thus, player A should use strategy 1 with probability $\frac{1}{3}$ and strategy 2 with probability $\frac{2}{3}$. Player B should use strategy 1 with probability $\frac{2}{3}$, strategy 2 with probability $\frac{1}{3}$, and should never use strategies 3 and 4. The value of the game is $-\frac{2}{9}$.

Chapter 10 Review Exercises

For Exercises 3-7, use the following payoff matrix.

$$
\begin{bmatrix}
-2 & 5 & -6 & 3 \\
0 & -1 & 7 & 5 \\
2 & 6 & -4 & 4
\end{bmatrix}
$$

3. The strategy $(1,1)$ means that player A chooses row 1 and player B chooses column 1. A negative number represents a payoff from A to B. The entry

at $(1,1)$ is -2, indicating that the payoff is 2 from A to B.

5. The entry at $(2,3)$ is 7. A positive number represents a payoff from B to A, indicating that the payoff is 7 from B to A.

7. Row 3 dominates row 1 and column 1 dominates column 4.

9. $\begin{bmatrix} -11 & 6 & 8 & 9 \\ -10 & -12 & 3 & 2 \end{bmatrix}$

Column 1 dominates both column 3 and column 4. Remove the dominated columns to obtain

$$
\begin{bmatrix}
-11 & 6 \\
-10 & -12
\end{bmatrix}.
$$

11. $\begin{bmatrix} -2 & 4 & 1 \\ 3 & 2 & 7 \\ -8 & 1 & 6 \\ 0 & 3 & 9 \end{bmatrix}$

Row 2 dominates row 3. Remove row 3 to obtain

$$
\begin{bmatrix}
-2 & 4 & 1 \\
3 & 2 & 7 \\
0 & 3 & 9
\end{bmatrix}.
$$

Column 1 dominates column 3. Remove column 3 to obtain

$$
\begin{bmatrix}
-2 & 4 \\
3 & 2 \\
0 & 3
\end{bmatrix}.
$$

13. $\begin{bmatrix} \boxed{\underline{-2}} & 3 \\ \underline{-4} & \boxed{5} \end{bmatrix}$

Underline the smallest number in each row, and box the largest number in each column. The -2 at $(1,1)$ is the smallest number in its row and the largest number in its column, so the saddle point is -2 at $(1,1)$. The value of the game is -2.

15. $\begin{bmatrix} \underline{-4} & -1 \\ 6 & \boxed{\underline{0}} \\ \boxed{8} & \underline{-3} \end{bmatrix}$

The 0 at $(2,2)$ is the smallest number in its row and the largest number in its column, so the saddle point is 0 at $(2,2)$. The value of the game is 0, so it is a fair game.

17. $\begin{bmatrix} \boxed{8} & 1 & \boxed{-7} & 2 \\ -1 & \boxed{4} & \boxed{-3} & 3 \end{bmatrix}$

The -3 at $(2,3)$ is the smallest number in its row and the largest number in its column, so the saddle point is -3 at $(2,3)$. The value of the game is -3.

19. $\begin{bmatrix} 1 & 0 \\ -2 & 3 \end{bmatrix}$

The optimum strategy for player A is

$$p_1 = \frac{3-(-2)}{1-(-2)-0+3} = \frac{5}{6},$$

$$p_2 = 1 - \frac{5}{6} = \frac{1}{6}.$$

The optimum strategy for player B is

$$q_1 = \frac{3-0}{1-(-2)-0+3} = \frac{3}{6} = \frac{1}{2},$$

$$q_2 = 1 - \frac{1}{2} = \frac{1}{2}.$$

The value of the game is

$$\frac{1(3)-0(-2)}{1-(-2)-0+3} = \frac{1}{2}.$$

21. $\begin{bmatrix} -3 & 5 \\ 1 & 0 \end{bmatrix}$

The optimum strategy for player A is

$$p_1 = \frac{0-1}{-3-1-5+0} = \frac{-1}{-9} = \frac{1}{9},$$

$$p_2 = 1 - \frac{1}{9} = \frac{8}{9}.$$

The optimum strategy for player B is

$$q_1 = \frac{0-5}{-3-1-5+0} = \frac{-5}{-9} = \frac{5}{9},$$

$$q_2 = 1 - \frac{5}{9} = \frac{4}{9}.$$

The value of the game is

$$\frac{-3(0)-5(1)}{-3-1-5+0} = \frac{-5}{-9} = \frac{5}{9}.$$

23. $\begin{bmatrix} -4 & 8 & 0 \\ -2 & 9 & -3 \end{bmatrix}$

Column 1 dominates column 2. Remove column 2 to obtain

$$\begin{bmatrix} -4 & 0 \\ -2 & -3 \end{bmatrix}.$$

The optimum strategy for player A is

$$p_1 = \frac{-3-(-2)}{-4-(-2)-0+(-3)} = \frac{-1}{-5} = \frac{1}{5},$$

$$p_2 = 1 - \frac{1}{5} = \frac{4}{5}.$$

The optimum strategy for player B is

$$q_1 = \frac{-3-0}{-4-(-2)-0+(-3)} = \frac{-3}{-5} = \frac{3}{5},$$

$q_2 = 0$ (column 2 was removed),

$$q_3 = 1 - \frac{3}{5} = \frac{2}{5}.$$

The value of the game is

$$\frac{-4(-3)-0(-2)}{-4-(-2)-0+(-3)} = \frac{12}{-5} = -\frac{12}{5}.$$

25. $\begin{bmatrix} 2 & -1 \\ -4 & 5 \\ -1 & -2 \end{bmatrix}$

Row 1 dominates row 3. Remove row 3 to obtain

$$\begin{bmatrix} 2 & -1 \\ -4 & 5 \end{bmatrix}.$$

The optimum strategy for player A is

$$p_1 = \frac{5-(-4)}{2-(-4)-(-1)+5} = \frac{9}{12} = \frac{3}{4},$$

$$p_2 = 1 - \frac{3}{4} = \frac{1}{4},$$

$p_3 = 0$ (row 3 was removed).

The optimum strategy for player B is

$$q_1 = \frac{5-(-1)}{2-(-4)-(-1)+5} = \frac{6}{12} = \frac{1}{2},$$

$$q_2 = 1 - \frac{1}{2} = \frac{1}{2}.$$

The value of the game is

$$\frac{2(5)-(-1)(-4)}{2-(-4)-(-1)+5} = \frac{6}{12} = \frac{1}{2}.$$

27. $\begin{bmatrix} -4 & 2 \\ 3 & -5 \end{bmatrix}$

Get rid of the negative numbers by adding 5 to each entry to obtain

$$\begin{bmatrix} 1 & 7 \\ 8 & 0 \end{bmatrix}.$$

To find the optimum strategy for player A,

Minimize $w = x + y$
subject to: $x + 8y \geq 1$
$7x \qquad \geq 1$
with $x \geq 0, y \geq 0$.

Solve this linear programming problem by the graphical method. Sketch the feasible region.

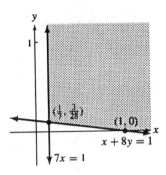

The region is unbounded, with corner points $\left(\frac{1}{7}, \frac{3}{28}\right)$ and $(1, 0)$.

Corner Point	Value of $w = x + y$
$\left(\frac{1}{7}, \frac{3}{28}\right)$	$\frac{1}{7} + \frac{3}{28} = \frac{1}{4}$
$(1, 0)$	$1 + 0 = 1$

The minimum value is $w = \frac{1}{4}$ at $\left(\frac{1}{7}, \frac{3}{28}\right)$. Thus, $g = \frac{1}{w} = 4$, and the optimum strategy for A is

$$p_1 = gx = 4\left(\frac{1}{7}\right) = \frac{4}{7},$$

$$p_2 = gy = 4\left(\frac{3}{28}\right) = \frac{3}{7}.$$

To find the optimum strategy for player B,

Maximize $z = x + y$
subject to: $x + 7y \leq 1$
$8x \qquad \leq 1$
with $x \geq 0, y \geq 0$.

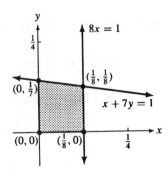

Corner Point	Value of $z = x + y$
$\left(0, \frac{1}{7}\right)$	$0 + \frac{1}{7} = \frac{1}{7}$
$\left(\frac{1}{8}, \frac{1}{8}\right)$	$\frac{1}{8} + \frac{1}{8} = \frac{1}{4}$
$\left(\frac{1}{8}, 0\right)$	$\frac{1}{8} + 0 = \frac{1}{8}$
$(0, 0)$	$0 + 0 = 0$

The maximum value is $z = \frac{1}{4}$ at $\left(\frac{1}{8}, \frac{1}{8}\right)$. The optimum strategy for B is

$$q_1 = gx = 4\left(\frac{1}{8}\right) = \frac{1}{2},$$

$$q_2 = gy = 4\left(\frac{1}{8}\right) = \frac{1}{2}.$$

The value of the game is $4 - 5 = -1$.

29. $\begin{bmatrix} 1 & 0 \\ -3 & 4 \end{bmatrix}$

Add 3 to each entry to obtain

$$\begin{bmatrix} 4 & 3 \\ 0 & 7 \end{bmatrix}.$$

To find the optimum strategy for player A,

Minimize $w = x + y$
subject to: $4x \qquad \geq 1$
$3x + 7y \geq 1$
with $x \geq 0, y \geq 0$.

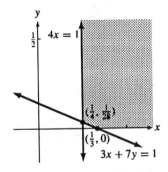

Corner Point	Value of $w = x + y$
$\left(\frac{1}{4}, \frac{1}{28}\right)$	$\frac{1}{4} + \frac{1}{28} = \frac{2}{7}$
$\left(\frac{1}{3}, 0\right)$	$\frac{1}{3} + 0 = \frac{1}{3}$

The minimum value is $w = \frac{2}{7}$ at $\left(\frac{1}{4}, \frac{1}{28}\right)$. Thus, $g = \frac{1}{w} = \frac{7}{2}$, and the optimum strategy for A is

$$p_1 = gx = \frac{7}{2}\left(\frac{1}{4}\right) = \frac{7}{8},$$

$$p_2 = gy = \frac{7}{2}\left(\frac{1}{28}\right) = \frac{1}{8}.$$

To find the optimum strategy for player B,

Maximize $z = x + y$

subject to: $4x + 3y \le 1$
$7y \le 1$

with $x \ge 0, y \ge 0$.

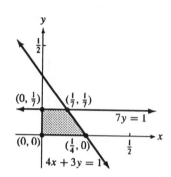

Corner Point	Value of $z = x + y$
$(0, \frac{1}{7})$	$0 + \frac{1}{7} = \frac{1}{7}$
$(\frac{1}{7}, \frac{1}{7})$	$\frac{1}{7} + \frac{1}{7} = \frac{2}{7}$
$(\frac{1}{4}, 0)$	$\frac{1}{4} + 0 = \frac{1}{4}$
$(0, 0)$	$0 + 0 = 0$

The maximum value is $z = \frac{2}{7}$ at $(\frac{1}{7}, \frac{1}{7})$. The optimum strategy for B is

$$q_1 = gx = \frac{7}{2}\left(\frac{1}{7}\right) = \frac{1}{2},$$

$$q_2 = gy = \frac{7}{2}\left(\frac{1}{7}\right) = \frac{1}{2}.$$

The value of the game is

$$\frac{7}{2} - 3 = \frac{1}{2}.$$

31. $\begin{bmatrix} 4 & 3 & 1 \\ -1 & 0 & 2 \end{bmatrix}$

Add 1 to each entry to obtain

$$\begin{bmatrix} 5 & 4 & 2 \\ 0 & 1 & 3 \end{bmatrix}.$$

The linear programming problem to be solved is as follows.

Maximize $z = x_1 + x_2 + x_3$
subject to: $5x_1 + 4x_2 + 2x_3 \le 1$
$x_2 + 3x_3 \le 1$
with $x_1 \ge 0, x_2 \ge 0, x_3 \ge 0$.

Use the simplex method to solve the problem. The initial tableau is

$$\left[\begin{array}{cccccc|c} x_1 & x_2 & x_3 & s_1 & s_2 & z & \\ 5 & \boxed{4} & 2 & 1 & 0 & 0 & 1 \\ 0 & 1 & 3 & 0 & 1 & 0 & 1 \\ \hline -1 & -1 & -1 & 0 & 0 & 1 & 0 \end{array}\right].$$

Pivot on each indicated entry.

$$\begin{array}{c} \\ \\ \\ R_1 + 5R_3 \rightarrow R_3 \end{array} \left[\begin{array}{cccccc|c} x_1 & x_2 & x_3 & s_1 & s_2 & z & \\ 5 & 4 & 2 & 1 & 0 & 0 & 1 \\ 0 & 1 & \boxed{3} & 0 & 1 & 0 & 1 \\ \hline 0 & -1 & -3 & 1 & 0 & 5 & 1 \end{array}\right]$$

$$\begin{array}{c} -2R_2 + 3R_1 \rightarrow R_1 \\ \\ R_2 + R_3 \rightarrow R_3 \end{array} \left[\begin{array}{cccccc|c} x_1 & x_2 & x_3 & s_1 & s_2 & z & \\ 15 & 10 & 0 & 3 & -2 & 0 & 1 \\ 0 & 1 & 3 & 0 & 1 & 0 & 1 \\ \hline 0 & 0 & 0 & 1 & 1 & 5 & 2 \end{array}\right]$$

Create a 1 in the columns corresponding to x_1, x_3, and z.

$$\begin{array}{c} \frac{1}{15}R_1 \rightarrow R_1 \\ \frac{1}{3}R_2 \rightarrow R_2 \\ \frac{1}{5}R_3 \rightarrow R_3 \end{array} \left[\begin{array}{cccccc|c} x_1 & x_2 & x_3 & s_1 & s_2 & z & \\ 1 & \frac{2}{3} & 0 & \frac{1}{5} & -\frac{2}{15} & 0 & \frac{1}{15} \\ 0 & \frac{1}{3} & 1 & 0 & \frac{1}{3} & 0 & \frac{1}{3} \\ \hline 0 & 0 & 0 & \frac{1}{5} & \frac{1}{5} & 1 & \frac{2}{5} \end{array}\right]$$

From this final tableau, we have

$$x_1 = \frac{1}{15}, x_2 = 0, x_3 = \frac{1}{3},$$

$$y_1 = \frac{1}{5}, y_2 = \frac{1}{5}, \text{ and } z = \frac{2}{5}.$$

Note that $g = \frac{1}{z} = \frac{5}{2}$. The optimum strategy for player A is

$$p_1 = gy_1 = \frac{5}{2}\left(\frac{1}{5}\right) = \frac{1}{2},$$

$$p_2 = gy_2 = \frac{5}{2}\left(\frac{1}{5}\right) = \frac{1}{2}.$$

The optimum strategy for player B is

$$q_1 = gx_1 = \frac{5}{2}\left(\frac{1}{15}\right) = \frac{1}{6},$$

$$q_2 = gx_2 = \frac{5}{2}(0) = 0,$$

$$q_3 = gx_3 = \frac{5}{2}\left(\frac{1}{3}\right) = \frac{5}{6}.$$

The value of the game is

$$\frac{5}{2} - 1 = \frac{3}{2}.$$

33. $\begin{bmatrix} -2 & 1 & 0 \\ 2 & 0 & -2 \\ 0 & -1 & 3 \end{bmatrix}$

Add 2 to each entry to obtain

$$\begin{bmatrix} 0 & 3 & 2 \\ 4 & 2 & 0 \\ 2 & 1 & 5 \end{bmatrix}.$$

The problem to be solved is as follows.

Maximize $\quad z = x_1 + x_2 + x_3$
subject to: $\qquad\qquad 3x_2 + 2x_3 \le 1$
$\qquad\qquad 4x_1 + 2x_2 \qquad \le 1$
$\qquad\qquad 2x_1 + x_2 + 5x_3 \le 1$
with $\qquad\qquad x_1 \ge 0, x_2 \ge 0, x_3 \ge 0.$

The initial simplex tableau is

$$\begin{array}{ccccccc}
x_1 & x_2 & x_3 & s_1 & s_2 & s_3 & z \\
\end{array}$$
$$\left[\begin{array}{ccccccc|c}
0 & 3 & 2 & 1 & 0 & 0 & 0 & 1 \\
\boxed{4} & 2 & 0 & 0 & 1 & 0 & 0 & 1 \\
2 & 1 & 5 & 0 & 0 & 1 & 0 & 1 \\
\hline
1 & -1 & -1 & 0 & 0 & 0 & 1 & 0
\end{array}\right].$$

Pivot on each indicated entry.

$$\begin{array}{ccccccc}
x_1 & x_2 & x_3 & s_1 & s_2 & s_3 & z \\
\end{array}$$
$$\begin{array}{r}
\\
\\
-R_2 + 2R_3 \to R_3 \\
R_2 + 4R_4 \to R_4
\end{array}
\left[\begin{array}{ccccccc|c}
0 & 3 & 2 & 1 & 0 & 0 & 0 & 1 \\
4 & 2 & 0 & 0 & 1 & 0 & 0 & 1 \\
0 & 0 & \boxed{10} & 0 & -1 & 2 & 0 & 1 \\
\hline
0 & -2 & -4 & 0 & 1 & 0 & 4 & 1
\end{array}\right]$$

$$\begin{array}{ccccccc}
x_1 & x_2 & x_3 & s_1 & s_2 & s_3 & z \\
\end{array}$$
$$\begin{array}{r}
-R_3 + 5R_1 \to R_1 \\
\\
\\
2R_3 + 5R_4 \to R_4
\end{array}
\left[\begin{array}{ccccccc|c}
0 & \boxed{15} & 0 & 5 & 1 & -2 & 0 & 4 \\
4 & 2 & 0 & 0 & 1 & 0 & 0 & 1 \\
0 & 0 & 10 & 0 & -1 & 2 & 0 & 1 \\
\hline
0 & -10 & 0 & 0 & 3 & 4 & 20 & 7
\end{array}\right]$$

$$\begin{array}{ccccccc}
x_1 & x_2 & x_3 & s_1 & s_2 & s_3 & z \\
\end{array}$$
$$\begin{array}{r}
\\
-2R_1 + 15R_2 \to R_2 \\
\\
2R_1 + 3R_4 \to R_4
\end{array}
\left[\begin{array}{ccccccc|c}
0 & 15 & 0 & 5 & 1 & -2 & 0 & 4 \\
60 & 0 & 0 & -10 & 13 & 4 & 0 & 7 \\
0 & 0 & 10 & 0 & -1 & 2 & 0 & 1 \\
\hline
0 & 0 & 0 & 10 & 11 & 8 & 60 & 29
\end{array}\right]$$

Create a 1 in the columns corresponding to x_1, x_2, x_3, and z.

$$\begin{array}{ccccccc}
x_1 & x_2 & x_3 & s_1 & s_2 & s_3 & z \\
\end{array}$$
$$\begin{array}{r}
\frac{1}{15}R_1 \to R_1 \\
\frac{1}{60}R_2 \to R_2 \\
\frac{1}{10}R_3 \to R_3 \\
\frac{1}{60}R_4 \to R_4
\end{array}
\left[\begin{array}{ccccccc|c}
0 & 1 & 0 & \frac{1}{3} & \frac{1}{15} & -\frac{2}{15} & 0 & \frac{4}{15} \\
1 & 0 & 0 & -\frac{1}{6} & \frac{13}{60} & \frac{1}{15} & 0 & \frac{7}{60} \\
0 & 0 & 1 & 0 & -\frac{1}{10} & \frac{1}{5} & 0 & \frac{1}{10} \\
\hline
0 & 0 & 0 & \frac{1}{6} & \frac{11}{60} & \frac{2}{15} & 1 & \frac{29}{60}
\end{array}\right]$$

From this final tableau, we have

$$x_1 = \frac{7}{60}, x_2 = \frac{4}{15}, x_3 = \frac{1}{10},$$
$$y_1 = \frac{1}{6}, y_2 = \frac{11}{60}, y_3 = \frac{2}{15}, \text{ and } z = \frac{29}{60}.$$

Note that $g = \frac{1}{z} = \frac{60}{29}$. The optimum strategy for player A is

$$p_1 = gy_1 = \frac{60}{29}\left(\frac{1}{6}\right) = \frac{10}{29},$$
$$p_2 = gy_2 = \frac{60}{29}\left(\frac{11}{60}\right) = \frac{11}{29},$$
$$p_3 = gy_3 = \frac{60}{29}\left(\frac{2}{15}\right) = \frac{8}{29}.$$

The optimum strategy for player B is

$$q_1 = gx_1 = \frac{60}{29}\left(\frac{7}{60}\right) = \frac{7}{29},$$
$$q_2 = gx_2 = \frac{60}{29}\left(\frac{4}{15}\right) = \frac{16}{29},$$
$$q_3 = gx_3 = \frac{60}{29}\left(\frac{1}{10}\right) = \frac{6}{29}.$$

The value of the game is

$$\frac{60}{29} - 2 = \frac{2}{29}.$$

In Exercises 37-41, use the following payoff matrix.

		Management	
		Friendly	Hostile
Labor	Friendly	\$600	\$800
	Hostile	\$400	\$950

37. Be hostile; then he has a chance at the \$950 wage increase, which is the largest possible increase.

39. If there is a .7 chance that the company will be hostile, then there is a .3 chance that it will be friendly. Find the expected payoff for each strategy.

Friendly: $.3(600) + .7(800) = \$740$
Hostile: $.3(400) + .7(950) = \$785$

Therefore, he should be hostile. The expected payoff is $785.

41. The 600 at $(1, 1)$ is the smallest number in its row and the largest number in its column, so it is a saddle point, and the game is strictly determined. Labor and management should both always be friendly, and the value of the game is 600.

43. $\begin{bmatrix} 2800 & 3200 \\ 5000 & -2000 \end{bmatrix}$

To guarantee that the value of the game is positive, we add 2000 to all entries in the matrix to obtain

$$\begin{bmatrix} 4800 & 5200 \\ 7000 & 0 \end{bmatrix}.$$

Let Hector choose row 1 with probability p_1 and row 2 with probability p_2. Then,

$$E_1 = 4800p_1 + 7000p_2$$
and $E_2 = 5200p_1$.

Let g represent the minimum of the expected gains, so that

$$E_1 = 4800p_1 + 7000p_2 \geq g$$
$$E_2 = 5200p_1 \qquad\quad \geq g.$$

Dividing by g yields

$$4800\left(\frac{p_1}{g}\right) + 7000\left(\frac{p_2}{g}\right) \geq 1$$
$$5200\left(\frac{p_1}{g}\right) \qquad\qquad \geq 1.$$

Let $x = \dfrac{p_1}{g}$ and $y = \dfrac{p_2}{g}$.

We have the following linear programming problem:

Minimize $w = x + y$
subject to: $4800x + 7000y \geq 1$
$\qquad\qquad 5200x \qquad\quad \geq 1$
with $x \geq 0, y \geq 0.$

Graph the feasible region.

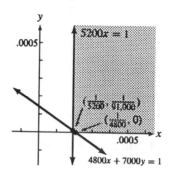

The corner points are $\left(\frac{1}{5200}, \frac{1}{91,000}\right)$ and $\left(\frac{1}{4800}, 0\right)$. The minimum value of $w = x + y$ is $\frac{37}{182,000}$ at $\left(\frac{1}{5200}, \frac{1}{91,000}\right)$. Thus, the value of the game is

$$g = \frac{1}{w} = \frac{182,000}{37} \approx 4918.92.$$

To find the value of the original game, we must subtract 2000:

$$4918.92 - 2000 = 2918.92.$$

The value of the original game is $2918.92.

The optimum strategy for Hector is

$$p_1 = gx = \frac{182,000}{37}\left(\frac{1}{5200}\right) = \frac{35}{37},$$
$$p_2 = gy = \frac{182,000}{37}\left(\frac{1}{91,000}\right) = \frac{2}{37}.$$

Hector should invest in blue-chip stocks with probability $\frac{35}{37}$ and growth stocks with probability $\frac{2}{37}$.

For Exercises 45-49, use the following payoff matrix.

		Opponent		
		Favors	Waffles	Opposes
	Favors	0	−1000	−4000
Candidate	Waffles	1000	0	−500
	Opposes	5000	2000	0

45. If the candidate is an optimist, she should oppose the new factory; then she has a chance at an additional 5000 votes.

47. Find the expected payoffs in vote changes under the three strategies.

Favors:

$$0(.40) + (-1000)(.35) + (-4000)(.25) = -1350$$

Waffles:

$$1000(.40) + 0(.35) + (-500)(.25) = 275$$

Opposes:

$$5000(.4) + 2000(.35) + 0(.25) = 2700$$

The candidate should oppose the new factory. He can expect a gain of 2700 votes.

49. This game has a saddle point of 0 at $(3, 3)$. The value of the game is 0. The strategy $(3, 3)$ means that the candidate opposes the factory and her opponent also opposes the factory. Therefore, each candidate should oppose the factory.

51.

Rontovia

		Attack 1	Attack 2
Ravogna	Defend 1	4	1
	Defend 2	3	4

To find the optimum strategy for Ravogna,

Minimize $w = x + y$

subject to: $4x + 3y \geq 1$

$x + 4y \geq 1$

with $x \geq 0, y \geq 0.$

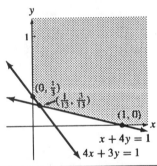

Corner Point	Value of $w = x + y$
$(0, \frac{1}{3})$	$0 + \frac{1}{3} = \frac{1}{3}$
$(\frac{1}{13}, \frac{3}{13})$	$\frac{1}{13} + \frac{3}{13} = \frac{4}{13}$
$(1, 0)$	$1 + 0 = 1$

The minimum value is $w = \frac{4}{13}$ at $(\frac{1}{13}, \frac{3}{13})$. Thus, $g = \frac{1}{w} = \frac{13}{4}$ is the value of the game. The optimum strategy for Ravogna is

$$p_1 = gx = \frac{13}{4}\left(\frac{1}{13}\right) = \frac{1}{4},$$

$$p_2 = gy = \frac{13}{4}\left(\frac{3}{13}\right) = \frac{3}{4}.$$

That is, Ravogna should defend installation #1 with probability $\frac{1}{4}$ and installation #2 with probability $\frac{3}{4}$.

To find the optimum strategy for Rontovia,

Maximize $z = x + y$

subject to: $4x + y \leq 1$

$3x + 4y \leq 1$

with $x \geq 0, y \geq 0.$

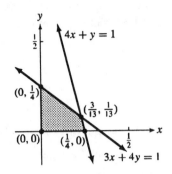

Corner Point	Value of $z = x + y$
$(0, \frac{1}{4})$	$0 + \frac{1}{4} = \frac{1}{4}$
$(\frac{3}{13}, \frac{1}{13})$	$\frac{3}{13} + \frac{1}{13} = \frac{4}{13}$
$(\frac{1}{4}, 0)$	$\frac{1}{4} + 0 = \frac{1}{4}$
$(0, 0)$	$0 + 0 = 0$

The maximum value is $z = \frac{4}{13}$ at $(\frac{3}{13}, \frac{1}{13})$. The optimum strategy for Rontovia is

$$q_1 = gx = \frac{13}{4}\left(\frac{3}{13}\right) = \frac{3}{4},$$

$$q_2 = gy = \frac{13}{4}\left(\frac{1}{13}\right) = \frac{1}{4}.$$

That is, Rontovia should attack installation #1 with probability $\frac{3}{4}$ and installation #2 with probability $\frac{1}{4}$.

Extended Application: Decision Making in the Military

1. Let the payoffs represent the number of days of bombing.

Japanese

		North	South
Kenney	North	2	2
	South	1	3

2. The saddle point is (North, North). Kenney should concentrate reconnaissance on the northern route. The Japanese should sail the northern route.

Chapter 10 Test

[10.1]

1. The Dean of Students at the local college must decide on the discipline rules for the upcoming year. She can choose to be Strict or Wimpy. Her success depends on whether the dorms are full of Rowdies or Studious folk.

$$\begin{array}{c} \\ \text{Strict} \\ \text{Wimpy} \end{array} \begin{array}{cc} \text{Rowdies} & \text{Studious} \\ \begin{bmatrix} 100 & -30 \\ -100 & 200 \end{bmatrix} \end{array}$$

(a) What course of action would an optimist take?

(b) What course of action would a pessimist take?

(c) What course of action would yield the largest expected payoff if the probability of Rowdies in the dorm is .6?

2. A politician is faced with three possible levels of opposition to his policies: Light, Medium, and Heavy. He must decide on three levels of advertising to combat this opposition: Low, Medium, or High. His payoff matrix is given below.

$$\begin{array}{c} \\ \text{Low} \\ \text{Medium} \\ \text{High} \end{array} \begin{array}{ccc} \text{Light} & \text{Medium} & \text{Heavy} \\ \begin{bmatrix} 100 & -30 & -60 \\ -10 & 75 & -70 \\ -50 & -20 & 50 \end{bmatrix} \end{array}$$

(a) What would an optimist do?

(b) What would a pessimist do?

(c) The probability of heavy opposition is .3, and of medium opposition is .4. What is the best course of action?

3. Remove any dominated strategies in the game with the following payoff matrix.

$$\begin{bmatrix} 2 & 3 & 0 & -1 \\ 1 & 2 & -2 & -5 \\ 0 & 3 & 1 & 0 \\ -1 & 0 & 0 & 1 \end{bmatrix}$$

4. For the following game, find the saddle point (if it exists), the strategies producing it, and the value of the game. (The entries represent dollar winnings.)

$$\begin{bmatrix} 1 & 20 & 3 \\ -1 & 1 & 0 \\ -3 & 0 & 1 \end{bmatrix}$$

[10.2]

5. Suppose a game has payoff matrix

$$M = \begin{bmatrix} 2 & -1 \\ 0 & 1 \end{bmatrix}.$$

(The entries represent dollar winnings.) Find the expected value of the game for the following strategies for players A and B.

$$A = \begin{bmatrix} .3 & .7 \end{bmatrix}; B = \begin{bmatrix} .75 \\ .25 \end{bmatrix}$$

6. For each game below, find the optimum strategy for each player and the value of the game.

(a) $\begin{bmatrix} 1 & -2 \\ 2 & 3 \end{bmatrix}$ (b) $\begin{bmatrix} 6 & -3 \\ 5 & 9 \end{bmatrix}$

[10.3]

7. Consider the game

$$\begin{bmatrix} -3 & 2 \\ 4 & -5 \end{bmatrix}.$$

(a) Find the optimum strategy for player A using the graphical method.

(b) Find the optimum strategy for player B using the simplex method.

(c) Find the value of the game.

8. Consider the game

$$\begin{bmatrix} 3 & 10 & 13 & 11 \\ 18 & 12 & 1 & 2 \end{bmatrix}.$$

Use the simplex method to find the optimum strategies for players A and B. Find the value of the game.

Chapter 10 Test Answers

1. **(a)** Wimpy   **(b)** Strict   **(c)** Strict

2. **(a)** Low advertising   **(b)** High advertising   **(c)** Medium advertising

3. $\begin{bmatrix} 2 & 0 & -1 \\ 0 & 1 & 0 \\ -1 & 0 & 1 \end{bmatrix}$

4. Saddle point: 1 at $(1,1)$; value of game: $1

5. $.55

6. **(a)** Player A: $p_1 = 0$, $p_2 = 1$ **(b)** Player A: $p_1 = \frac{4}{13}$, $p_2 = \frac{9}{13}$
   Player B: $q_1 = 1$, $q_2 = 0$ Player B: $q_1 = \frac{12}{13}$, $q_2 = \frac{1}{13}$
   Value of game $= 2$ Value of game $= \frac{69}{13}$

7. **(a)** Player A: $p_1 = \frac{9}{14}$, $p_2 = \frac{5}{14}$ **(b)** Player B: $q_1 = \frac{1}{2}$, $q_2 = \frac{1}{2}$ **(c)** Value of game $= -\frac{1}{2}$

8. Player A: $p_1 = \frac{2}{3}$, $p_2 = \frac{1}{3}$; player B: $q_1 = \frac{3}{8}$, $q_2 = 0$, $q_3 = 0$, $q_4 = \frac{5}{8}$; value of game: 8

Explorations in Finite Mathematics Software

This software enhances the understanding and visualization of many of the topics covered in the textbook. The self-documented software consists of two programs called FINITE1 and FINITE2.

Part 1 (Execute by entering FINITE1)

Part 1 covers finite math topics involving linear equations, linear inequalities, and matrices. In particular, it contains routines to perform Gaussian elimination, matrix operations, matrix inversion, simplex method, graphical solution of linear programming problems, the method of least-squares, and regular and absorbing Markov chains. The software follows the conventions of the text.

Part 2 (Execute by entering FINITE2)

Part 2 contains routines for Venn diagrams, Venn diagram counting problems, Venn diagram probability problems, computation of combinations, permutations, and factorials, statistical analysis of data, Galton board, binomial distribution, areas under normal curve, simple interest, compound interest, loan analysis, annuity analysis, and finance tables.

Requirements

The software runs on *any* IBM compatible computer having at least 384 K of memory and a graphics adapter, that is, CGA, EGA, MCGA, VGA, or Hercules.

Technical Support

If you have questions about the software, contact David I. Schneider by e-mail at dis@math.umd.edu.

Matrices On Diskette

About fifty matrices appearing in examples from the text have been saved on the diskette with either suggestive names or names of the form CxSxEx. For instance, the matrix appearing in Chapter 8, Section 3, Example 7 has the name C8S3E7. The list of saved matrices appropriate to the current routine is displayed on the screen whenever you request "Load a matrix saved on disk."

To Invoke Explorations in Finite Mathematics from DOS

1. Place the diskette in a drive, say drive A.
2. Type A: and press the Enter key.
3. Type FINITE1 or FINITE2 and press the Enter key.
(*Note:* If a Hercules adapter is used, the program MSHERC.COM must be run before FINITE1 or FINITE2 is entered.)

To Invoke Explorations in Finite Mathematics from Windows 3.1

1. Place the diskette in a drive, say drive A.
2. From Program Manager, double-click on the DOS icon in the Main program group. Or, exit Windows.
3. Type A: and press the Enter key.
4. Type FINITE1 or FINITE2 and press the Enter key.

To Invoke Explorations in Finite Mathematics from Windows 95 (or later version)

1. Place the diskette in a drive, say drive A.
2. Click the Start button, point to Programs, and click MS-DOS Prompt.
3. If the DOS window does not fill the screen, hold down the Alt key and press the Enter key to enlarge the window.
4. Type A: and press the Enter key.
5. Type FINITE1 or FINITE2 and press the Enter key.